Quark Model and High Energy Collisions

To the memory of

Vladimir Mikhailovitch Shekhter

PREFACE

This book treats one of the directions in elementary particle physics, namely, the application of the quark model to high energy hadron interactions. The main topics under consideration are binary processes, multiparticle production and hadron-nucleus collisions. Since we are considering a relatively restricted subject matter, we have the possibility of presenting a detailed exposition of the results as well as the required calculational techniques. We have made an effort to elucidate the material to such an extent that a novice in the field should be able to make his own calculations after the book is studied. The audience for whom the book is written is assumed to be familiar with quantum field theory, including Feynman diagrams, and also to be acquainted with the eperimental situation in high energy physics.

The subjects considered are inevitably connected with some other domains of elementary particle physics. In our effort to make our book as self-contained as possible we have discussed the necessary background material from these adjacent domains. Chapter 2 gives a concise account of the main experimental results in high energy strong interactions; the foundations of the Regge-pole phenomenology and multiperipheral dynamics are also briefly described there. In Chapter 3 the methods for treating composite systems are explained, while Chapter 4 discusses the theory of multiple rescattering at high energies. The present status of the quark model is reviewed in Chapter 5.

The material covered in Chapters 2 and 5 are described in more detail in books like

J.C. Polkinghorne, *Models of High Energy Physics*, Cambridge Univ. Press, Cambridge (1975).

P.D.B. Collins, *An Introduction to Regge Theory and High Energy Physics*, Cambridge Univ. Press, Cambridge (1975).

K. Huang, *Quarks, Leptons & Gauge Fields*, World Scientific, Singapore (1982).

F.E. Close, *An Introduction to Quarks and Partons*, Academic Press, New York (1979).

Some auxiliary material has been relegated to the Appendices and could probably aid the reader in making various explicit calculations.

We would like to thank A.A. Anselm, I.M. Dremin, S.S. Gerstein, V.G. Grishin, E.M. Levin, N.N. Nikolaev, M.G. Ryskin, E.V. Shuryak for extremely helpful discussions of the problems considered in the book. We are also deeply indebted to V.N. Gribov: his comments were extremely helpful for us. We are happy to thank our friends and colleagues V.M. Braun, L.G. Dakhno, A.K. Likhoded and P.E. Volkovitski: some papers written in collaboration with them are reflected in this work. The authors are grateful to K.K. Phua for his encouragement in writing this book. Two of us (V.V.A. and Yu.M.S.) are much obliged to O.I. Sumbaev for the permanent support of the investigations presented here: the book could hardly be accomplished without it.

We are specially indebted to our deceased friend V.M. Shekhter. Our collaboration with him has provided much material which is being passed on in this book. This book is dedicated to his memory.

Gatchina, USSR
Moscow, USSR
Budapest, Hungary

V.V. Anisovich
M.N. Kobrinsky
J. Nyiri
Yu.M. Shabelski

CONTENTS

Quark Model and High Energy Collisions

Chapter 1
INTRODUCTION

1.1 The Quark-Gluon Structure of Hadrons

It was assumed long ago that the observed elementary particles are not elementary at all. Already in the late forties Fermi and Yang[1] suggested that the pion is a composite system of a nucleon and an antinucleon. In the early fifties the discovery of the K-mesons and hyperons gave rise to different models, in which some particles were considered as fundamental ones, others as composite systems. The best known model of this kind was that of Sakata[2] in which the proton, the neutron and the Λ-hyperon were chosen as fundamental particles. The Sakata model and the scheme of unitary symmetry SU(3) which was built up on the basis of the fundamental p, n, Λ fields led to a proper classification of the pseudoscalar and vector mesons, but faced difficulties in the description of baryons. The Eightfold Way which was suggested by Gell-Mann[3] and Ne'eman[4] provided a possibility to describe both the mesons and the baryons. A splendid verification of this symmetry was the experimental discovery of the Ω^- hyperon.

The idea of the quark structure of hadrons appeared first in the papers of Gell-Mann[5] and Zweig[6]. It was shown, that the SU(3) octet symmetry may be realized on the basis of a fundamental triplet of some *hypothetical particles, called quarks by Gell-Mann, carrying fractional* electric charges. The quantum numbers of these quarks q(u,d,s) are presented in Table 1.1.

By now it is well known that u, d and s are not the only quarks: there exists also a group of heavy quarks. Experimentally two sorts (flavours) of these heavy quarks were discovered: the c-quark and the b-quark. The existence of a third quark, t, was also expected: after long unsuccessful searches it seems to have been found in recent experiments. The masses of heavy quarks are considerably larger than those of the light quarks. Because of this, dealing with soft processes (i.e. with processes with small momentum transfers) we can restrict ourselves to the consideration of the quarks (u,d,s) which realize the lowest representation of the $[SU(3)]_{flavour}$ group.

Table 1.1 Quantum numbers of light quarks

flavour	charge	isospin		strangeness	baryon charge
u	2/3		I_3=1/2	0	1/3
		I=1/2			
d	-1/3		I_3=-1/2	0	1/3
s	-1/3	I = 0		- 1	1/3

In the quark picture of hadrons, the meson consists of a quark-antiquark pair, while the baryon is a system of three quarks:

$$M = q\bar{q}, \qquad B = qqq \tag{1.1}$$

In general, one may say that in quark dynamics the only existing states are quark states with zero triality (i.e. when $|N(q) - N(\bar{q})|_{\text{mod } 3} = 0$).

In the non-relativistic quark model the light quarks realize the fundamental representation of a larger symmetry group SU(6) which acts in the space of spins and flavours:

$$q = (u^{\uparrow}, u^{\downarrow}, d^{\uparrow}, d^{\downarrow}, s^{\uparrow}, s^{\downarrow}). \tag{1.2}$$

The indices $\uparrow$ and $\downarrow$ denote the quark spin projection $s_z = \pm 1/2$ on the quantization axis. The introduction of the group

$[SU(6)]_{flavour \times spin}$ instead of the group $[SU(3)]_{flavour}$ was suggested by Gürsey and Radicati[7] and Sakita[8]. The analogy which motivated this step was the successful transition from $[SU(2)]_{isospin}$ to $[SU(4)]_{isospin \times spin}$ in nuclear physics, suggested by Wigner[9].

The SU(6) symmetry might be interpreted in terms of the wave functions of the composite quark systems; in doing so, the introduction of new quark quantum numbers is required[10-12]. The recent picture of coloured quarks was formulated by Gell-Mann[13]. In this picture, each quark possesses a new quantum number, colour, which can have three values:

$$q_i \qquad i = 1,2,3 \text{ (or red, yellow, green)} . \tag{1.3}$$

The coloured quarks realize the lowest representation of the colour group $[SU(3)]_{colour}$. Instead of the requirement of zero triality for the quark bound states here it is postulated, that the observable hadrons are singlets of the $[SU(3)]_{colour}$ group, i.e. they are white states. For the simplest mesons and baryons this means

$$M = \sum q_i \bar{q}_i$$

$$B = \sum_{i,k,\ell} \varepsilon_{ik\ell}\, q_i q_k q_\ell \quad . \tag{1.4}$$

Here the sum runs over the quark colours; $\varepsilon_{ik\ell}$ is the unit totally antisymmetric tensor. A more detailed picture of the history of the quark models is given in many review papers[14-20].

An important step in the development of the hadron quark structure is found in the parton hypothesis[21,22]. According to this, a fast hadron consists of point-like particles called partons. In deep inelastic collisions the photon or the weak intermediate boson interacts with the parton with large momentum transfer; hence in these processes the parton structure can be observed. There is experimental evidence showing that interacting partons are particles carrying the quantum numbers of quarks. The nucleon consists of three valence quarks-partons and a large number (a "sea") of quark-antiquark pairs and gluons. Investigating the momentum distributions one can conclude that the quark-partons carry about half of the momentum of the fast nucleon. The remaining half of the momentum is carried away by gluons which do not interact with the electro-weak field. The investigation of other hard processes such as $\mu^+\mu^-$ production in hadron collisions, hadron production with large p_T and, in particular, e^+e^- annihilation, confirmed the hypothesis of the quark-parton hadron structure and gave serious arguments in favour of the existence of gluons; besides, in these processes new heavy quarks were discovered. The existence of gluons led to three-jet events in e^+e^- annihilation: there must be gluon jets as well as quark jets. In fact such events are observed experimentally and the analysis of these events shows that there are jets which are initiated by particles with $J^P = 1^-$. (The quark-parton hypothesis and the experimental status of the hard processes are described in detail, e.g., in Refs. 23 to 27.

Our present understanding of hadrons as bound states of quarks is based on the theory of quantum chromodynamics. This is a non-Abelian gauge theory based on the theory of Yang-Mills fields[28)] (for a discussion of non-abelian gauge theory we refer, e.g., to the textbook[29)]). In QCD the interaction of quarks is due to the exchange of massless gluons just as the electromagnetic interaction arises from photon exchange. The main process in QCD is the emission of gluons, that in QED is the emission of photons, (Fig. 1.1a). In contrast to QED, where, along with the electron, one neutral photon exists, in QCD three types of quarks (three colours) are assumed and each of them can transform into another via the emission of 8 possible coloured gluons. Colour charge of gluons leads to the consequence that not only quarks emit gluons, but gluon emission by gluons and gluon-gluon scattering are also taking place (Fig. 1.1b and c). The requirement of three colours determines the theory unambiguously.

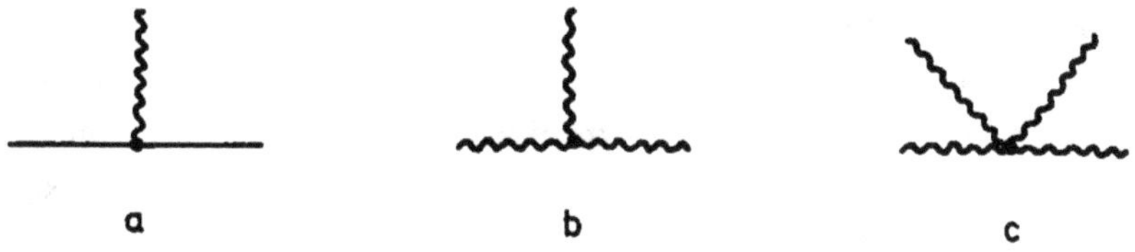

Fig. 1.1 QCD interactions vertices: gluon emission by the quark (a) or by the gluon (b); gluon-gluon scattering (c).

The most striking difference between QED and QCD is that, while electrons and photons are observed in nature, quarks and gluons are not seen as free particles. The hope that in QCD the confinement of coloured objects does take place is based on the increase of the effective charge at large distances. At the same time the decrease of the effective charge at large momenta[30-32] leads to a very important consequence, namely quarks and gluons are asymptotically free at small distances. The asymptotic freedom gives a theoretical justification of the parton model: quarks considered in QCD are quark-partons. At the same time perturbative QCD predicts some small deviations from the results of the naive parton model for hard processes. The detailed test of perturbative QCD calculations is of extreme interest; however the possibilities for the direct comparison of these calculations with the experiment are rather poor. (One can get introduced to the technique of the QCD calculations and its characteristic results in review papers, e.g. see Refs. 33 and 34.

According to our present understanding, QCD is an adequate theory of hadronic matter interactions both at small and large distances. However because of the growth of the effective charge in QCD at large distances perturbative methods cannot be applied to soft processes. A possible way out seems to be phenomenology based on QCD, which returns us to the quark model on a new level of understanding. Indeed, by stating that the colour forces allow the existence of colourless free states only, one implies that these forces are sufficiently strong.

For strong forces it would be natural to lead to a complicated structure of the hadrons as composite systems of quarks and gluons. On the other hand, many years of experimental analysis proved that hadrons can be well described as systems consisting of three quarks (baryons: $B = qqq$) or a quark and an antiquark (mesons: $M = q\bar{q}$) with a relatively small admixture of quark-antiquark pairs. Thus we see that the quarks of the quark model are rather complicated systems themselves. These constituent quarks are the objects of consideration for strong-interaction phenomenology.

1.2 Dressed Quarks

The attempts to combine the results of the quark model with the quark-parton picture have led to the assumption of double hadron structure long ago[35-38]. According to this assumption, practically all the quark-gluon matter inside the hadron is contained in two (mesons) or three (baryons) spatially separated clusters. Every cluster is a dressed (or constituent) quark; it consists of a valence QCD quark (or antiquark), surrounded by quark-antiquark pairs and gluons (see Fig. 1.2). This hypothesis of the double hadron structure states that the hadrons are built similarly to the light nuclei like deuterium or tritium, which consist of two or three nucleons. Such an analogy allows one to call this hadron structure "quasinuclear" (the name "quasinuclear" was introduced by Lipkin[39]). The underlying structure of the dressed quarks manifests itself in the deep inelastic scattering processes.

The quantum numbers of the dressed quark and that of the QCD quarks coincide (it is evident from hadron spectroscopy data — see Chap. 5); however, the other properties of these objects differ rather significantly. Let us first of all consider quark masses. The masses of QCD quarks may be estimated within the framework of current algebra with the use of the partial conservation of axial current hypothesis (PCAC). These estimates lead to comparatively low mass values of the light QCD quarks[40-43]; they are presented in Table 1.2.

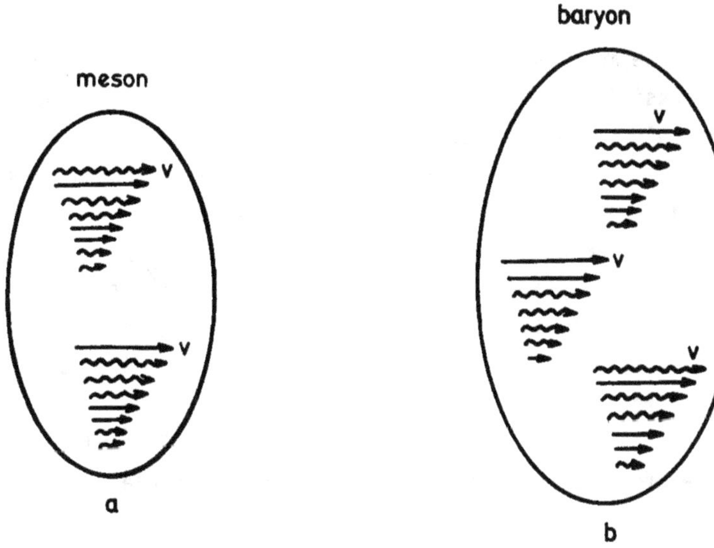

Fig. 1.2 Parton structure of the meson (a) and of the nucleon (b). Nucleon consists of three (meson of two) dressed quarks; each dressed quark (antiquark) consists of the valence quark-parton (straight arrow, marked by index V), sea gluons (wavy arrows) and sea quark-partons (straight arrows).

The masses of the dressed quarks m_q (q = u,d,s), which are estimated on the basis of hadron spectroscopy data (see Chap. 5), are considerably larger than those of the QCD quarks. The values of these masses allow one to calculate the magnetic moments of the baryons and the radiation widths of the vector mesons (Sec. 5.2) in the framework of the quasinuclear quark model. The results of such calculations are in reasonable agreement with the experimental data, though there are also some discrepancies between the predicted values with the observed ones.

As one may see from Table 1.2, the values of masses of the bare (QCD) and dressed quarks are essentially different. For the u-quark the ratio of the dressed to current quark masses approaches two orders of magnitude. The question arises whether such a large difference is reasonable. One must take into account that the absence of free quarks makes it impossible to give an exact definition of the quark mass; it is possible only to consider an effective value of the quark mass. According to Ref. 44, it seems to be natural to define the quark mass through the inverse propagator $\hat{S}_q^{-1}(k) = \hat{k} - m_q(k^2)$, where k is the quark momentum. The quark mass must depend significantly on its virtuality. The variation of the $m(k^2)$-value when k^2 decreases from the large values, typical for the hard processes, to the small virtualities considered in the quark model, may be estimated starting from the perturbation theory. It was found in Ref. 45 that the increase of $m(k^2)$ by one to two orders of magnitude while k^2 tends to 0.1-0.5 GeV^2, typical for the quark model, looks quite reasonable.

Table 1.2 Quark masses

light quarks			heavy quarks		
q	QCD (current) quark masses	Constituent (dressed) quark masses m_q	q	heavy quark masses	
u	4 MeV	$m_u \simeq m_d \simeq 300$	c	~ 1.5	GeV
d	7 MeV	- 400 MeV	b	~ 4.5	GeV
s	150 MeV	$m_s \simeq 3/2\ m_u$	t	30-50	GeV

The existence of the dressed quarks — quark-gluon clusters inside the hadron — must lead to a specific picture of high energy hadron interactions. Due to the quasinuclear hadron structure this interaction must obey quark additivity. The additive quark model (AQM) of hadron-hadron collisions predicts the ratio of the total nucleon-nucleon and pion-nucleon cross sections and is in good agreement with experiment[46,47]

$$\sigma_{tot}(NN)/\sigma_{tot}(\pi N) = \frac{3}{2} \ . \tag{1.5}$$

Measurements of the cross sections of the strange particle (kaons and hyperons) interactions also agree with the AQM predictions (see Sec. 6.2).

Impressive arguments in favour of the quasinuclear hadron structure are found in hadron-nucleus high energy collisions. The probability for two or three quarks of the projectile hadron to interact in nuclear matter is rather large. For a hypothetical super-heavy nucleus ($A \to \infty$) all dressed quarks of the projectile interact practically with unit probability; in this case the ratio of the secondary multiplicities in the central region of high energy NA and πA collisions must be equal to the ratio of the number of dressed quarks in the projectiles[48]:

$$\langle n_{ch} \rangle_{NA} / \langle n_{ch} \rangle_{\pi A} = \frac{3}{2} \qquad \text{as} \qquad A \to \infty \, . \tag{1.6}$$

For real nuclei this ratio is less than 3/2 and depends on the probability of the dressed quark inelastic interaction with the nucleus. This probability may be calculated using the value of quark-nucleon inelastic cross section ($\sigma_{inel}(qN) \simeq 1/2\ \sigma_{inel}(\pi N) \simeq 1/3\ \sigma_{inel}(NN)$) and the nucleon-density distribution inside the nucleus. The same probability determines the inclusive cross section of secondary hadron production in the beam-fragmentation region, since the fragmentational particles are formed by the quark-spectators, which did not interact with the target. Numerous relations for the inclusive production in the central and fragmentation regions, obtained in Chap. 8 for hadron-nucleus collisions, are in a reasonable agreement with the experimental data.

A crucial point regarding hadron structure is the ratio of the size of the dressed quark (r_q) and that of the hadron (R_h). The values of r_q estimated from the data on hadron-hadron collisions belong to the interval

$$\frac{1}{30} \le \frac{r_q^2}{R_N^2} \lesssim \frac{1}{5} \tag{1.7}$$

i.e. according to all such estimates the size of the dressed quark is appreciably less than that of the hadron. The ratio r_q^2/R_N^2 was estimated also from studies of the hard processes (e.g. from the average transverse momenta of the partons $\langle k_T \rangle$ and appeared to be about $1/5 \div 1$ (see e.g. Refs. 49 to 52). However, these estimates are model dependent. Model-independent estimates obtained from the quark-parton correlation-function calculations in the framework of QCD[53]), gives the value $r_q^2/R_N^2 \simeq 1/5$.

The success of the quark model in the description of the static hadron features and in the soft collision physics is undoubted. However, this cannot hide the fact that some predictions of the quark model are in an evident disagreement with the data. As stated by Glashow in Ref. 54: "Here is a subject full of mystery ... The mystery is why the non-relativistic quark model works so well, and sometimes, why it does not". An example of a disagreement between model and the data may be easily found when the relation (1.5) is considered in more detail. The difference $\sigma_{tot}(NN)/\sigma_{tot}(\pi N) - 3/2$, if it is determined by the Glauber screening corrections, must be negative; however, the experiment gives a positive value. The discrepancies of the quark-model predictions with the data for hyperon magnetic moments were also widely discussed. One may note that all these discrepancies are rather small. The shadow corrections in the ratio $\sigma_{tot}(NN)/\sigma_{tot}(\pi N)$ are about 10-20%, the predicted hyperon magnetic moments also differ from the measured values by approximately 20%. The situation is similar for other quark-model results. The whole picture may be summarized as follows: the quark model, which is based on the assumption of the existence of dressed quarks, is in agreement with the experiment within an accuracy of 20%. On the other hand, the existing 20% discrepancies are to be taken rather seriously. They probably show that the assumed quasinuclear hadron structure is inadequate for the description of some probably inherent hadron features. The understanding of the reasons for such failures of the quark model seems to us to be extremely important for the future development of the soft interaction theory.

There are at least two evident directions of studies which may extend our knowledge of hadron structure. One of them is the expansion of the field of application of the quark model, which would undoubtedly give some new understanding of the problem. The consideration of the multihadron production processes (Chaps. 7 and 8) is a step in this direction, it allows one to capture some features of the mechanism of soft colour neutralization and quark confinement. We shall discuss these problems later once more; here we turn our attention to the other direction of studies, which is of ultimate importance from our point of view.

We now consider the problem of the reconciliation of the quark model with QCD. This problem would be solved finally if one could derive the main assumptions of the model (AQM) from the fundamental principles of the theory (QCD). However the theoretical level of QCD-understanding of the soft processes is by now far from this hypothetical aim. We may only consider a rather simplified question: Is the assumption of the dressed quarks compatible with QCD? In other words, are there any possibilities in QCD of finding two characteristic sizes in soft processes — the radius of the hadron and that of the dressed quark? Up to now, we have several preliminary versions of the QCD scenarios with two fundamental sizes.

It was assumed[55)] that the smallness of the dressed quark size is related with the existence of the heavy glueball states with $M_{glueball} \simeq 1.5$-2 GeV: the radius of the dressed quark may be caused by the t-channel exchanges by the colourless gluonic states. In this case $r_q \sim M^{-1}_{glueball}$.

Spontaneous or dynamical breaking of the gauge symmetry may generate nonzero gluon mass. Such a possibility was discussed e.g. in Refs. 56 and 57. Probably the phase of the quark-gluon matter where the dressed quarks exist includes also comparatively heavy gluons; *then* $r_q \sim m^{-1}_{gluon}$. *Some arguments in favour of the heavy gluons are* considered in Ref. 58; the results of the calculations in Ref. 59, where the reggeization of the pomeron channel in the framework of QCD was made, have shown that gluons are effectively heavy in hadronic processes.

Another assumption was suggested in Ref. 60, where it was claimed that QCD vacuum has "granular" structure due to existence of instanton fluctuations of the typical size r_q. According to Ref. 60, the suppression of such a fluctuation by the valence quark-parton leads to the formation of the dressed quark. The solution with instanton-like fluctuations of the r_q-scale in the gluodynamic vacuum was obtained in Ref. 61 with the help of Feynman's variational method[62].

Thus although now we cannot state exactly why two fundamental sizes exist, we think that the small radius of the dressed quark is probably related to the small effective radius of the purely gluonic interaction, while the large size of hadron is caused by the size of the quark-interaction region. The large mass of the glueballs or the large effective mass of gluons may reflect the fact that gluons are confined at appreciably smaller distances than quarks are. This problem will probably be clarified soon, and then we shall get a definite answer.

Multiple production processes provide additional indications of the effective short-ranged gluon interactions. The main characteristics of these processes like average secondary multiplicities and inclusive spectra are described under the assumption of the quasinuclear hadron structure quite reasonably within the usual quark-model accuracy. Two essential features of this description should be underlined (see Chap. 7):

i) Hadron-production probabilities obey the rules of quark statistics;

ii) Quark hadronization proceeds in a "soft" way.

The fulfillment of the quark statistical rules indicates that some equilibrium "cloud" of dressed quarks is produced in the multiple production process. The subsequent hadronization of these quarks proceeds with the same probability for any quark flavour or spin projections, so that the statistical proportions in the quark cloud are not destroyed in the hadronization process — this is called "soft hadronization". This assumption is supported by the fact, that the x-distributions of the fragmentation particles in hadron-hadron and hadron-nucleus collisions repeat with good accuracy those of quark-spectators. It means that the transition "dressed quarks → hadrons" proceeds without a significant momentum change. Such a soft transition shows that quark confinement does not manifest itself as some obstacle which prevents quarks from leaving the interaction region (or the excited hadron state). Quarks may fly out easily, but they also easily produce additional quark-antiquark pairs and fly out together with the newly produced quarks (or quarks) so that the observed final state is always colourless.

It is necessary to emphasize here, that the last stage of hadronization proceeds without the explicit participation of gluons. The short-range nature of the gluonic interactions (or small radius of the gluons confinement) assumes that the gluonic degrees of freedom are effectively "included" into the dressed quarks.

Summarizing the previous considerations, we may conclude that the quark model goes through a qualitative transformation and its difficulties are probably "the difficulties of growth". In many aspects the quark model is motivated by QCD, but nowadays it cannot be derived from its first principles. The model describes quantitatively a broad class of the soft interaction phenomena, but these calculations are rather rough and for the present there is no consistent scheme which would allow us to reach a higher level of accuracy.

Studies of the dressed quark formation mechanism in the framework of QCD will probably give some deeper understanding of the quark structure of hadrons, which lies beyond the considered quasinuclear quark model and provides the possibility of more precise calculations. It is, however, just a hope; by now some crucial points of the quark model are guided by experimental data. In other words, the recent quark model is essentially a semi-phenomenological construction.

Our book is devoted mainly to high energy interactions, and we are confronted with the necessity of choosing an adequate language for the description of these interactions. Our choice is Regge-pole theory and the corresponding multiperipheral picture for multiple production processes (see Chap. 2). This interaction picture was discovered in the early sixties[63,64]; reformulated by R. Feynman[21] it became famous as the parton model. Regge-pole theory has a remarkable property: the structure of the collision process may be considered in the framework of this theory independent of the type of the colliding particles. Thus, Regge-pole theory, which was originally formulated for hadron collision processes, probably may be used for the description of quark collisions as well.

References

1. E. Fermi and C.N. Yang, Phys. Rev. 76, 1739 (1949).
2. S. Sakata, Prog. Theor. Phys. 16, 686 (1956).
3. M. Gell-Mann, "The Eightfold Way: a Theory of Strong Interaction Symmetry", CIT Rept. CTSL-20 (1961).
4. Y. Ne'eman, Nucl. Phys. 26, 222 (1961).
5. M. Gell-Mann, Phys. Lett. 8, 214 (1964).
6. G. Zweig, "An SU(3) Model of Strong Interaction Symmetry and its Breaking", CERN Rept. No. 8182/TH401 (1964).
7. F. Gürsey and L. Radicati, Phys. Rev. Lett. 13, 173 (1964).
8. B Sakita, Phys. Rev. 136B, 1756 (1964).
9. E. Wigner, Phys. Rev. 51, 106 (1937).
10. O.W. Greenberg, Phys. Rev. Lett 13, 598 (1964).
11. N.N. Bogoliubov, B.V. Struminski and A.N. Tavkhelidze, "To the Composite Models", Preprint JINR D-1968, 1964, (in Russian).
12. M. Han and Y. Nambu, Phys. Rev. 139, B1006 (1965).
13. M. Gell-Mann, Acta Phys. Austriaca (Schladming Lectures) Suppl. IX, 773 (1972).
14. N.N. Bogoliubov, "Theory of elementary particle symmetry", In High Energy Physics and Elementary Particle Physics, V.P. Shelest, ed. "Naukova Dumka", Kiev, 1967, (in Russian).

15. J.J.J. Kokkedee, The Quark Model, W.A. Benjamin, New York, 1969.

16. H.J. Lipkin, Phys. Rep., 8C, 173 (1973).

17. O.W. Greenberg and C.A. Nelson, Phys. Rep. 32C 71 (1977).

18. G. Zweig, "Origins of the Quark Model", CALT Rept. No. 68-805 (1980).

19. Y. Ne'eman, "Patterns, Structure and then Dynamics: Discovering Unitary Symmetry and Concerning Quarks", TAUP Rept. No. 142-81 (1981).

20. R.H. Dalitz, Prog. Part. Nucl. Phys. 8, 7 (1982).

21. R. Feynman, Photon-Hadron Interactions, W.A. Benjamin, New York, 1972.

22. J.D. Bjorken and E. Paschos, Phys. Rev. 186, 1975 (1969).

23. R.P. Feynman, "Partons", in Selected Lectures from the Hawaii Conference, S. Pakvasa and S.F. Tuan, eds. World Scientific, Singapore, 1983.

24. J.D. Bjorken, "Some Topics in Weak and Electromagnetic Interactions", *ibid.*

25. R. Kögerler and D. Schildknecht, eds., Electroweak Interactions at High Energy, Proc. of the 1982 DESY Workshop, World Scientific, Singapore, 1983.

26. Th. Papadopoulos and N.D. Tracas, eds., First Hellenic School on Elementary Particle Physics, World Scientific, Singapore, 1983.

27. H. Galić, B. Guberina, and D. Tadic, eds., Phenomenology of Unified Theories, Proc. of the Topical Conf., Dubrovnik, 1983, World Scientific, Singapore, 1984.

28. C.N. Yang and R.L. Mills, Phys. Rev. 96, 191 (1954).

29. K. Huang, Quarks, Leptons and Gauge Fields, World Scientific, Singapore, 1983.
K. Moriyasu, An Elementary Primer for Gauge Theory, World Scientific, Singapore, 1983.

30. I.B. Khriplovich, Yad. Fiz. 10, 409 (1969); Sov. J. Nucl. Phys. 10, 235 (1970).

31. H.D. Politzer, Phys. Rev. Lett. 30, 1346 (1973).

32. D.J. Gross and F. Wilczek, Phys. Rev. Lett. 30, 1343 (1973).

33. W. Marciano and H. Pagels, Phys. Rep. 36C, 137 (1978).

34. Yu. L. Dokshitzer, D.T. Dyakonov and S.I. Troyan, Phys. Rep. 58C, 269 (1980).

35. V.V. Anisovich, "Strong Interactions at High Energies and the Quark-Parton Model", in Proceedings of the IX-th LNPI Winter School, Vol. 3, p. 106, Yu. N. Novikov, ed. (Leningrad), 1974 (in Russian).

36. G. Altarelli, N. Cabibbo, L. Maiani and R. Petronzio, Nucl. Phys. B69 531 (1974).

37. T. Kanki, Prog. Theor. Phys. 56, 1885 (1976).

38. R.C. Hwa, Phys. Rev. D22, 759, 1593 (1980).

39. H.J. Lipkin, "A Quasinuclear Quark Model for Hadron", Preprint FERMILAB-Conf-78/73-THY (1978).

40. H. Leutwyler, Phys. Lett. 48B, 45 (1974).

41. M. Gell-Mann, Oppenheimer Lectures, Preprint IAS, Princeton, 1975.

42. S. Weinberg, Harvard Preprints HUTP-77/A057, HUPT-77/A061 (1977).

43. B.L. Ioffe, Yad. Fiz. 29, 1611 (1979); Sov. J. Nucl. Phys. 29, 827 (1979).

44. H. Georgi and H.D. Politzer, Phys. Rev. D14, 1829 (1976).

45. V.M. Shekhter, Yad. Fiz. 33, 817 (1981); Sov. J. Nucl. Phys. 33, 426 (1981).

46. E.M. Levin and L.L. Frankfurt, Pis'ma v ZhETF 2, 105 (1965); JETP Lett. 2, 65 (1965).

47. H.J. Lipkin and F. Sheck, Phys. Rev. Lett. 16, 71 (1966).

48. V.V. Anisovich, Phys. Lett. 57B, 87 (1975).

49. P. Chiapetta and M. Greco, Phys. Lett. 106B, 219 (1981).

50. P. Allen *et al.*, Nucl. Phys. B188, 1 (1981).

51. J.P. Berge *et al.*, Nucl. Phys. B203, 16 (1982).

52. B. Cox, "A Comparison of Dimuon Production by 125 GeV/c $\bar{p}$ and π with Predictions of the Drell-Yan Model", FERMILAB-Conf. 83/13 EXP (1983).

53. E.V. Shuryak and A.I. Vainstein, Nucl. Phys. B199, 451 (1982).

54. S.L. Glashow, "Particle Physics far from High Energy Frontier", Preprint HUTP-80/A089 (1980).

55. V.N. Gribov, Talk at the Autumn Session of the Academy of Science of the USSR (1979).

56. R.N. Mohapatra, J.S. Pati and A. Salam, Phys. Rev. D13, 1733 (1976).

57. E. Ma, Phys. Rev. D17, 623 (1978).

58. G. Parisi and R. Petronzio, Phys. Lett. 94B, 51 (1980).

59. L.N. Lipatov, "Conformal properties of the pomeron in QCD". In Group Theory Methods in Physics, Moscow, Nauka, 1983 (in Russian).

60. E.V. Shuryak, Nucl. Phys. B203, 93, 116, 140 (1982).

61. D.I. Dyakonov and V. Yu. Petrov, "Instanton-based vacuum from Feynman variational principle", Leningrad Preprint, LNPI-900 (1983).

62. R.P. Feynman, A.R. Hibbs, Quantum Mechanics and Path Integrals, McGraw-Hill, N.Y., 1965.

63. V.N. Gribov, ZhETF, 41, 667 (1961); Sov. Phys. JETP, 14, 478 (1962).

64. D. Amati, S. Fubini and A. Stanghellini, Nuovo Cim. 26, 896 (1962).

Chapter 2

HIGH ENERGY HADRON INTERACTIONS

2.1 General Picture of the Energy Dependence of Hadron Interactions

High energy accelerator experiments in the past several decades produced a rather large amount of data. There are now hundreds of known hadron processes observed at different energies. The experimental situation is regularly discussed at the International Conferences on High Energy Physics and the the International Symposia on Multiparticle Dynamics the materials of which present a detail information (see, e.g., Refs. 1-4).

The best known characteristics of hadron-collision processes at high energies are the total cross sections (σ_{tot}), the elastic cross sections (σ_{el}) and the cross sections of diffraction-dissociation processes (σ_{DD}). For multiparticle production processes well-studied quantities are the total inelastic cross sections $(\sigma_{inel} = \sigma_{tot} - \sigma_{el})$, average and partial multiplicities, inclusive cross sections and correlation functions. The total cross sections for the entire energy interval considered are given in Fig. 2.1. On the basis of the under-lying physics of rather different character the considered interaction processes as functions of the energy can be grouped under the following regions:

1) low energy region: $p_{lab} \lesssim (3\text{-}5)$ GeV/c,
2) intermediate energy region: 3-5 GeV/c $\lesssim p_{lab} \lesssim$ 30-80 GeV/c,
3) high energy region: $(3\text{-}8) \times 10$ GeV/c $\lesssim p_{lab} \lesssim (3\text{-}5) \times 10^3$ GeV/c,
4) region of superhigh energies: $p_{lab} \gtrsim (3\text{-}5) \times 10^3$ GeV/c.

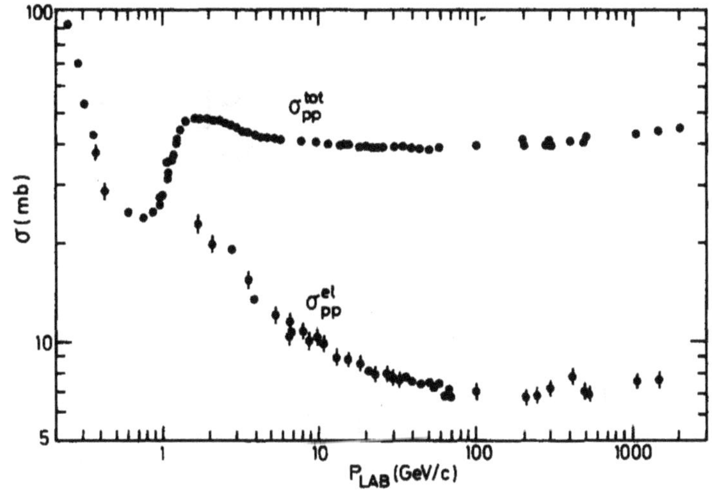

Fig. 2.1a Energy dependence of the total and elastic cross sections of pp interactions.

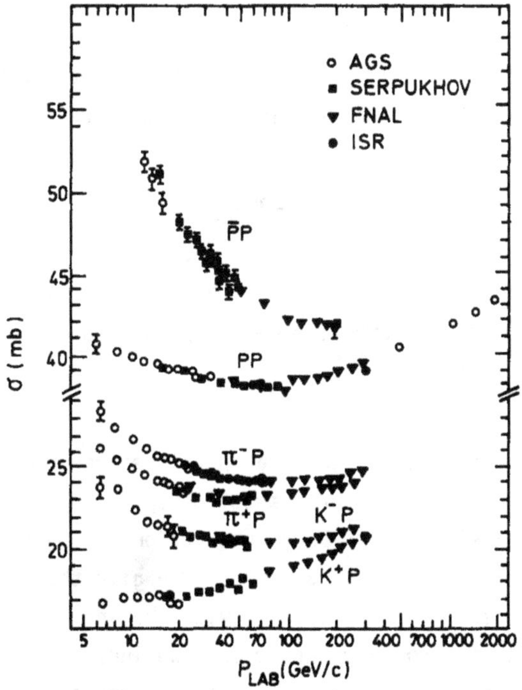

Fig. 2.1b Energy dependence of the total $\pi^{\pm}p$, $K^{\pm}p$, pp, and $p\bar{p}$ cross sections.

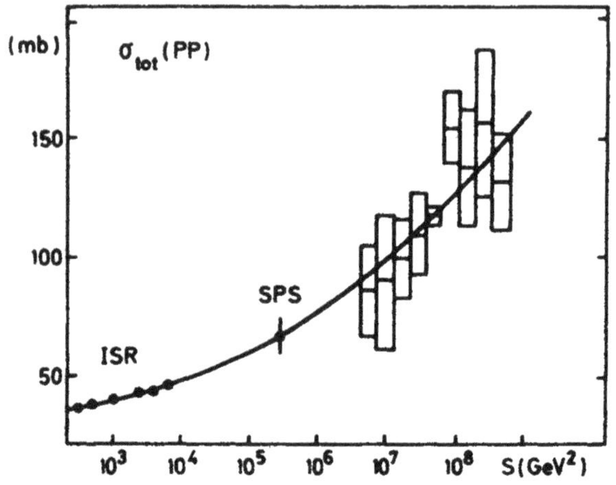

Fig. 2.1c Experimental data for $\sigma_{tot}(pp)$, extracted from the analysis of atmospheric showers, initiated by cosmic rays.

The region of relatively low energies (up to a few GeV in the c.m.s.) is dominated by resonance-formation processes (Fig. 2.2).

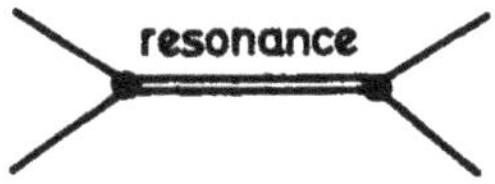

Fig. 2.2 Resonance formation — a typical process of low-energy collision.

Partial-wave analysis of the data in this energy region gives the main part of information for hadron spectroscopy. Approaching intermediate energies, inelastic two-body interactions being to play a role in the collision process (Fig. 2.3a, b): they give the main contributions to σ_{tot} up to $p \sim 10$-15 GeV/c. With the increase of the energy the multiplicities of secondary particles also increase. However, most of the observed particles are decay products of resonances like Δ, ρ, ω etc. The estimated average multiplicity of prompt particles (i.e.

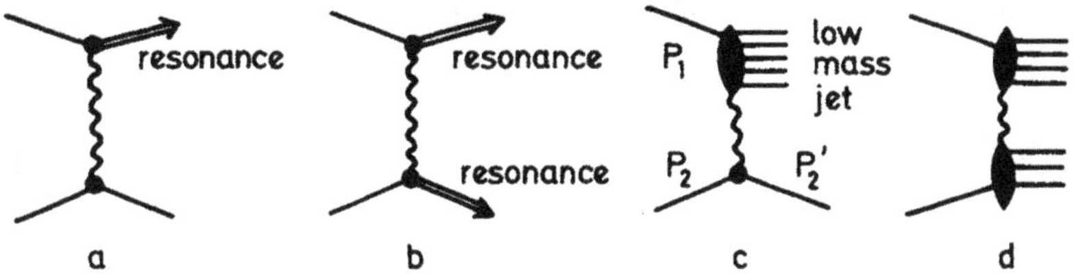

Fig. 2.3 Characteristic inelastic process in the intermediate energy region: two-body (a, b) and quasi-two-body (c, d) particle production. For the process (c) $s = (p_1 + p_2)^2$, $t = (p_2 - p_2')^2$, $M^2 = (p_1 + p_2 - p_2')^2$.

of those which are not from resonance decays) is about 3-4 at $p \simeq 50$ GeV/c.

The stretch covering the region 10 GeV/c $\lesssim p \lesssim$ 50 GeV/c was studied using various phenomenological models, based on the Regge-pole-exchange approach. A number of successful descriptions of the experimental data were obtained here. They revealed the general features of collision processes in this region: the cross sections of the binary (i.e. two-body and quasi-two-body (Figs. 2.3c, d)) reactions as functions of the momentum transferred squared t are of evidently diffractive nature.

When $p \gtrsim 50$ GeV/c the total cross sections and the elastic ones become nearly constant. The multiparticle production processes which begin to play a significant role here, are governed by the multiperipheral dynamics (see Section 2.2). The elastic cross sections are about 10-20 percent of the total cross sections. At sufficiently high energies (in pp-scattering this is $\sqrt{s} \gtrsim 20$ GeV) a diffractional minimum appears at $|t| \simeq 1.4$ GeV/c^2. In the region of small t (at $|t| \lesssim 0.1$ GeV^2/c^2) the elastic cross section is usually parametrized in the form

$$\frac{d\sigma}{dt} = \frac{\sigma_{tot}^2}{16\pi}(1+\rho^2)e^{b(s)t} \tag{2.1}$$

The proportionality to the total cross section squared is due to the connection between the imaginary part of the scattering amplitude and the total cross section (see Appendix B). The parameter ρ is determined by

$$\rho = \frac{\mathrm{Re}A(s,\,0)}{\mathrm{Im}A(s,\,0)} \quad . \tag{2.2}$$

At high energies ρ is small. In Section 6.4 some data for ρ values at different energies are presented together with the Regge-pole fit.

The slope $b(s)$ is usually parametrized in logarithmic form

$$b(s) = a_0 + 2\alpha' \ln s/s_0 \quad , \tag{2.3}$$

where $s_0 = 1\ \mathrm{GeV}^2$. With the increase of the energy the diffractional cone shrinks, i.e. the transverse dimension of the interaction region of hadrons increases. At high energies the shrinkage parameter is roughly the same for all elastic processes (this phenomenon was predicted in the Regge-pole theory[5]). The universal value of α', according to Ref. 6, is

$$\alpha' = 0.14 \pm 0.02\ (\mathrm{GeV}/c)^2 \quad . \tag{2.4}$$

The diffraction-dissociation processes involve two-body processes and quasi-two-body processes, as shown in Fig. 2.3. A single diffraction dissociation in defined by processes Figs. 2.3a, c, a double one by the processes of Figs. 2.3b, d. These processes are considered in Section 6.6, where the corresponding experimental data are presented.

The value of the cross section $\sigma_{DD}^{(1)}$ of the single diffraction dissociation in pp-scattering is 6-8 mb, i.e. about 15% of the total cross section. The double diffraction-dissociation cross section $\sigma_{DD}^{(2)}$ is not well known; existing estimates give here 2-10% of the total cross section. Note that the sum $\sigma_{el} + \sigma_{DD}$ gives about 35%.

In the high energy region the Pomeranchuk theorem[7], which states that the total cross sections of particles and antiparticles are equal, is approximately fulfilled. This can be especially well observed if one considers the differences between the total cross sections. From the point of view of the Regge-pole approach this means that at high energies the main contribution to the scattering amplitude is given by exchange states with vacuum-quantum numbers. In Section 2.3 the differences of the total cross sections are considered in the framework of Regge phenomenology; the corresponding data are also presented.

The region $p > (3\text{-}5) \times 10^3$ GeV/c is called, rather conditionally, the region of superhigh energies. Here the increase of $\sigma_{tot}(\bar{p}p)$ is observed. The character of this increase is not yet understood. Approximations like

$$\sigma_{tot} = \sigma_0 + \gamma \ln^n s/s_0 \quad , \tag{2.5}$$

where σ_0, γ, n and s_0 are phenomenological parameters, lead to small γ values ($\gamma/\sigma_0 \sim 1/40$) and to $1 \lesssim n \lesssim 2$ (see for instance Refs. 6, 8-10). Cosmic ray data (see Fig. 1.1c) agree with the maximum increase of the total cross section; however, these cosmic data are not reliable enough to allow the determination of n in Eq. (2.5) with certainty.

2.2 Multiparticle Production and Inclusive Processes

2.2.1 *Average multiplicity and limited transverse momenta*

For high energies $p_{lab} \gtrsim 50$ GeV, more than 80% of the total cross section is associated with the processes in which new particles are produced. These processes are called multiparticle production (or simply multiple production) processes.

As the beam energy increases, the average number of the produced secondaries per inelastic event $\langle n \rangle$ also increases. However, this growth is appreciably slower than the fastest allowed by the available phase space: in the latter case one would expect $\langle n \rangle \sim \sqrt{s}$, while experimental data in the available accelerator beam energy range $E_{lab} \sim 10\text{-}10^5$ GeV show that[a] $\langle n \rangle \sim \ln s$ or $\ln^2 s$. (The detailed

---

[a] Data at higher energies come from cosmic ray experiments; these data are less reliable. In the following we shall restrict our considerations to the available accelerator energies.

considerations of the total and average secondary multiplicities in hadron-hadron collisions are given in Section 7.4 where one may find relevant experimental data.) This means that the average value of energy (or momentum) carried by every secondary particle grows with the beam energy. The most fundamental property of the multiparticle production processes is that the momenta of secondaries grow only in the longitudinal direction, i.e. along the beam axis. The average value of the transverse components of the secondary momenta remains less than 0.7 GeV/c while the initial energy changes by several orders of magnitude (Fig. 2.4).

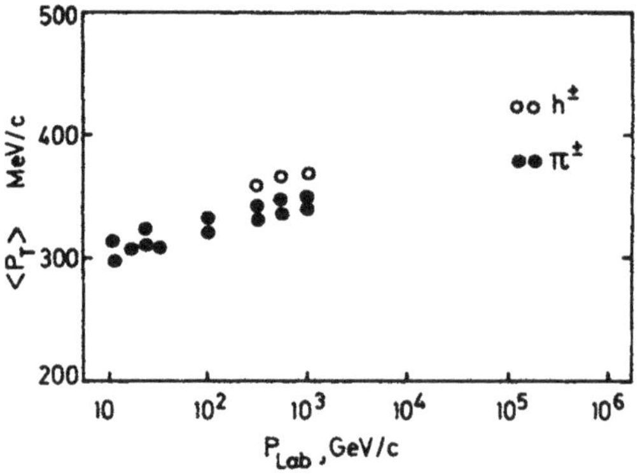

Fig. 2.4 Average transverse momentum $\langle p_T \rangle$ versus p_{lab} ($h^\pm = \pi^\pm + K^\pm + p^\pm$).

Let us consider the secondary particle distribution in the inelastic collision of hadrons a and b. In Fig. 2.5 this distribution is presented on the Peyrou plot[11]. The abscissa axis on this plane of secondary particle momentum is the longitudinal component of the momentum in c.m. frame p_L^* (hereafter all starred values refer to c.m. frame); the ordinate axis is the absolute value of the transverse momentum p_T. The physical region for a secondary particle is $p_L^{*2} + p_T^{*2} \le p_{max}^{*2}$ and $p_T > 0$. Every secondary particle gives a point on this plot.

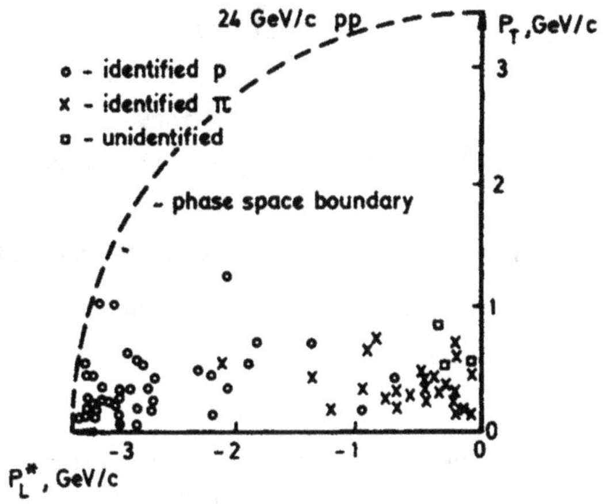

Fig. 2.5a Experimental distribution of the secondary protons (circles) and pions (crosses) in 24 GeV/c pp collision on the Peyrou plot[11]. Only the backward hemisphere is shown, since pp collision is symmetric in the c.m. frame.

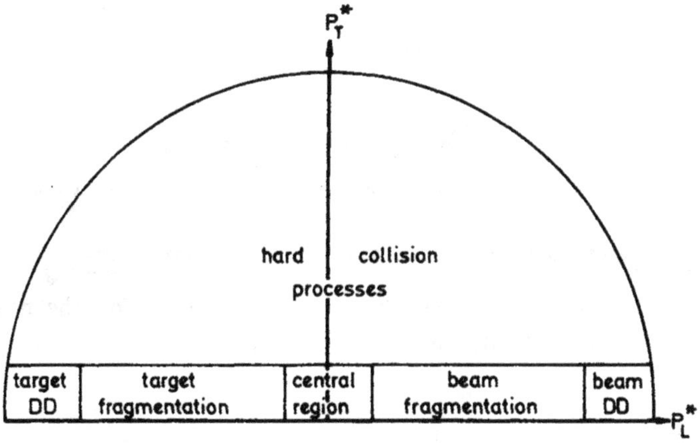

Fig. 2.5b Schematic definitions of different regions on the Peyrou plot for high-energy ab collision.

Figure 2.5a presents experimental distributions of the secondary pions and protons in pp collisions (only the backward hemisphere is shown, since pp collisions are backward-forward symmetric in c.m. frame). It is clearly seen that nearly all the secondaries are concentrated in a narrow strip $p_T^* \lesssim 1$ GeV/c. Figure 2.5b gives a schematic definition of the different regions in hadron-hadron collisions.

The region of small $|p_T^*| \lesssim 1$ GeV/c is the region of soft particle production, while particles with large p_T^* are produced in hard collisions.

For high initial energies most of the secondaries are concentrated in the region of the Peyrou plot where $|\vec{p}_L^*| << |\vec{p}_a^*|$. This region is called pionization (slow particles are mostly pions) or central region; we prefer the latter name and shall use it in the following. The regions of $|p_T^*| < 1$ GeV/c and $|p_L^*| \sim |p_a^*|$ are known as the fragmentation regions[12)] of the projectile and of the target. The particle with the largest $|p_L^*|$ value in these regions usually carries away about half of the initial momentum ($|\vec{p}_a^*|$ or $|\vec{p}_b^*|$ respectively); often it is the particle of the same sort as the projectile (target) particle. This property is known as leading particle effect; it was discovered long ago in cosmic ray experiments[13)]. In Fig. 2.5a it is clearly seen that secondary particles with large p_L^* are mainly protons, while those with small $|p_L^*|$ are mainly pions.

The regions of $|p_L^*| \simeq |\vec{p}_a^*|$ and $|p_L^*| \simeq |\vec{p}_b^*|$ are dominated by diffraction-dissociation processes of the beam or the target (or both).

2.2.2 *Kinematic variables for the secondary particles*

Different kinematic variables are convenient for the description of secondary particle distributions in different regions of the Peyrou plot.

Secondaries produced in the central region are responsible for the growth of the average multiplicity $\langle n \rangle$ with energy increase. At moderately high energies, $s \sim 10^2$-10^3 GeV^2, $\langle n \rangle \sim \ln s$. Thus the convenient variable for the central region is that for which the size of the physical region is also proportional to $\ln s$; one may think that secondary particles in the central region will be distributed approximately uniformly according to this variable. The latter is known as rapidity:

$$y = \frac{1}{2} \ln \frac{E + p_L}{E - p_L} = \ln (E + p_L)/m_T \quad , \tag{2.6}$$

where $m_T^2 = \vec{p}_T^{\,2} + m^2$. Rapidity transforms additively under Lorentz boosts with velocity β along the beam axis:

$$y' = y - \frac{1}{2} \ln \frac{1 + \beta}{1 - \beta} \quad . \tag{2.7}$$

In particular, for the transformation from the lab. frame (y) to the c.m. frame (y^*) $\beta = |\vec{p}_a|/(E_a + m_b)$ and at $|\vec{p}_a| >> m_a, m_b$, $y^* \simeq y - \frac{1}{2} \ln s/m_b^2$. Phase-space boundaries for rapidity variables at high energies are:

$$\ln m_T/m_b < y < \ln s/m_T m_b$$

$$- \frac{1}{2} \ln s/m_T^2 < y^* < \frac{1}{2} \ln s/m_T^2 \quad . \tag{2.8}$$

The size of the physical region for rapidity is a Lorentz invariant

$$y_{max} - y_{min} = y^*_{max} - y^*_{min} \simeq \ln s/m_T^2 \tag{2.9}$$

and grows logarithmically with the energy increase.

In some high energy experiments only angular distributions of the secondaries are measured, while energy and momentum are not measurable at all (e.g. nuclear emulsion experiments). In these cases y cannot be determined; a useful variable for the central region is pseudorapidity:

$$\eta^* = - \ln \mathrm{tg}(\theta^*/2) \quad , \tag{2.10}$$

where θ^* is the angle between the momentum of the secondary particle

$\vec{p}^*$ in c.m. frame and the beam axis:

$$\mathrm{tg}\,\theta^* = |\vec{p}_T^*|/p_L^* \quad . \tag{2.11}$$

Let us obtain the relation between η^* and y^*. Since $0 \leq \theta^* \leq \pi$, we may write

$$\mathrm{tg}\,\theta^*/2 = |\vec{p}_T|/(|\vec{p}^*| + p_L^*) \tag{2.12}$$

so

$$\eta^* = -\ln \mathrm{tg}\,\theta^*/2 = \frac{1}{2} \ln \frac{|\vec{p}^*| + p_L^*}{|\vec{p}^*| - p_L^*} \quad . \tag{2.13}$$

Comparing (2.13) and (2.6) we see that if $|\vec{p}^*| \gg m$, $\eta^* \simeq y^*$, i.e. for ultrarelativistic particles rapidity and pseudorapidity approximately coincide.

Fragmentational particles carry away a finite fraction of the maximal possible longitudinal momentum $p_{L_{max}}$. Thus the convenient variable for these particles is Feynman's scaling variable

$$x = p_L/p_{L_{max}} \quad . \tag{2.14}$$

It is easy to connect x and y, since $p_L = m_T \mathrm{sh}\, y$, $E = m_T \mathrm{ch}\, y$. In the laboratory frame at high energies $p_{L_{max}} \simeq |\vec{p}_a|$, while in c.m.s. $|p^*_{L_{max}}| \simeq \sqrt{s}/2$ and thus

$$x = \frac{m_T \mathrm{sh}\, y}{|\vec{p}_a|}$$

$$x^* = \frac{2m_T \mathrm{sh}\, y^*}{\sqrt{s}} \quad . \tag{2.15}$$

In the beam-fragmentation region, when $p_L^* \gg m_T$, the variables x and

x^* approximately coincide:

$$x \simeq x^* \left(1 + \frac{m_T^2}{s \cdot x^{*2}}\right) \quad . \tag{2.16}$$

The invariant phase-space element may be expressed in terms of the variables $y(y^*)$, η^* and x (or x^*):

$$\frac{d^3p}{2p_0(2\pi)^3} = \frac{dy\, dp_T^2}{2(2\pi)^3} = \frac{p_T \mathrm{ch}\, \eta^*}{\sqrt{m^2 + \vec{p}_T^{\,2} \mathrm{ch}^2 \eta^*}} \cdot \frac{d\eta^* dp_T^2}{2(2\pi)^3} = \frac{\sqrt{s}}{p_0} \cdot \frac{dx^* dp_T^2}{(2\pi)^3} \simeq \frac{dx^* dp_T^2}{2x^*(2\pi)^3}$$

(2.17)

2.2.3 *Secondary particle distributions in the central and the fragmentational regions*

Inclusive cross section for particle production (see Appendix B, Eqs. (B.11), (B.12)) may be expressed in terms of rapidity or pseudo-rapidity variables:

$$\frac{d^3\sigma}{d^3p^*}(a + b \to h + X) = \frac{1}{\pi} \frac{d\sigma}{dy^* d^2p_T} = \sqrt{1 + \frac{m^2}{p_T^2 \cdot \mathrm{ch}^2\eta^*}} \frac{d\sigma}{d\eta^* d^2p_T}$$

(2.18)

As was already mentioned, in the energy region where $\langle n \rangle \sim \ln s$ one may assume that the distributions $d\sigma/dy^*$ in the central region will be approximately constant. Experimental data show that $d\sigma/dy^*$ is approximately constant indeed; however, the variations of $d\sigma/dy^*\big|_{y^*=0}$ with the energy increase are quite noticeable. For example, when $\sqrt{s}$ grows from the ISR region $\sqrt{s} \simeq 60$ GeV to SPS region $\sqrt{s} = 540$ GeV, the value of $d\sigma/d\eta^*\big|_{\eta^*=0}$ increases by the factor of 1.5.

For high energies in the fragmentation region the invariant cross section

$$p_0 \frac{d\sigma}{d^3p} \simeq \frac{x}{\pi} \frac{d\sigma}{dx dp_T^2} \tag{2.19}$$

is approximately independent of s, i.e. it is function of x and

p_T^2 only when $x \gg 2m_T/\sqrt{s}$ (Feynman scaling law[14]). This scaling is observed for some secondary particles (e.g. K_S^0 or p in pp collisions) starting from $s \sim 10^2$ GeV; for some other secondaries (such as K^- or $\bar{p}$ in pp collisions) it is significantly broken. It seems that scaling in the fragmentation region is best fulfilled for those secondaries, which may be considered as fragments of the projectile (target) hadron in the framework of the quark model (see Chapter 7 for details). This model predicts also scale breaking at superhigh energies (Section 7.6); however at the SPS energy $\sqrt{s}$ = 540 GeV the predicted breaking is rather small, about 10-15%.

Experimental data for similar inclusive reactions are often presented in the original papers as functions of different considered variables. Some characteristic properties of the measured values may be hidden as a result of unappropriate choice of the independent variable (or, conversely, proper choice of the variable may emphasize some important features of the data). In order to illustrate this, let us consider an example of artificial "experimental" distributions, presented in Fig. 2.6. We assume that secondary pions and protons in pp collisions at $s = 10^3$ GeV^2 are produced with the following cross sections:

$$p_0^* \cdot \frac{d\sigma}{dp_L^* dp_T^2}(pp \to hX) = f(p_L^*, p_T)\exp(-7p_T^2) \quad , \tag{2.20}$$

where $h = p$ or π and

$$f_h(p_L, p_T) = \begin{cases} C_h, & |p_L^*| \le p_{L_{max}}^*/2 \\ 2C_h(1 - p_L^*/p_{L_{max}}^*), & p_L^* \ge p_{L_{max}}^*/2 \end{cases} \quad . \tag{2.21}$$

In Eq. (2.21) the only p_T-dependent factor is $p_{L_{max}}^*$; for simplicity we assume $C_\pi = C_p$. In Figs. 2.6a, e the distributions $2p_0^*/\sqrt{s} \cdot d\sigma/dx^* \simeq x^* \cdot d\sigma/dx^*$ are presented; they have similar x-dependences. The same "experimental data" are presented in Figs. 2.6b, f as distributions

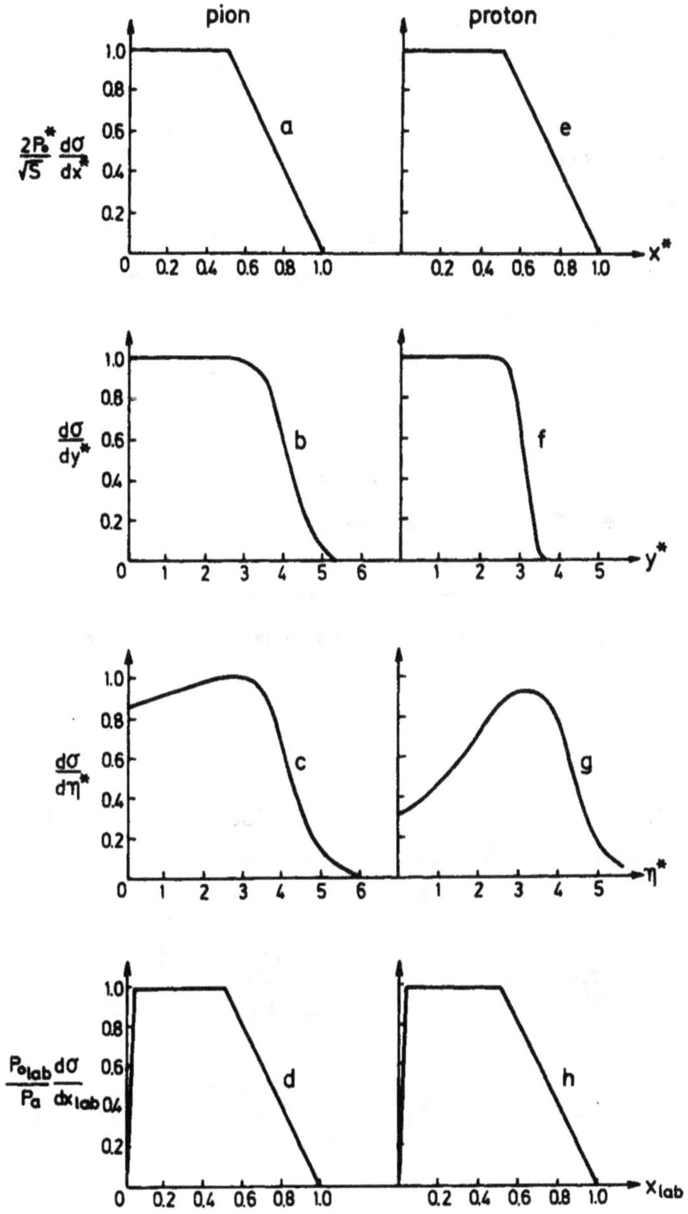

Fig. 2.6 Artificial distributions of secondary pions (a-d) and protons (e-h) which present the same "experimental data" as functions of different variables: x^*(a, e), y^*(b, f), η^*(c, g), and x_{lab}(d, h). The "experimental data" are determined by the Eqs. (2.20), (2.21).

$d\sigma/dy^*$, which appear to be approximately constant for $y^* \lesssim 3.5$ for pions and in a somewhat narrower region $y^* \lesssim 3$ for protons. The distributions $d\sigma/d\eta^*$ (Figs. 2.5c, g) have a characteristic dip at small η^* values, which is due to the factor $\sqrt{1 + m_T^2/(p_T^2 \mathrm{ch}^2 \eta^*)}$ in (2.18); this dip is deeper for the heavier particle (proton). As a result, the $d\sigma/d\eta^*$ distribution for protons has no central plateau at all. The distributions $(p_{0_{lab}}/p_a)(d\sigma/dx_{lab}) \approx x_{lab}(d\sigma/dx_{lab})$ in the laboratory frame are shown in Figs. 2.5d, h. They practically coincide with $x^* d\sigma/dx^*$ everywhere except the region of $x_{lab} \lesssim 0.02$.

2.2.4 *Multiperipheral collisions*

The considered important properties of the multiparticle production are inherent features of the multiperipheral collision picture. Let us consider the process of N-particle production in a high energy ab collision (Fig. 2.7a) in the rest frame of the b particle. All the secondaries may be enumerated in decreasing order of their longitudinal momenta:

$$p_{1L} \geq p_{2L} \geq \cdots \geq p_{NL} \quad . \tag{2.22}$$

The ladder obtained in Fig. 2.7a is called multiperipheral if all squares of the momentum transferred along the ladder (t_i) and squares of the adjacent invariant masses $s_{i,i+1}$ are limited at $s \to \infty$:

$$t_i = (p_a - \sum_{k=1}^{i} p_k)^2 = \left(1 - \sum_{k=1}^{i} x_i\right)\left(m_a^2 - \sum_{j=1}^{i} \frac{m_{jT}^2}{x_j}\right) - \left(\sum_{j=1}^{i} \vec{p}_{jT}\right)^2$$

$$s_{i,i+1} = (p_i + p_{i+1})^2 = (x_i + x_{i+1})\left(\frac{m_{iT}^2}{x_i} + \frac{m_{(i+1)T}^2}{x_{i+1}}\right) - (\vec{p}_{iT} + \vec{p}_{(i+1)T})^2$$

(2.23)

The fast decrease of the cross section with the transferred momentum growth is a well-known property of binary reactions at high energies (e.g. elastic scattering). It is usually connected with the peripheral

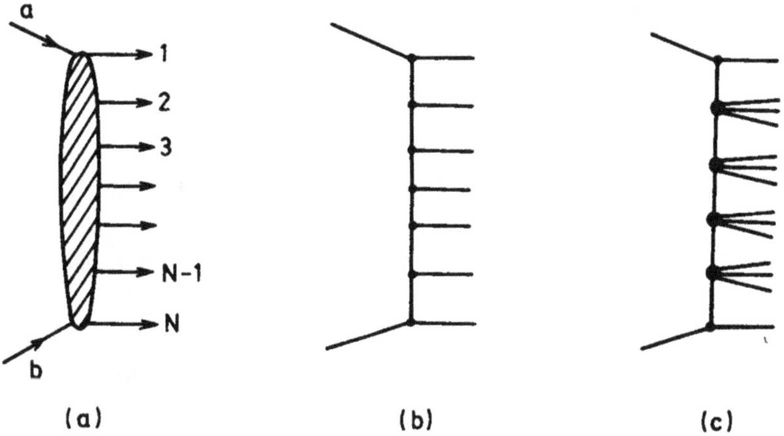

Fig. 2.7 (a) Multiperipheral N-particle production process.
(b, c) Different types of multiperipheral diagrams: cut-ladder diagram (b); diagram with cluster production (c).

nature of the interaction, governed by t-channel light-state exchanges. Similarly, a multiperipheral production process may be considered as particle production via t-channel light-state exchanges.

The boundedness of t_i and s_i means that average secondary transverse momenta $\langle p_{iT} \rangle$ are limited by some characteristic mass of t-channel exchanged states:

$$\langle p_{iT}^2 \rangle \lesssim \mu^2 \quad . \tag{2.24}$$

Besides, the ratios of the longitudinal momenta of the adjacent particles in the ladder are of the same order of magnitude:

$$x_1 \sim x_2/x_1 \sim x_3/x_2 \; \ldots \sim x_N/x_{N-1} \quad . \tag{2.25}$$

In other words, rapidity gaps between the adjacent particles are approximately equal to each other:

$$y_1 - y_2 \simeq y_2 - y_3 \simeq \dots \simeq y_{n-1} - y_N \quad .$$

There exist a lot of multiperipheral models; all of them obey the general properties considered but differ in their details of the production mechanism. The simplest versions[15,16] are based on the cut-ladder diagrams of the Fig. 2.5b type, while more complicated schemes include particle-cluster production (Fig. 2.5c). Various comparisons of multiperipheral models with the experimental data, as well as more detailed considerations of the production dynamics may be found in Refs. 17-19 and references given therein.

At high energies multiperipheral production processes give the main contribution to the imaginary part of the elastic scattering amplitude (see Appendix B, Eq. (B.20)). This contribution has the same asymptotic energy behaviour as the amplitude of Regge-pole exchange (see the next Section).

2.3 Regge-Pole Phenomenology

Regge-pole exchange theory gives a useful tool to the description of many quantitative features of high energy hadronic collisions. Since its first introduction into particle physics in the early sixties[5], a great number of review papers and books have appeared in this field and we refer the readers to them for the full list of references as well as for more details on the subject (see e.g. Refs. 20 to 25). The main aim of this section is only to give a guide for the basic results of Regge phenomenology; it will probably help in the understanding of some of the foregoing topics.

2.3.1 *Binary reactions*

Regge-pole expansion of the binary amplitude $12 \to 34$ (see Fig. 2.8a) assumes a power behaviour for the latter as a function of the energy variable $s = (p_1 + p_2)^2$ for fixed small-$|t|$ values. In the complex s-plane this amplitude has two cuts — the left and the right one

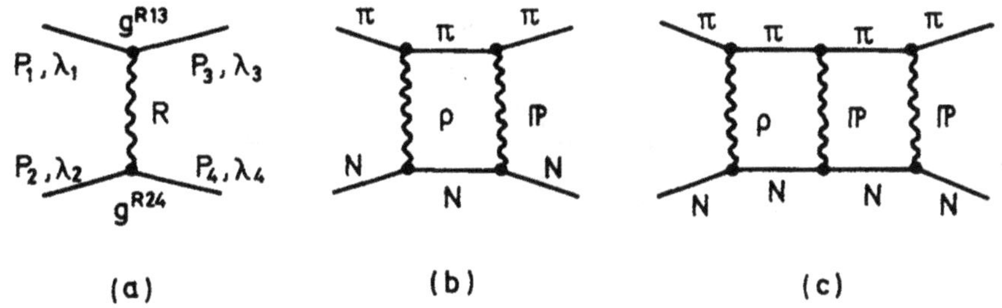

Fig. 2.8 Regge-pole theory diagrams:
(a) single pole exchange in the binary process $1+2 \to 3+4$,
(b) double $\rho\mathbb{P}$, (c) triple $\rho\mathbb{P}\mathbb{P}$ pole exchanges in πN elastic scattering which may be effectively accounted by the renormalization of ρ-pole parameters.

(see Fig. 2.9). The physical region of the direct process $1+2 \to 3+4$ is on the upper lip of the right cut (s-channel), while that of crossed process $1+\bar{4} \to \bar{2}+3$ is on the lower lip of the left cut (u-channel). The crossing symmetric amplitude may be considered as a function of the crossing symmetric energy variable, which is usually chosen as $\nu = (s-u)/2$. Regge parametrization of the amplitude is used at high energies and small momentum transferred. Since $s+t+u = \sum_{i=1}^{4} m_i^2$ for $s \to \infty$ and small $|t|$ we may neglect t and particle masses. Then in the direct channel $\nu \simeq s$, while in the crossed channel $\nu \simeq -u$.

The contribution of a single t-channel Regge pole R to an s-channel helicity amplitude of the process in Fig. 2.8a is given by:

$$A^R_{\lambda_1\lambda_3\lambda_2\lambda_4}(\nu, t) = R(\nu, t)\tau^R_{\lambda_1\lambda_3\lambda_2\lambda_4}(t)$$

$$R(\nu, t) = \frac{1 + \xi_R \exp[-i\pi\alpha_R(t)]}{\sin[\pi\alpha_R(t)]} \cdot \left(\frac{\nu}{\nu_0}\right)^{\alpha_R(t)} \qquad (2.26)$$

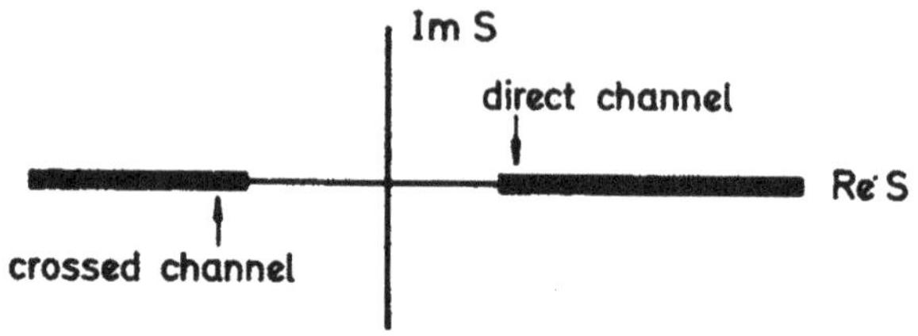

Fig. 2.9 Physical regions of the direct $1+2\to 3+4$ and crossed $1+\bar{4}\to\bar{2}+3$ channels of the reaction.

Here λ_i are the external particle helicities, $\xi_R = \pm 1$ is the Reggeon signature, $\tau^R_{\lambda_1\lambda_3\lambda_2\lambda_4}$ the residue function, $\alpha_R(t)$ the pole trajectory and ν_o is a constant scale factor. The representation (2.26) is assumed to be valid in the asymptotic limit $\nu\to\infty$, t fixed up to leading order in $1/\nu$. Let us consider the most essential properties of the Reggeon exchange amplitude.

i. All energy dependence of $A(\nu, t)$ comes purely from the factor $(\nu/\nu_o)^{\alpha_R(t)}$.

ii. $A(\nu, t)$ is an analytic function of ν, its phase is uniquely determined. $A^R(\nu, t)$ is even (odd) under the crossing transformation $s \leftrightarrows u (\nu \leftrightarrows -\nu)$ if the signature $\xi_R = +1$ (-1).

iii. The most significant property of the residue function $\tau^R_{\lambda_1\lambda_3\lambda_2\lambda_4}$ is the factorization property[26,27]: $\tau^R_{\lambda_1\lambda_3\lambda_2\lambda_4}(t) = g^R_{\lambda_1\lambda_3}(t)\cdot g^R_{\lambda_2\lambda_4}(t)$.

In the elastic scattering process the vertex functions due to angular momentum conservation must behave for $t\to\infty$ as $g^R_{\lambda_1\lambda_3}(t) \sim (-t)^{|\lambda_3-\lambda_1|/2}$. The presence of the factorization provides an analogy between Reggeon (like Fig. 2.8) and Feynman diagrams. $R(\nu, t)$ plays the role of the Reggeon propagator, while residue functions stand for the usual vertex couplings. This analogy is still deeper because $R(\nu, t)$ has poles at t values where $\alpha_R(t)$ converts into an integer number. Indeed, for $\xi_R = +1$ near the point t_{2j}, where

$$\alpha_R(t_{2j}) = 2j \tag{2.27}$$

the signature factor is

$$\frac{1 + \exp[-i\pi\alpha_R(t)]}{\sin[\pi\alpha_R(t)]} \simeq \frac{2}{\pi\alpha'(t_{2j})(t - t_{2j})} \tag{2.28}$$

where $\alpha'(t)$ is the slope of the Reggeon trajectory: $\alpha' = d\alpha/dt$. Similarly $R(\nu, t)$ has a pole for $\xi_R = -1$ and $\alpha_R(t_{2j+1}) = 2j + 1$. Thus we may say that a signature-even Reggeon corresponds to the t-channel exchange of particles with even angular momentum (spin) while a signature-odd Reggeon with that of odd spin.

A remarkable property of the Regge trajectories is that any of the well established hadrons lies on some trajectory at $t > 0$.

We shall begin with meson trajectories, which are studied in more detail; the list of the leading trajectories is given in Table 2.1. Usually the notation of the trajectory is the name of the lightest particle on it (e.g. ρ-trajectory, etc.). The only exception is the Pomeranchuk (alias pomeron or vacuum) trajectory **P**; sometimes the f-trajectory is referred as **P**'.

The experimental data show that all meson trajectories are approximately linear in t in a rather broad t interval $|t| \lesssim 2\ (\mathrm{GeV}/c)^2$:

$$\alpha_R(t) = \alpha_R(0) + \alpha'_R \cdot t \quad , \tag{2.29}$$

where $\alpha_R(0)$ (intercept) and α'_R (slope) are some real numbers. From the Table 2.1 we can see that all meson trajectories, except the Pomeron one, have approximately equal slopes $\alpha' \simeq 0.8$; the slope $\alpha'_{\mathbf{P}}$ is prominently less. The mass of the lightest particle that could lie on the Pomeron trajectory may be estimated with the help of $\alpha'_{\mathbf{P}}$; its

Table 2.1 Meson trajectories and their quantum numbers

$\alpha(t)$	Trajectory Name	Particles on the trajectory	Quantum numbers ξ_R	I	$P_R\xi_R$(naturality)
$1+0.15t$	$\mathbb{P}$(Pomeron)		+1	0	+1
$0.5+0.85t$	f or $\mathbb{P}'$	f(1270), h(2040)	+1	0	+1
	ω	ω(783), ω(1670)	-1	0	+1
	ρ	ρ(770), g(1690)	-1	1	+1
	A_2	A_2(1320), δ(2030)	+1	1	+1
$0.15+0.8t$	Φ	Φ(1020)	-1	0	+1
	f'	f'(1515)	+1	0	+1
$0+0.7t$	π	π, A_3(1680)	+1	1	-1
	B	B(1235)	-1	1	-1
	A_1	A_1(1270), π(2050)	-1	1	-1
$0.3+0.85t$	K^*	K^*(892), K^*(1780)	-1	1/2	+1
	K^{**}	K^*(1430), K^*(2060)	+1	1/2	+1
$-0.25+0.85t$	K	K, L(1580)	+1	1/2	-1
	Q_1	Q_1(1280)	-1	1/2	-1
	Q_2	Q_2(1400)	-1	1/2	-1

quantum numbers should be $I^G = 0^+$, $J^P = 2^+$ and the mass about 2.7 GeV. Such a heavy hadron may be probably associated with some glueball state (see Introduction); then the Pomeron is mainly a gluonic object.

A prominent feature of the meson trajectories is the approximate equality of the trajectories with opposite signature (e.g. $\rho - A_2$, $f - \omega$, $\pi - A_1$ etc.). The analysis of the experimental data shows that not only the trajectories, but also the couplings are roughly independent of the signature. This property is called exchange degeneracy and may be easily understood in terms of the quark model as absence (or smallness) of the exchange forces in the qq (crossed) channel as compared with the direct $q\bar{q}$ channel.

The vertex functions $g^R_{\lambda_1\lambda_3}(t)$ are usually assumed to be smooth functions of t apart from the trivial kinematic factor. The simplest parametrization, compatible with angular momentum conservation is

$$g^R_{\lambda_1\lambda_3}(t) = \beta^{R13}_{\lambda_1\lambda_3}(t) \cdot (-t'/4M^2)^{|\lambda_1 - \lambda_3|/2} \tag{2.30}$$

In (2.30), $t' = t - t_{min}$, M is some scale factor (like nucleon mass). Strictly speaking, the vertex function should only be a function of t, not t' (t' depends on s). However, parametrization (2.30) is often used in practical calculations. It is probably equivalent to some effective account of some additional j-plane singularities like daughter trajectories. $\beta(t)$ is usually parametrized as an exponent in t or t' and may be interpreted as particle-Reggeon form factor. Then $\beta \equiv \beta(0)$ (or $\beta(t_{min})$) is a vertex coupling constant, which characterizes the relative strength of particle-Reggeon coupling just like the electric charge characterizes the strength of the particle-photon coupling. The list of useful couplings is given in Table 2.2

Strong interactions are assumed to obey approximate SU(3) symmetry. This means that the couplings of different Reggeons and particles from the same SU(3) multiplets are related to each other through SU(3) Clebsh-Gordan coefficients. For example, $\beta^{\rho_0 K^+K^+} = -\frac{1}{2}\beta^{\rho^0\pi^-\pi^-} =$

Table 2.2 The basic coupling "strengths" $\beta^{R13}_{\lambda_1\lambda_3}(0)$ for the leading meson trajectories[24].

Exchange	Coupling	Value	Exchange	Coupling	Value
	$\beta^{\rho\pi^0\pi^-}_{00}$	8.0		$\beta^{\pi\rho^0\pi}_{00}$	4.4
	$\beta^{\rho\omega\pi^-}_{10}$	13.0		$\beta^{\pi\rho^0\pi^-}_{10}$	15.18
	$\beta^{\rho np}_{1/2\ 1/2}$	2.3		$\beta^{\pi np}_{1/2\ -1/2}$	35.7
ρ	$\beta^{\rho np}_{1/2\ -1/2}$	18.4	π	$\beta^{\pi np}_{1/2\ 1/2}$	0
	$\beta^{\rho\Delta^{++}p}_{1/2\ -1/2}$	12.5		$\beta^{\pi\Delta^{++}p}_{1/2\ 1/2}$	-5.3
	$\beta^{\rho\Delta^{++}p}_{3/2\ 1/2}$	$\sqrt{3}\cdot 12.5$		$\beta^{\pi\Delta^{++}p}_{1/2\ -1/2}$	22.6
	$\beta^{\rho\Delta^{++}p}_{1/2\ 1/2}$	0		$\beta^{\pi\Delta^{++}p}_{3/2\ 1/2}$	31.3
	$\beta^{\rho\Delta^{++}p}_{3/2\ -1/2}$	0		$\beta^{\pi\Delta^{++}p}_{3/2\ -1/2}$	-27.1
	$\beta^{\omega B^+\pi^+}_{00}$	3.0		$\beta^{A_1\rho^0\pi}_{00}$	6.76
ω	$\beta^{\omega B^+\pi^+}_{10}$	6.6		$\beta^{A_1\rho\pi}_{10}$	5.0
			A_1	$\beta^{A_1 np}_{1/2\ 1/2}$	5.0
				$\beta^{A_1 np}_{1/2\ -1/2}$	0

$\frac{1}{\sqrt{2}}\beta^{\bar{K}^{*0}K^{+}\pi^{+}}$ etc. The exchange degeneracy and SU(3) symmetry for the residues reduce the number of independent couplings to a few number of the "prototypes". The list of the prototype couplings is given in Table 2.3.

Up to now we considered only meson Regge trajectories. Baryon trajectories and their contributions are studied in much less detail. The established nonstrange baryons may be classified into a nucleon trajectory $\alpha_N(t) \simeq -0.3 + 0.9t$ ($N_{1/2^+}(939)$, $N_{5/2^+}(1680)$, $N_{3/2^+}(2220)$) and a Δ trajectory $\alpha_\Delta(t) \simeq 0.0 + 0.9t$ ($\Delta_{3/2^+}(1232)$, $\Delta_{7/2^+}(1950)$, $\Delta_{11/2^+}(2420)$). Λ and Σ baryons also lie on the appropriate Regge trajectories. The experimental evidence for baryon exchange comes from meson-baryon backward scattering and flavour-exchange reactions; the corresponding cross sections are very small and it is difficult to study them with the proper accuracy. In addition, from the point of view of the quark model baryon exchange is forbidden in the impulse approximation (see Chapter 6).

Helicity amplitudes used in the representation (2.26) are very convenient for the explicit calculations of the cross sections, density matrices etc. However, sometimes usual invariant amplitudes[28,29] are preferable (e.g. when absorptive corrections or some other rescattering effects are calculated). Invariant amplitudes for higher spins should be written with some caution, taking into account that all equations are valid in the leading order of s only. All difficulties can be avoided by constructing invariants in terms of the transverse components of particle momenta and polarization vectors. However, in the simplest case of spin 1/2 particles this is not relevant; here the vertex function may be written as

$$g^{R13}_{\lambda_1\lambda_3}(t) = \bar{u}^{\lambda_3}(p_3)\hat{C}_\alpha(t)u^{\lambda_1}(p_1) \tag{2.31}$$

assuming that only leading-order-in-s terms are accounted for. The valence subscript α of the covariant vertex function $\hat{C}_\alpha(t)$ should

Table 2.3 "Prototype" couplings for binary reactions

Process 12 → 34 $J^P(1), J^P(2) \to J^P(3), J^P(4)$	Prototype coupling R(13) or R(24)	
	Natural parity exchange	Unnatural parity exchange
$0^- \; 0^- \to 0^- \; 0^-$	$\rho(\pi\pi)$	-
$\frac{1}{2}^+ \frac{1}{2}^+ \to \frac{1}{2}^+ \frac{1}{2}^+$	$\rho(NN)$	$\pi(NN)$
$0^- \frac{1}{2}^+ \to 0^- \frac{1}{2}^+$	$\rho(\pi\pi)$, $\rho(NN)$	-
$0^- \frac{1}{2}^+ \to 0^- \frac{3}{2}^+$	$\rho(\pi\pi)$, $\rho(\Delta N)$	-
$0^- \frac{1}{2}^+ \to 1^- \frac{1}{2}^+$	$\rho(\omega\pi)$, $\rho(NN)$	$\pi(\rho\pi)$, $\pi(NN)$
$1^- \frac{1}{2}^+ \to 0^- \frac{1}{2}^+$		$A_1(\rho\pi)$, $A_1(NN)$
$0^- \frac{1}{2}^+ \to 1^- \frac{3}{2}^+$	$\rho(\omega\pi)$, $\rho(\Delta N)$	$\pi(\rho\pi)$, $\pi(\Delta N)$
$1^- \frac{1}{2}^+ \to 0^- \frac{3}{2}^+$		
$\frac{1}{2}^+ \frac{1}{2}^+ \to \frac{1}{2}^+ \frac{3}{2}^+$	$\rho(NN)$, $\rho(\Delta N)$	$\pi(NN)$, $\pi(\Delta N)$
$\frac{1}{2}^+ \frac{1}{2}^+ \to \frac{3}{2}^+ \frac{3}{2}^+$	$\rho(\Delta N)$	$\pi(\Delta N)$
$0^- \frac{1}{2}^+ \to 1^+ \frac{1}{2}^+$	$\omega(B\pi)$, $\rho(NN)$	-
$0^- \frac{1}{2}^+ \to 1^+ \frac{3}{2}^+$	$\omega(B\pi)$, $\rho(\Delta N)$	-

be coupled with the covariant part of the Reggeon propagator, which coincides with the customary covariant part of the propagator in the leading order in s. Using the explicit form of the wave functions $u^\lambda(p)$ for fermions with a definite helicity value λ (see Appendix A) one may easily connect the covariant vertex function with that for helicity amplitude. Thus, for natural parity exchange $(P_R = \xi_R)$

$$\hat{C}_\alpha(t) = g_1(t)(p_1 + p_3)_\alpha \hat{I} + g_2(t)\gamma_{\dot{\alpha}} \quad , \tag{2.32}$$

while for unnatural parity exchange

$$\hat{C}_\alpha(t) = f_1(t)\gamma_5(p_1 + p_3) + f_2(t)\gamma_5\gamma_\alpha \quad . \tag{2.33}$$

Here invariant couplings g_i and f_i may be expressed in terms of helicity couplings

$$\begin{aligned} \beta_{1/2\ 1/2} &= 2\left(\frac{m_1 + m_3}{2} \cdot g_1 + g_2\right) \\ \beta_{1/2\ -1/2} &= 1/2 \cdot g_1 \end{aligned} \tag{2.34}$$

(natural parity exchange) and

$$\begin{aligned} \beta_{1/2\ 1/2} &= 2i\left(-\frac{m_1 + m_3}{2} \cdot f_1 + f_2\right) \\ \beta_{-1/2\ 1/2} &= -1/2 \cdot f_1 \end{aligned} \tag{2.35}$$

(unnatural parity exchange). The additional constraints of C-invariance may reduce the number of independent couplings: thus, for A_1-exchange $f_1 = 0$, while for π-exchange $\beta_{1/2\ 1/2} = 0$. The detailed consideration of the procedure of covariant reggeization may be found in Refs. 30, 31, where all necessary expressions are given for higher spin particles.
In the phenomenological analysis of the total cross sections it is usually assumed, that only single Regge-pole exchanges contribute, while multiple exchanges are negligible. In this case the amplitudes

of the different elastic scatterings may be symbolically written as

$$\begin{aligned}
A(\pi^+p) &= \mathbf{P} + f + \rho \\
A(\pi^-p) &= \mathbf{P} + f - \rho \\
A(K^+p) &= \mathbf{P} + f + \omega + A_2 + \rho \\
A(K^-p) &= \mathbf{P} + f - \omega + A_2 - \rho \\
A(pp) &= \mathbf{P} + f + \omega + A_2 + \rho \\
A(\bar{p}p) &= \mathbf{P} + f - \omega + A_2 - \rho
\end{aligned} \tag{2.36}$$

where only leading trajectories are taken into account. Here the Reggeon symbol stands for the corresponding contribution of the type (2.36). In the linear combinations of the total cross sections the contributions of some trajectories cancel, and we may isolate the contributions of a single Reggeon, e.g.

$$\begin{aligned}
\Delta\sigma(\pi^{\pm}p) &= \sigma_{tot}(\pi^+p;\ s) - \sigma_{tot}(\pi^-p;\ s) \\
&= 2\beta^{\rho\pi\pi}\beta^{\rho NN}\left(\frac{s}{s_0}\right)^{\alpha_\rho(0) - 1}
\end{aligned} \tag{2.37}$$

The parameters of Reggeon intercepts $\alpha_R(0)$ and couplings β^R for the leading non-vacuum trajectories (ρ, ω, A_2, etc.) are usually obtained via the fit of cross section differences by expressions like (2.37); such fits are shown in Fig. 2.10 by the solid lines.

In fact the relation (2.37) determines the contribution of the "effective" ρ-Reggeon, in which the ρ-pole exchange is only a Born term, the next terms being $\rho\mathbf{P}$, $\rho\mathbf{P}\,\mathbf{P}$, etc. exchanges (see Figs. 2.8b, c). The multiple exchanges correspond to the cuts in the complex J-plane; their contributions to the scattering amplitude behave at large s as $s^{\alpha}\ln^{-n}s$.

In most cases accounting for Regge cuts is not necessary from the phenomenological point of view. However, there are some situations

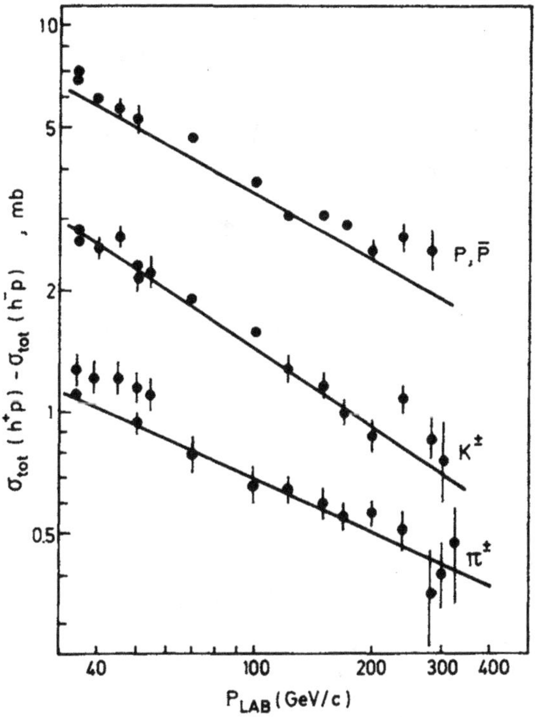

Fig. 2.10 The differences of the total cross sections of $\pi^{\pm}p$, $K^{\pm}p$, and $p^{\pm}p$ collisions. The curves correspond to the single pole exchange fit according to the Eqs. (2.36), (2.37).

where cut contributions are dominating or at least significant. For example, the differential cross section of pn charge-exchange reaction $pn \to np$ vanishes like t (or faster) as $t \to 0$ if only Regge poles contribute, while the experimental data are peaked forward. Another famous example is the polarization in πN charge-exchange collisions: it would be identically zero, if the ρ-pole dominated both helicity amplitudes.

The contributions of the Regge cuts $\mathbf{P\,P}$, $\mathbf{P\,P\,P}$ *etc. lead to* the preasymptotic growth of the total cross section even if $\alpha_{\mathbf{P}}(0) = 1$. However, all attempts to relate this growth with the observed behaviour of the cross section seem to be unsuccessful: the experimental growth is significantly faster. Some recent models describe the behaviour of the total cross section assuming $\alpha_{\mathbb{P}}(0) > 1$ (in Ref. 32 it was taken as $\alpha_{\mathbb{P}}(0) = 1.07$). In this case only the account of Regge cuts leads to the correct asymptotic behaviour $\sigma_{tot} \sim \ln^2 s$, compatible with the Froissart bound.

2.3.2 *Inclusive reactions*

The inclusive cross sections are connected with the discontinuities of the $3 \to 3$ forward elastic amplitudes by the generalized optical theorem[33] (see Appendix B, Eq. (B.24) and Fig. B.4). When $s_1 = (p_1 + k)^2$ and $s_2 = (p_2 + k)^2$ are high enough, we may write the $3 \to 3$ amplitude via the Regge-pole exchange diagram of Fig. 2.11a. Due to the factorization of Reggeons, the upper and lower vertices in Fig. 2.11a are the same as that in the related elastic $2 \to 2$ amplitude; the third vertex where two particles are coupled with two Reggeons is a new type of coupling, and the corresponding vertex functions should be determined from the experimental data. For the invariant cross section of the process $a + b \to h + X$ we may write:

$$F(p^*) = p_0^* \frac{d^3\sigma}{d\vec{p}^*} \simeq 1/s \sum_{R_1 R_2} \left(\frac{s_1}{s_0}\right)^{\alpha_{R_1}(0)} \cdot \left(\frac{s_2}{s_0}\right)^{\alpha_{R_2}(0)} \cdot f_{hh}^{R_1 R_2}(p_T^2) \quad , \tag{2.38}$$

where the sum is taken over all possible Reggeons in the ah and bh channels; $\beta^{R_1 aa}$ and $\beta^{R_2 bb}$ vertices in (2.38) are incorporated into $f_{hh}^{R_1 R_2}$ vertex. s_1 and s_2 may be expressed in terms of the particle h momentum:

$$s_1 \simeq \sqrt{s}\, m_T\, e^{-y^*} = (s/2) x_-$$

$$s_2 \simeq \sqrt{s}\, m_T\, e^{y^*} = (s/2) x_+$$

$$x_\pm = \sqrt{x^2 + 4m_T^2/s} \pm x \quad . \tag{2.39}$$

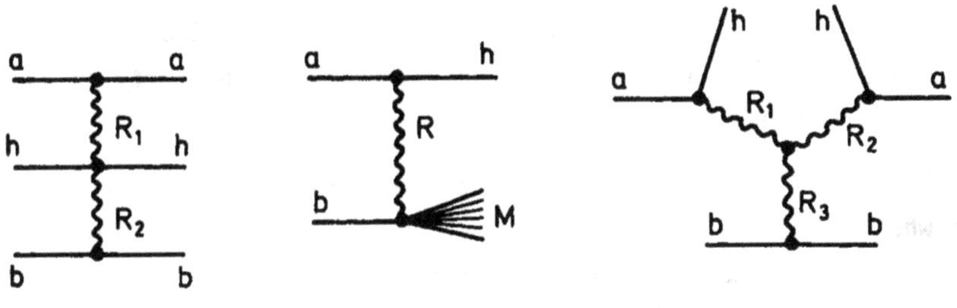

Fig. 2.11 (a) Double-Reggeon diagram for $3 \to 3$ elastic scattering amplitude;
(b) Reggeon diagram for inclusive reaction $a + b \to h + X$ at $M^2 \ll s$;
(c) Triple-Reggeon diagram for the inclusive cross section $a + b \to h + X$ (see Fig. b) in the region $m_h^2, m_a^2, m_b^2 \ll M^2 \ll s$.

In (2.39) $s_1, s_2 \gg m_a^2, m_b^2, m_h^2$ is assumed. Substituting these expressions into Eq. (2.38) we get[34]:

$$F(y^*, p_T^2) = \frac{1}{s} \sum_{R_1R_2} f_{hh}^{R_1R_2}(p_T^2)\left(\frac{sm_T^2}{s_0^2}\right)^{[\alpha_{R_1}(0) + \alpha_{R_2}(0)]/2}$$

$$\times \exp[(\alpha_{R_2}(0) - \alpha_{R_1}(0))y^*] \quad . \tag{2.40}$$

For $x \gg 2m_T/\sqrt{s}$ one obtains from (2.40)

$$F(x, p_T^2) = \frac{1}{s} \sum_{R_1R_2} f_{hh}^{R_1R_2}(p_T^2) \cdot \left(\frac{s}{s_0}\right)^{\alpha_{R_2}(0)} \left(\frac{m_T^2}{s_0}\right)^{\alpha_{R_1}(0)} x^{\alpha_{R_2}(0) - \alpha_{R_1}(0)} \quad . \tag{2.41}$$

In Reference 35 the secondary pion spectra in pp collisions were fitted using Eqs. (2.40), (2.41). The values of $f_{hh}^{R_1R_2}(p_T^2)$, obtained in this fit, allowed for calculations of the pion spectra in πp and Kp collisions[36]. This confirms the factorization property of double-Reggeon vertex functions.

Now we shall consider the inclusive reaction $a+b\to h+X$ in the case when particle h carries away almost the whole initial momentum, i.e. the case $1-x \ll 1$. In this case the amplitude of the inelastic reaction $a+b\to h+X$, where X is some cluster of particles with the total mass M, may be presented as the Reggeon exchange of Fig. 2.11b. If M^2 is large enough, the lower blob in Fig. 2.11b should be a multiperipheral ladder again; thus the inclusive cross section in this case is related to the discontinuity of the triple-Reggeon graph of Fig. 2.11c[34]

$$\frac{d^2\sigma}{dtdM^2} = \sum f^{R_1R_2R_3}(t)\left(\frac{M^2}{m_b^2}\right)^{\alpha_{R_2}(0)}\left(\frac{s}{M^2}\right)^{2\alpha_{R_1}(t)-1} \tag{2.42}$$

The triple-Reggeon vertex $f^{R_1R_2R_3}$ is an unknown function of t; in (2.42) it incorporates the factorized vertices β^{R_1ac}, β^{R_2ac} and β^{R_3bb}. The right-hand side of (2.42) may be easily expressed in terms of x and p_T since

$$x \simeq 1 - M^2/s$$

$$t \simeq t_{min} - p_T^2/x$$

$$t_{min} = (1-x)(m_a^2 - m_h^2/x) \quad . \tag{2.43}$$

Thus for the triple-Pomeron contribution $(R_1 = R_2 = R_3 = \mathbf{P})$ we get

$$x\frac{d^2\sigma}{dxdp_T^2} \sim (1-x)^{-1} \quad . \tag{2.44}$$

Such a behaviour was indeed observed in experiments (see Fig. 2.12).

The amplitudes of different triple-Pomeron vertices were deter-

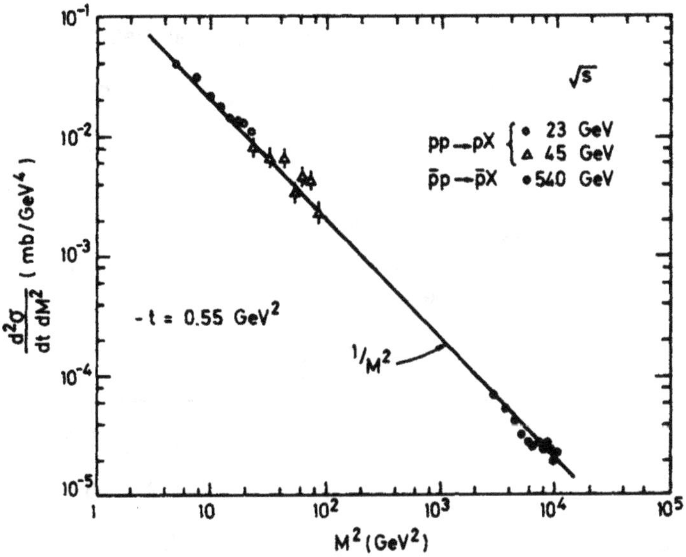

Fig. 2.12 Experimental spectra in the reactions $pp \to pX$ and $\bar{p}p \to \bar{p}X$ at $x \to 1$; the curve corresponds to triple-Pomeron behaviour (Eq. (2.44)): at large s, $M^2 (1-x) \simeq M^2/s$, thus $1/M^2 \sim (1-x)^{-1}$.

mined in the analysis of the inclusive cross section as $x \to 1$[37-39]. The values obtained in Ref. 39 are presented in Table 2.4.

2.3.3 *The AGK cutting rules*

We have already mentioned that in phenomenological studies absorption corrections to the single Regge-pole exchange amplitudes for binary reactions are usually accounted for effectively by the renormalization of Reggeon parameters. Sometimes, however, it is necessary to account for these corrections explicitly — the corresponding examples were also considered briefly. The question is, how to include absorption corrections (if any) into the Regge-pole description of the inclusive processes.

This problem may be solved exactly with the help of the Abramovsky-Gribov-Kancheli (AGK) rules[40], which are based on the unitarity condition.

Table 2.4 Triple-Reggeon vertices in the parametrization[39] $f^{R_1R_2R_3}(t) = G_{R_1R_2R_3} \exp(b_{R_1R_2R_3} t)$. As compared to the Eq. (2.42) (see text), $f^{R_1R_2R_3}(t)$ in the Table does not incorporate Reggeon-particle vertices β^{Rac}, β^{Rbb}.

Vertex $R_1R_2R_3$	$G_{R_1R_2R_3}$ mb/GeV^2	$b_{R_1R_2R_3}$ (GeV/c)$^{-2}$
P P P	3.24 ± 0.35	4.25 ± 0.24
R*) **P P**	7.2 ± 1.9	-1.2 ± 0.5
2ReR**P P**	6.9 ± 1.1	8.5 ± 3.7
P P R	3.2 ± 0.6	1.7 ± 0.4
RRR	51.9 ± 7.8	0
2Re**P**RR	-9.3 ± 2.2	0

*) In Ref. 39 R stands for the exchange-degenerate Reggeon with $\alpha_R(0) = 0.5$ (ρ, ω, A_2 etc.).

The unitarity condition (Eq. (B.20)) relates the imaginary part of the elastic scattering amplitude with the sum of cross sections of all possible processes in the intermediate state. Let us consider a Regge-pole diagram with one-Pomeron exchange (see Fig. 2.13a). The main intermediate states are here multiperipheral ladders with a number of particles proportional to $\ln s$. Graphically this can be described as shown in Fig. 2.13b-d: the imaginary part of the one-Pomeron exchange amplitude (the corresponding cut is shown in Fig. 2.13b, c by dotted lines) is determined by the modulus squared of the multiperipheral ladder-formation amplitude.

In the case of a diagram with the exchange of two Pemerons there

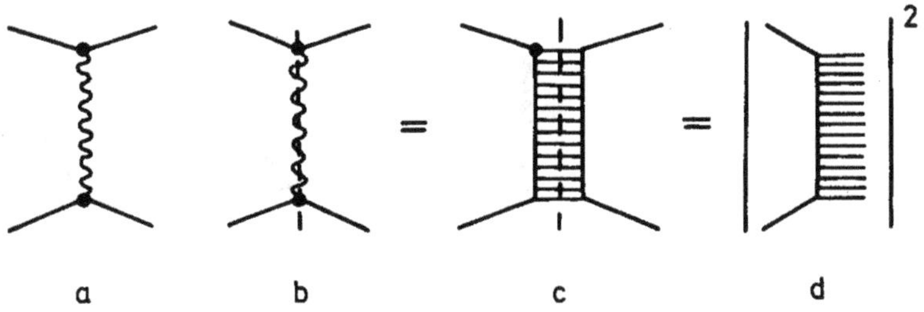

Fig. 2.13 (a) Reggeon diagram of single-Pomeron exchange and (b-d) its cuts.

are there classes of intermediate states, i.e. three types of cuts shown in Fig. 2.14 are possible. The cut between Reggeons corresponds to the cross section of the elastic scattering or of the diffraction dissociation (Fig. 2.14a). The one-Pomeron cut (Fig. 2.14b) corresponds to the production of one multiperipheral ladder similar to the diagram of Fig. 2.13. The contributions of the diagrams in Figs. 2.13b and 2.14b are of different signs, i.e. the diagram in Fig. 2.14b is the first absorptive correction to the cross section of the process in Fig. 2.13b. Finally, a cut is possible which slices through two Pomerons. This corresponds to the production of two multiperipheral ladders simultaneously[41)]. The contributions of other types of cuts (e.g. Fig. 2.14d) turn out to be asymptotically small in the theory with $\alpha_{\mathbb{P}}(0) = 1$.

It is shown by AGK that differently-sliced Reggeon diagrams are connected by simple combinatorial coefficients, and all of them contain the same vertex functions[40)]. If the contribution of diagrams with the exchange of ν Pomerons to the total cross section (i.e. the imaginary part of the elastic scattering amplitude) is S_ν, the contribution to the cross section which is connected with the cut of $n \leq \nu$ Pomerons

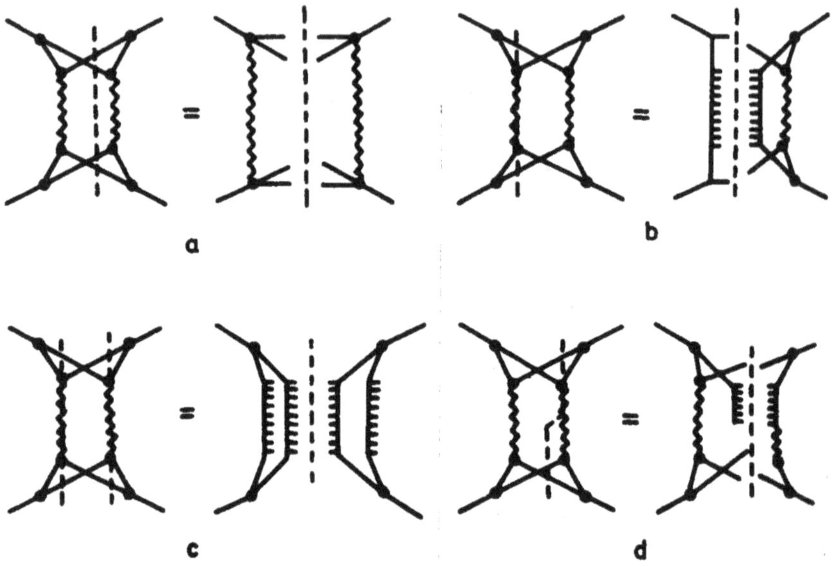

Fig. 2.14 Different cuts of double-Pomeron diagrams. The contributions of the diagrams (d) are asymptotically small in the theory with $\alpha_{\mathbb{P}}(0) = 1$.

equals

$$\sigma^{(\nu,\,n)} = (-1)^{n-1} \frac{\nu!}{n!(\nu-n)!} 2^{\nu-1} S_\nu \quad , \quad n \geq \nu$$

$$\sigma^{(\nu,\,0)} = (1 - 2^{\nu-1}) S_\nu \quad , \quad n = 0 \quad . \tag{2.45}$$

These relations are the AGK cutting rules. The sum of all contributions, naturally, equals S_ν

$$\sigma^{(\nu,\,0)} + \sum_{n=1}^{\nu} \sigma^{(\nu,\,n)} = S_\nu \tag{2.46}$$

for each $\nu \geq 1$ value.

The AGK cutting rules lead to several important consequences. The cross section of the production of n multiperipheral ladders[40,42] turns out to be, taking into account all absorption corrections,

$$\sigma^{(n)} = \sum_{\nu=n}^{\infty} \sigma^{(\nu,\, n)} = (-1)^{n-1} \sum_{\nu=n}^{\infty} \frac{\nu!}{n!(\nu-n)!} 2^{\nu-1} s_{\nu} \quad . \tag{2.47}$$

Let us consider the inclusive spectrum of an arbitrary hadron h in the central region. If the energy is sufficiently high, each multiperipheral ladder gives to this spectrum the same contribution $f_{hh}^{\mathbf{P}\,\mathbf{P}}$. The total contribution is

$$f(h) = \sum_{n=1}^{\nu} (-1)^{n-1} \frac{\nu!}{n!(\nu-n)!} 2^{\nu-1} f_{hh}^{\mathbf{P}\,\mathbf{P}} = f_{hh}^{\mathbf{P}\,\mathbf{P}} \tag{2.48}$$

i.e. the inclusive cross sections in the central region are determined by the simplest Mueller-Kancheli diagram only (Fig. 2.11a). The contributions of all absorption corrections cancel.

References

1. P. Petiau and M. Porneuf, eds. Proc. of the XXI Int. Conf. on High Energy Physics, Paris, 1982.

2. Proc. of the XXII Int. Conf. on High Energy Physics, Leipzig, 1984.

3. W. Kittel, W. Metzger and A. Stergiou, eds. Proc. of the XIII Int. Symp. on Multiparticle Dynamics, Volendam, 1982, World Scientific, Singapore, 1983.

4. P. Yager and J.F. Gunion, eds. Proc. of the XIV Int. Symp. on Multiparticle Dynamics, Lake Tahoe, 1983, World Scientific, Singapore, 1984.

5. V.N. Gribov, ZhETF 41, 667 (1961); Sov. Phys. JETP 14, 478 (1962); G.F. Chew and S.C. Frautschi, Phys. Rev. Lett. 7, 394 (1961).

6. J.P. Burq et al., Phys. Lett. 109B, 124 (1982); Nucl. Phys. B217 (1983).

7. I.Ya. Pomeranchuk, ZhETF 34, 725 (1958); Sov. Phys. JETP 7, 499 (1958).

8. F. Takagi, "An Estimation of Total Proton-Nucleon Cross Sections between 10^3 and 10^6 TeV from Air-Shower Data", Preprint TU/83/265 (1983).

9. V.M. Braun, "Shadow Corrections in Additive Quark Model", Preprint LNPI-854, Leningrad, 1983.

10. I.M. Dremin and M.I. Nazirov, Pis'ma v ZhETF 37, 163 (1983); JETP Letters 37, 198 (1983).

11. Ch. Peyrou, Proc. of the Int. Conf. on Elementary Particles. Aix-en-Provence, 1961, v. II, p. 1031.

12. J. Benecke et al., Phys. Rev. 188, 2159 (1969).

13. G.I. Zatzepin, ZhETF 19, 1104 (1949) (in Russian).

14. R.P. Feynman, Phys. Rev. Lett. 23, 1415 (1969).

15. L. Bertocchi, S. Fubini and M. Tonin, Nuovo Cim. 25, 625 (1962).

16. D. Amati, A. Stanghellini and S. Fubini, Nuovo Cim. 26, 896 (1962).

17. I.M. Dremin et al., ZhETF 48, 952 (1965); Sov. Phys. JETP 21, 633 (1965).

18. K.G. Boreskov et al., Yad. Fiz. 15, 361 (1972); 17, 1285 (1973); Sov. J. Nucl. Phys. 15, 203 (1972); 17, 669 (1974).

19. E.M. Levin and M.G. Ryskin, Yad. Fiz. 17, 388 (1973); Sov. J. Nucl. Phys. 17, 199 (1973).

20. V.N. Gribov, ZhETF 53, 654 (1967); Sov. Phys. JETP 26, 414 (1968).

21. G.C. Fox and C. Quigg, Ann. Rev. Nucl. Sci. 23, 219 (1973).

22. P.D.B. Collins and E.J. Squires, Regge Poles in Particle Physics, Springer, Berlin, 1968.

23. P.D.B. Collins, An Introduction to Regge Theory and High Energy Physics, Cambridge U.P., 1976.

24. A.C. Irving and R.P. Worden, Phys. Reports 34C, 117 (1977).

25. M. Baker and K.A. Ter-Martirosyan, Phys. Reports 28C, 3 (1976).

26. M. Gell-Mann, Phys. Rev. Lett. 8, 263 (1962).

27. V.N. Gribov and I.Ya. Pomeranchuk, ZhETF 42, 1141 (1962); Sov. Phys. JETP 15, 788 (1962); Phys. Rev. Lett. 8, 343 (1962).

28. V.N. Gribov, L.B. Okun and I.Ya. Pomeranchuk, ZhETF 45, 1114 (1963); Sov. Phys. JETP 18, 769 (1964).

29. V.N. Gribov and D.V. Volkov, ZhETF 44, 1068 (1963); Sov. Phys. JETP 17, 720 (1963).

30. V.A. Kudryavtsev, E.M. Levin and A.A. Shchipakin, Yad. Fiz. 9, 1274 (1969); 10, 148 (1969); Sov. J. Nucl. Phys. 9, 742 (1969); 10, 86 (1969).

31. P.D.B. Collins and F.D. Gault, Nucl. Phys. B112, 483 (1976).

32. M.S. Dubovikov and K.A. Ter-Martirosyan, Nucl. Phys. B124, 163 (1977); B.Z. Kopeliovich, L.I. Lapidus. ZhETF 71, 61 (1976); Sov. Phys. JETP 44, 31 (1976).

33. A.H. Mueller, Phys. Rev. D2, 2963 (1976); O.V. Kancheli, Pis'ma v ZhETF 11, 397 (1970); JETP Letters, 11, 267 (1970).

34. V.A. Abramovsky, O.V. Kancheli and T.D. Mandzavidze, Yad. Fiz. 13, 1102 (1971); Sov. J. Nucl. Phys. 13, 630 (1971).

35. M.N. Kobrinsky, A.K. Likhoded and A.N. Tolstenkov, Yad. Fiz. 20, 775 (1974); Sov. J. Nucl. Phys. 20, 414 (1974).

36. A.K. Likhoded and A.N. Tolstenkov, Preprint IFVE-74-51, Serpukhov, 1974 (in Russian).

37. D.P. Roy and R.G. Roberts, Nucl. Phys. B77, 240 (1974).

38. R.D. Field and G.F. Fox, Nucl. Phys. B80, 367 (1976).

39. Yu.M. Kazarinov et al., ZhETF 70, 1152 (1976); Sov. Phys. JETP, 43, 598 (1976).

40. V.A. Abramovsky, V.N. Gribov and O.V. Kancheli, Yad. Fiz. 18, 595 (1979); Sov. J. Nucl. Phys. 18, 308 (1979).

41. V.A. Abramovsky and O.V. Kancheli, Pis'ma v ZhETF 15, 559 (1972); JETP Letters 15, 397 (1972).

42. K.A. Ter-Martirosyan, Phys. Lett. 44B, 377 (1973).

Chapter 3
COMPOSITE SYSTEMS

From now on we are going to consider different composite systems: hadrons, consisting of dressed quarks and nuclei, which are composite systems of nucleons. In the present chapter we will describe elements of the techniques to handle such systems. The results given concern mainly non-relativistic composite systems, i.e. those in which the constituents are non-relativistic particles in the rest frame. A significant point in the description of such systems is the possibility of considering them in any arbitrarily chosen reference frame including the cases when they moves with high velocity.

The presented technique is based on the formalism of the Feynman integrals.

3.1 Non-Relativistic Composite Systems

Let us start with the simple but fundamental problem of the interaction of two spinless non-relativistic particles which form a composite system.

We will consider the amplitude of two particles with equal masses m interacting via the exchange of some light particle with a mass $\mu \ll m$. This amplitude is determined by the sum of ladder diagrams (Fig. 3.1).

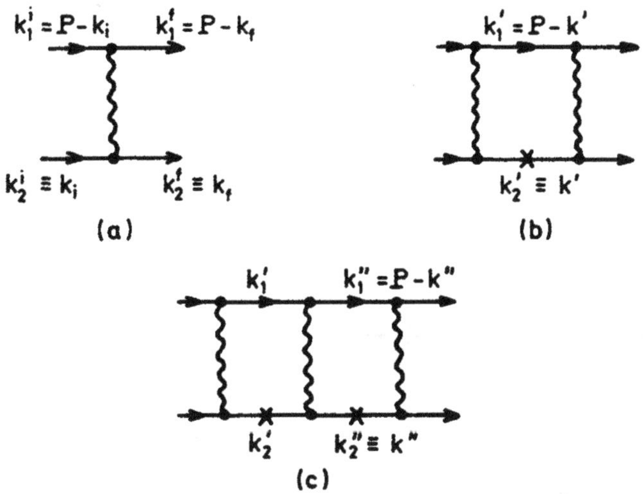

Fig. 3.1 Ladder-type diagrams which determine the scattering amplitude for two non-relativistic particles. The crossed propagator in the Feynman integrals are taken on the mass shell.

The 4-momenta of the initial particles are

$$(k_{10}^i, \vec{k}_1^i) \simeq (m + \frac{\vec{k}_1^{i2}}{2m}, \vec{k}_1^i)$$

$$(k_{20}^i, \vec{k}_2^i) \simeq (m + \frac{\vec{k}_2^{i2}}{2m}, \vec{k}_2^i) \tag{3.1}$$

Denoting

$$E_{1i} = \frac{\vec{k}_1^{i2}}{2m} \quad , \quad E_{2i} = \frac{\vec{k}_2^{i2}}{2m} \tag{3.2}$$

we introduce the total momentum and the total kinetic energy:

$$E = E_{1i} + E_{2i} \quad , \quad \vec{P} = \vec{k}_1^i + \vec{k}_2^i \quad . \tag{3.3}$$

The momenta of the final particles differ from Eqs. (3.1)-(3.3) by the subscript replacement $i \to f$.

Hereafter we shall also use the notation

$$k_2^i \equiv k_i \quad , \quad k_2^f \equiv k_f \tag{3.4}$$

The amplitude of the diagram Fig. 3.1a is

$$A^{(1)} = \frac{g^2}{\mu^2 - (k_i - k_f)^2} \simeq \frac{g^2}{\mu^2 + (\vec{k}_i - \vec{k}_f)^2} \tag{3.5}$$

In Eq. (3.5) we took note that in the non-relativistic limit $(E_i - E_f)^2 << (\vec{k}_i - \vec{k}_f)^2$.

In order to calculate the diagram of Fig. 3.1b let us introduce the non-relativistic integration variable

$$E' = k'_{10} - m \tag{3.6}$$

Non-relativistic limit for the particle propagator is:

$$\frac{1}{m^2 - k'^2 - i0} \longrightarrow \frac{1}{-2mE' + \vec{k}'^2 - i0} \tag{3.7}$$

Hence, the amplitude of the diagram Fig. 3.1b may be written as:

$$A^{(2)} = \int \frac{dE'd^3k'}{i(2\pi)^4} \cdot \frac{g^2}{\mu^2 + (\vec{k}_i - \vec{k}')^2} \cdot \frac{1}{(-2mE' + \vec{k}'^2 - i0)}$$
$$\times \frac{1}{[-2m(E - E') + (\vec{P} - \vec{k}')^2 - i0]} \cdot \frac{g^2}{\mu^2 + (\vec{k}' - \vec{k}_f)^2} \quad . \tag{3.8}$$

Integration over E' may be carried out explicitly. Here and for all the diagrams of Fig. 3.1 we enclose the pole of the lower particle propagator (crossed in Fig. 3.1b) by the integration contour. In the considered diagram 3.1b this is equivalent to the replacement:

$$(-2mE' + \vec{k}'^2 - i0)^{-1} \rightarrow 2\pi i\delta(-2mE' + \vec{k}'^2) \quad . \tag{3.9}$$

As a result we get

$$A^{(2)} = \int \frac{d^3k'}{2m(2\pi)^3} \cdot \frac{g^2}{\mu^2 + (\vec{k}_i - \vec{k}')^2} \cdot \frac{1}{[-2mE + \vec{k}'^2 + (\vec{P} - \vec{k}')^2 - i0]} \times \frac{g^2}{\mu^2 + (\vec{k}' - \vec{k}_f)^2} \quad . \tag{3.10}$$

The right-hand side of the Eq. (3.10) depends on two variables: $4mE - \vec{P}^2$ and $(\vec{k}_i - \vec{k}_f)^2$, which are the non-relativistic limits of the invariant variables s and t:

$$s \simeq 4m^2 + 4mE - \vec{P}^2 \quad .$$

$$t \simeq -(\vec{k}_i - \vec{k}_f)^2 \quad . \tag{3.11}$$

All the other diagrams of Fig. 3.1 depend on the same variables.

Similar calculations are valid for the diagram with triple rescattering (Fig. 3.1c):

$$A^{(3)} = \int \frac{d^3k'}{2m(2\pi)^3} \cdot \frac{d^3k''}{2m(2\pi)^3} \cdot \frac{g^2}{\mu^2 + (\vec{k}_i - \vec{k}')^2} \cdot \frac{1}{-2mE + \vec{k}'^2 + (\vec{P} - \vec{k}')^2 - i0} \times \frac{g^2}{\mu^2 + (\vec{k}' - \vec{k}'')^2} \cdot \frac{1}{-2mE + \vec{k}''^2 + (\vec{P} - \vec{k}'')^2 - i0} \cdot \frac{g^2}{\mu^2 + (\vec{k}'' - \vec{k}_f)^2} \tag{3.12}$$

The total amplitude is a sum over the number of scatterings:

$$A(s, t) = \sum_{n=1}^{\infty} A^{(n)}(s, t) \quad . \tag{3.13}$$

Let us consider the analytic properties of the amplitude. It has singularities in t at the values

$$t = (n\mu)^2 \quad , \quad n = 1,2,3,\ldots \quad . \tag{3.14}$$

The amplitude $A^{(1)}$ has a pole at $t = \mu^2$ (one-particle exchange), $A^{(2)}$ a singularity at $t = 4\mu^2$ (two-particle exchange) and so on.

In the s-plane the amplitude (3.13) has only a threshold singularity at

$$s = 4m^2 \tag{3.15}$$

which corresponds to the possibility that both constituents are on the mass shell. If the non-relativistic limit of the Eq. (3.7) is not taken in the ladder diagrams, they will have also singularities at $s = (2m+\mu)^2$, $(2m+2\mu)^2$ etc., corresponding to the production of particles with the mass μ. An example of a ladder-diagram cut, which gives a singularity at $s = (2m+\mu)^2$, is shown in Fig. 3.2. These singularities are absent in the non-relativistic limit, and the amplitude (3.13) does not describe the processes of particle production.

Let us make this clear by an example of the diagram of Fig. 3.2. It is easy to see that the cut shown in Fig. 3.2 gives zero contribution, for it contains the real decay of the particle with momentum k'' into the particles with momenta k' and $(k'' - k')$ (all the lower propagators are taken on the mass shell). Such a decay is evidently impossible.

If a bound state is produced as a result of the interaction, the amplitude (3.13) has also a pole singularity at

$$s = M^2 \tag{3.16}$$

where M is a mass of the bound state.

The position of the singularities in the complex s plane looks most simple in the case when the bound state occurs at the zero value of the orbital momentum $\ell = 0$. The partial amplitude with $\ell = 0$ is

$$A_0(s) = \int_{-1}^{1} \frac{dz}{2} A(s, t) \tag{3.17}$$

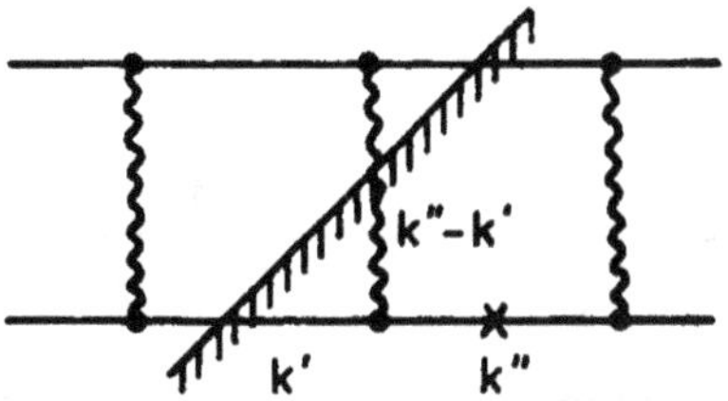

Fig. 3.2 An example of a cut-ladder diagram with three particles in the intermediate state.

where z is the cosine of the angle between the initial and final momenta $\vec{k}_i$, $\vec{k}_f$ in the c.m. reference frame

$$t = -2(\frac{s}{4} - m^2)(1 - z) \tag{3.18}$$

The partial amplitude has the same right half-plane singularities as the total amplitude $A(s, t)$: the threshold singularity at $s = 4m^2$ and a pole at $s = M^2$ (see Fig. 3.3). The left half-plane singularities of the partial amplitude correspond to the t-channel singularities of the total amplitude $A(s, t)$. They occur when the lower integration limit in the Eq. (3.17) $(z = -1)$ coincides with the t-channel singularity position $t = (n\mu)^2$, i.e. at

$$-s + 4m^2 = (n\mu)^2 \quad , \quad n = 1,2 \ldots \quad . \tag{3.19}$$

Since $\mu/m \ll 1$, the left half-plane singularities (3.19) at small n values are close to the point $s = 4m^2$, i.e. in the region where the non-relativistic approximation is valid (see Fig. 3.3). At large n the singularities of $A_0^{(n)}(s, t)$ are far from the $s = 4m^2$ point; they practically lie in this case in the region where a non-relativistic approximation is not valid. Thus the large n items of the series (3.13) have no first sheet (i.e. physical) singularities in the neighbourhood of $s \sim 4m^2$. The question is whether it is necessary to account for the amplitudes $A^{(n)}$ with large n in Eq. (3.13). All the amplitudes $A^{(n)}$ have singularities on the second (unphysical) sheet of the complex s-plane, under the cut from the threshold singularity.

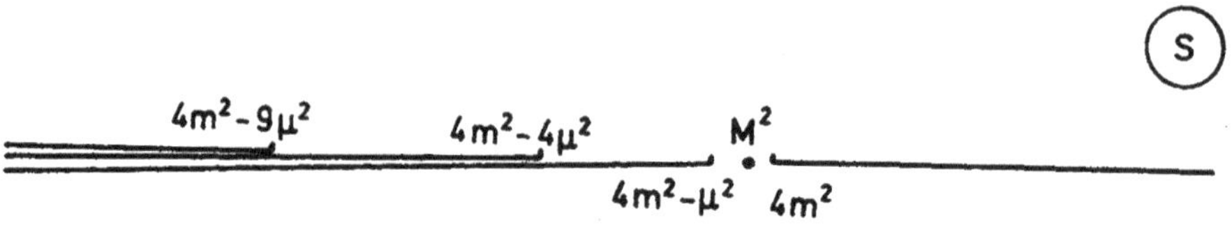

Fig. 3.3 The positions of the elastic partial amplitude singularities on the first (physical) sheet of the complex s-plane: threshold singularity at $s = 4m^2$, pole at $s = M^2$ and left cuts at the points $s = s_n = 4m^2 - (n\mu)^2$.

These singularities are always, independently of n, close to the point $s = 4m^2$. Thus all the terms of the series (3.13) are significant; the evident exception is the case of small g values, when perturbation theory is applicable. However, taking account the singularities (3.19) at large n, we would in fact be stretching the accuracy of the non-relativistic approximation.

In this approximation the scattering amplitude is determined by its singularities in the neighbourhood of the threshold point $s = 4m^2$ (both on the first and on the second sheets of the complex s-plane). If consideration of these nearest singularities is sufficient for the present problem, the non-relativistic approximation is closed and self-consistent. When the distant singularities contribute considerably, the non-relativistic approximation becomes invalid and the relativistic effects become significant.

Let us now consider the form factor and the disintegration amplitude of the composite system.

The partial amplitude near the point $s = M^2$ is well approximated by the one-pole contribution

$$A_0(s) \simeq \frac{G^2}{M^2 - s} \,. \tag{3.20}$$

G is the vertex of the transition "composite system ⟶ constituents" (see Fig. 3.4a). This vertex determines the amplitudes of the composite-

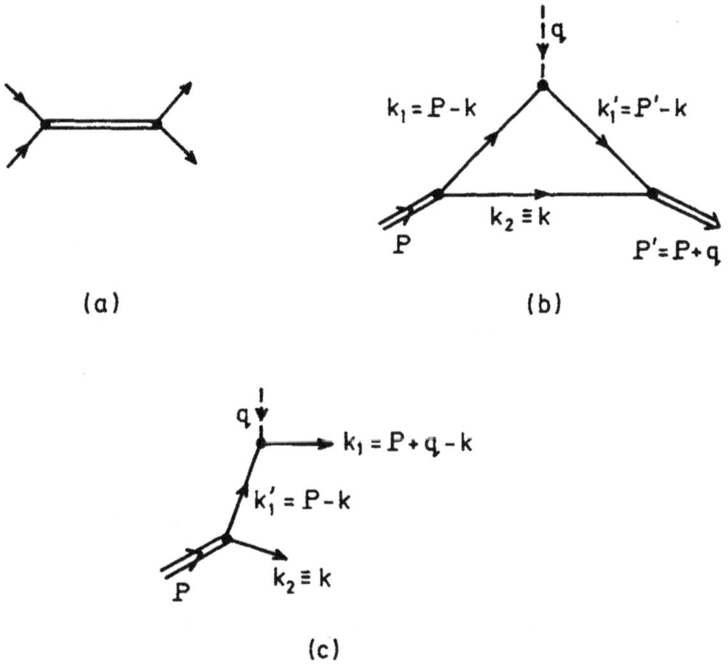

Fig. 3.4 (a) Pole diagram which corresponds to the virtual transitions of the scattering constituents into the bound state;
(b) Form factor of the bound state;
(c) Disintegration of the bound state.

system elastic scattering (Fig. 3.4b) and disintegration (Fig. 3.4c) by a definite (for example, electromagnetic) field. If the effective Lagrangian of the transition "composite system → constituents" can be written as

$$\mathscr{L}_{eff} = G(\Phi\varphi_1\varphi_2 + \Phi\varphi_1^+\varphi_2^+) \tag{3.21}$$

(where Φ is a composite system field, φ_1, φ_2 the constituents), the amplitude of the process in Fig. 3.4b may be calculated with the help of the standard Feynman rules:

$$A(\vec{q}^{\,2}) = \int \frac{dEd^3k}{i(2\pi)^4} \cdot \frac{eG^2(k_{1\mu} + k'_{1\mu})\varepsilon_\mu}{(-2mE_1 + \vec{k}_1^2 - i0)(-2mE'_1 + \vec{k}_1'^2 - i0)(-2mE_2 + \vec{k}_2^2 - i0)} \quad . \tag{3.22}$$

The momentum notations are shown in Fig. 3.4b; ε_μ is the photon polarization vector. In the Eq. (3.22) we used the non-relativistic approximation (3.7) from the very beginning. The 4-momentum of the composite system P is:

$$P = (2m + E_0, \vec{P}) = (2m - \epsilon + \frac{\vec{P}^2}{4m}, \vec{P}) \tag{3.23}$$

where ϵ is the binding energy $\epsilon = 2m - M$. The 4-momentum P' may be expressed through $E'_0 = -\epsilon + \frac{\vec{P}'^2}{4m}$ and $\vec{P}' = \vec{P} + \vec{q}$ in a similar way: in Eq. (3.23) it is necessary to make the substitution $E_0 \to E'_0$, $\vec{P} \to \vec{P}'$.

Let us assume that the external field is Coulomb-like. In the non-relativistic limit $(k_{1\mu} + k'_{1\mu})\varepsilon_\mu \simeq 2m$. In this case $A(\vec{q}^{\,2})$ is equal to

$$A(\vec{q}^{\,2}) = \int \frac{dEd^3k}{i(2\pi)^4} \cdot \frac{G^2 \cdot 2me}{[-2m(E_0 - E) + (\vec{P} - \vec{k})^2 - i0]}$$

$$\times \frac{1}{[-2m(E'_0 - E) + (\vec{P}' - \vec{k})^2 - i0] \cdot (-2mE + \vec{k}^2 - i0)} \quad . \tag{3.24}$$

Enclosing the pole singularity $(-2mE + \vec{k}^2 - i0)^{-1}$ by the integration contour in Eq. (3.24), we get:

$$A(\vec{q}^{\,2}) = 4me \int \frac{d^3k}{(2\pi)^3} \cdot \frac{\frac{1}{16m^3} \cdot G^2}{\left[-E_0 + \frac{\vec{k}^2}{2m} + \frac{(\vec{P} - \vec{k})^2}{2m}\right]\left[-E'_0 + \frac{\vec{k}^2}{2m} + \frac{(\vec{P}' - \vec{k})^2}{2m}\right]} \quad . \tag{3.25}$$

The integrand in the Eq. (3.25) is the composite system form factor $F(\vec{q}^2)$:

$$A(\vec{q}^2) = 4meF(\vec{q}^2) \quad , \quad F(0) = 1 \quad . \tag{3.26}$$

It may be expressed in terms of the relative momenta of the constituents $\frac{1}{4}(\vec{k}_2 - \vec{k}_1)^2 = (\vec{k} - \frac{1}{2}\vec{P})^2$ and $\frac{1}{4}(\vec{k}_2 - \vec{k}_1')^2 = (\vec{k} - \frac{1}{2}\vec{P}')^2$,

$$-E_0 + \frac{\vec{k}^2}{2m} + \frac{(\vec{P} - \vec{k})^2}{2m} = \epsilon + \frac{(\vec{k} - \frac{1}{2}\vec{P})^2}{m} . \tag{3.27}$$

The vertex function G in fact depends on the relative momentum squared $(\vec{k} - \frac{1}{2}\vec{P})^2$. It includes all possible rescatterings of the constituents (Fig. 3.5a) and satisfies the equation which is presented in a graphic form in Fig. 3.5b. The same calculations as those used for the Eq. (3.8) give for Fig. 3.5b:

$$G[(\vec{k} - \frac{1}{2}\vec{P})^2] = \frac{1}{(2m)^2}\int \frac{d^3k'}{(2\pi)^3} \cdot \frac{g^2}{\mu^2 + (\vec{k} - \vec{k}')^2} \cdot \frac{G[(\vec{k}' - \frac{1}{2}\vec{P})^2]}{\left\{\epsilon + \frac{(\vec{k}' - \frac{1}{2}\vec{P})^2}{m}\right\}} . \tag{3.28}$$

In the right-hand side of the Eq. (3.28) the factor $g^2/[\mu^2 + (\vec{k} - \vec{k}')^2]$ is in fact the constituent-interaction potential, which was previously chosen in a Yukawa-like form. In the general case of an arbitrary potential $V[(\vec{k} - \vec{k}')^2]$ Eq. (3.28) will have the form:

$$G[(\vec{k} - \frac{1}{2}\vec{P})^2] = \frac{1}{(2m)^2}\int \frac{d^3k'}{(2\pi)^3} V[(\vec{k} - \vec{k}')^2] \cdot \frac{G[(\vec{k}' - \frac{1}{2}\vec{P})^2]}{\epsilon + \frac{(\vec{k}' - \frac{1}{2}\vec{P})^2}{m}} . \tag{3.28a}$$

Thus in the integrand in the right-hand side of Eq. (3.25) G^2 should

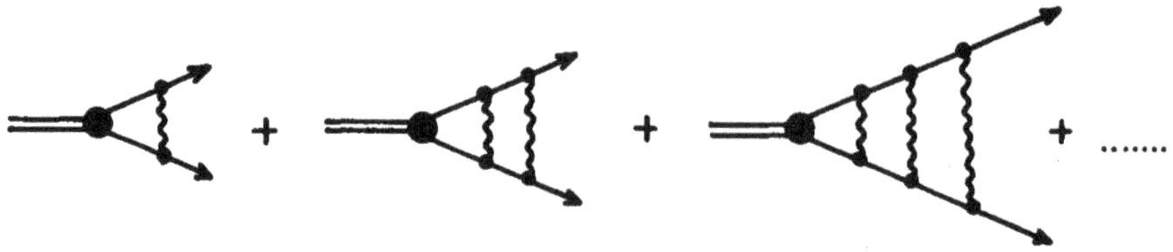

(a)

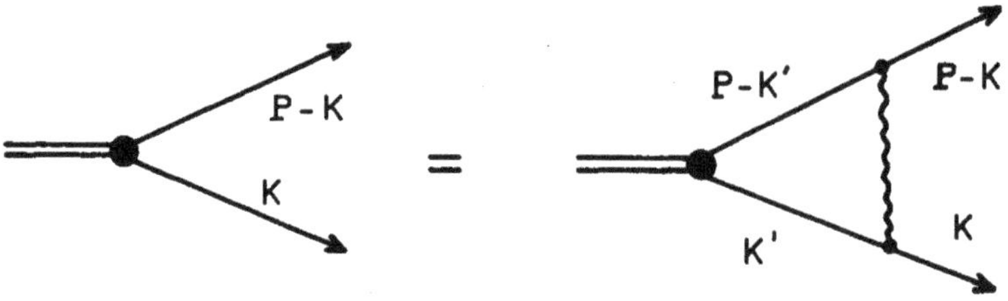

(b)

Fig. 3.5 (a) Diagrams with rescatterings of the constituents, which are to be included into the total vertex function;
(b) Graphical representation of the Eq. (3.28) for the vertex function.

be replaced by

$$G^2 \to G[(\vec{k} - \frac{1}{2}\vec{P})^2]\ G[(\vec{k} - \frac{1}{2}\vec{P}')^2] \quad . \tag{3.29}$$

The form factor may be also expressed in terms of the bound-state wave function:

$$F(\vec{q}^{\,2}) = \int \frac{d^3k}{(2\pi)^3} \Psi[(\vec{k} - \frac{1}{2}\vec{P})^2]\ \Psi[(\vec{k} - \frac{1}{2}\vec{P}')^2] \ . \tag{3.30}$$

The latter is

$$\Psi[(\vec{k}-\frac{1}{2}\vec{P})^2] = \frac{1}{4m^{3/2}}\cdot\frac{G[(\vec{k}-\frac{1}{2}\vec{P})^2]}{\epsilon+\frac{(\vec{k}-\frac{1}{2}\vec{P})^2}{m}}$$

Equation (3.28a) is in fact a Schrödinger equation for the wave function ψ. The integral equation for the scattering amplitude $A(s, t)$, which is described by the sum of the diagrams in Fig. 3.1, may also be written in a form similar to Eq. (3.28a). Using the non-relativistic variables (3.11), we should get:

$$A[2mE-\vec{P}^2, (\vec{k}_i-\vec{k}_f)^2]$$
$$= V[(\vec{k}_i-\vec{k}_f)^2] + \int\frac{d^3k'}{(2\pi)^3}\cdot\frac{V[(\vec{k}_f-\vec{k}')^2]}{\frac{\vec{k}'^2}{2m}+\frac{(\vec{P}-\vec{k}')^2}{2m}-E}A[2mE-\vec{P}^2, (\vec{k}'-\vec{k}_i)^2] \ . \tag{3.31}$$

Expanding both parts of the Eq. (3.31) into partial waves, we may rewrite (3.31) as equations for the partial amplitudes A_l. The amplitude of the process in Fig. 3.4c in the non-relativistic approximation is equal to the wave function of the composite system up to a normalization factor. If the coupling constant of the constituent interaction with the external field is equal to unity, this amplitude is

$$A_{wf} = \frac{G[(\vec{k}-\frac{1}{2}\vec{P})^2]}{-2m(E_0-E)+(\vec{P}-\vec{k})^2} = \frac{G[(\vec{k}-\frac{1}{2}\vec{P})^2]}{2m\left(\epsilon+\frac{(\vec{k}-\frac{1}{2}\vec{P})^2}{m}\right)} = 2\sqrt{m}\,\Psi((\vec{k}-\frac{1}{2}\vec{P})^2) \ . \tag{3.32}$$

The momenta are shown in Fig. 3.4c.

These results may be generalized for an N-body composite system. We assume for definiteness that all the constituents are different particles with equal masses m. The effective interaction which describes the disintegration of the system has the form

$$\mathscr{L}_{eff} = G(\Phi^{+}\varphi_1\varphi_2,\ldots,\varphi_N + \Phi\varphi_1^{+}\varphi_2^{+},\ldots,\varphi_N^{+}) \quad . \tag{3.33}$$

The form factor of such a system in the non-relativistic approximation is determined by the Feynman integral:

$$F(\vec{q}^{\,2}) = \frac{1}{N}\int \prod_{j=2}^{N} \frac{dE_j d^3k_j}{i(2\pi)^4} \cdot \frac{1}{(-2mE_j + k_j^2 - i0)}$$

$$\times \frac{G(\vec{k}_{12}^{\,2}, \vec{k}_{23}^{\,2},\ldots)G(\vec{k}_{12}^{\,\prime 2}, \vec{k}_{23}^{\,\prime 2},\ldots)}{(-2mE_1 + \vec{k}_1^{\,2} - i0)(-2mE_1' + \vec{k}_1^{\,\prime 2} - i0)} \quad . \tag{3.34}$$

The momentum notations are clear from the Fig. 3.6. The energy-momentum conservation relations for Eq. (3.34) are:

$$E_1 = E_0 - \sum_{j=2}^{N} E_j \quad , \qquad \vec{k}_1 = \vec{P} - \sum_{j=2}^{N} \vec{k}_j \quad ,$$

$$E_0 = P_0 - Nm = -\epsilon + \frac{\vec{P}^2}{2Nm} \tag{3.35}$$

and similar expressions with the replacement $\vec{P} \to \vec{P}' = \vec{P} + \vec{q}$ for E_1' and $\vec{k}_1'$. The vertex functions G depend on the different relative momenta $\vec{k}_{12} = \frac{1}{2}(\vec{k}_1 - \vec{k}_2)$, $\vec{k}_{23} = \frac{1}{2}(\vec{k}_2 - \vec{k}_3)$ and so on.

The integrations over E_j $(j \geq 2)$ in the Eq. (3.34) are carried out explicitly, and as a result we obtain

$$F(\vec{q}^{\,2}) = \int \prod_{j=1}^{N} \frac{d^3k_j}{(2\pi)^3}\, \psi(\vec{k}_1, \vec{k}_2,\ldots,\vec{k}_N)\Psi^*(\vec{k}_1 + \vec{q}, \vec{k}_2,\ldots,\vec{k}_N)$$

$$\times (2\pi)^3\delta^3(\sum_{j=1}^{N} \vec{k}_j - \vec{P}) \tag{3.36}$$

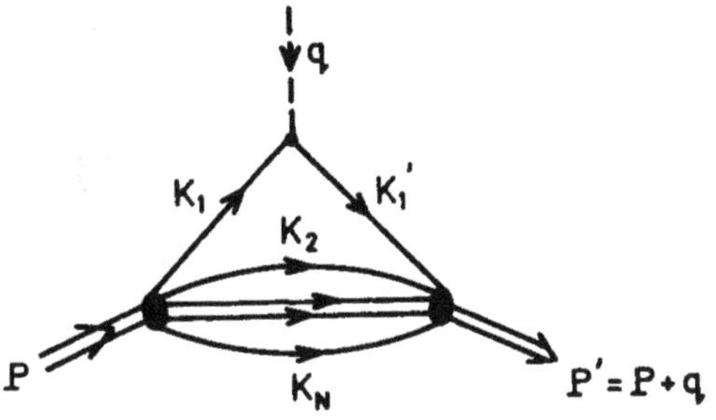

Fig. 3.6 Form factor of the N-particle composite system.

where

$$\Psi(\vec{k}_1, \vec{k}_2, \ldots, \vec{k}_N) = \frac{1}{\sqrt{N \cdot (2m)^{N+1}}} \cdot \frac{G(\vec{k}_{12}^2, \vec{k}_{23}^2, \ldots)}{E_0 - \sum_{j=1}^{N} \frac{\vec{k}_j^2}{2m}} \quad . \tag{3.37}$$

Here this form of the wave function is more convenient.

3.2 Lorentz-Invariant Description of Non-Relativistic Systems

In the foregoing parts of the book we often consider collisions of two non-relativistic composite systems for high energies and small momentum transfer. Typical examples are shown in Fig. 3.7. In a situation when the transferred momentum in the collision of the constituents is small, they exchange with each other by practically purely space-component momenta, while the transferred energy is negligible. Thus any composite particles in the diagrams of Fig. 3.7 may be described in the non-relativistic approximation of the previous section in its rest frame. However, the other composite system moves fast in this frame, and in order to describe both colliding composite particles simultaneously it is necessary to consider a non-relativistic composite system in a fast moving reference frame. In other words, it is necessary to present the expressions obtained in terms of the Lorentz-invariant

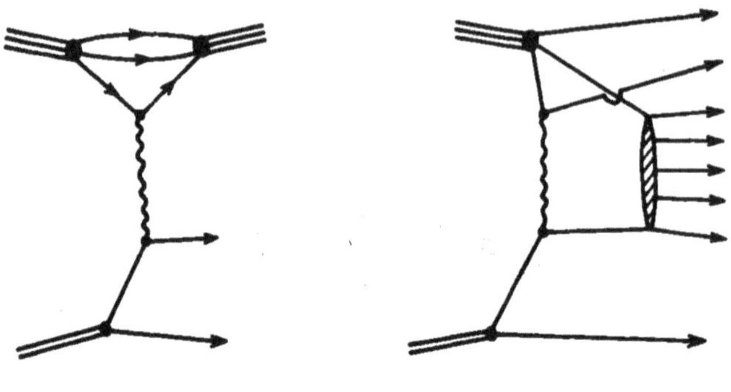

Fig. 3.7 Examples of high-energy collisions of composite systems which are to be considered within the relativistic situation.

variables. Such a procedure may be called "trivial relativization"; we shall consider it in the present section.

Let us start with the elastic scattering amplitude (3.13). If we were able to calculate all the diagrams of the series (3.13) explicitly, the trivial relativization would be carried out in a straightforward way: the diagrams $A^{(n)}$ depend only on $4mE - \vec{P}^2$ and $(\vec{k}_i - \vec{k}_f)^2$, and we replace them by the invariant variables s and t according to Eq. (3.11) we get the expression for $A^{(n)}$ which is valid in any reference frame. However, such an explicit calculation is usually impossible, and we must make a transformation to the Lorentz-invariant variables in the integrals. Let us consider, for example, Eq. (3.10).

The amplitude $A^{(2)}$ may be written in the form:

$$A^{(2)} = \int \frac{d^3k_1'}{2m(2\pi)^3} \cdot \frac{d^3k_2'}{2m(2\pi)^3} \cdot \frac{g^2}{\mu^2 + (\vec{k}_i - \vec{k}_2')^2} \cdot \frac{(2\pi)^3 \delta(\vec{k}_1' + \vec{k}_2' - \vec{P})}{E' - E - i0}$$

$$\times \frac{g^2}{\mu^2 + (\vec{k}_f - \vec{k}_2')^2} \quad . \tag{3.38}$$

Here $E' = \frac{\vec{k}_1'^2}{2m} + \frac{\vec{k}_2'^2}{2m}$; the energy-momentum 4-vector of the initial state is $P = (2m + E, \vec{P})$, where $E = \frac{\vec{k}_1^2}{2m} + \frac{\vec{k}_2^2}{2m}$. Let us introduce the 4-vector of energy-momentum of the intermediate state:

$$\tilde{P} = (2m + E', \vec{P}) \quad . \tag{3.39}$$

The total kinetic energies of the initial and final states in the c.m. frame are

$$E_{CM} = E - \frac{\vec{P}^2}{4m} \quad , \quad E'_{CM} = E' - \frac{\vec{P}^2}{4m}$$

Thus the Eq. (3.38) may be identically rewritten as

$$A^{(2)} = \frac{1}{\pi}\int_0^\infty dE'_{CM}\int \frac{d^3\vec{k}_1'}{2m(2\pi)^3}\cdot\frac{d^3\vec{k}_2'}{2m(2\pi)^3}\cdot\frac{g^2}{\mu^2 + (\vec{k}_i - \vec{k}_2')^2}$$

$$\times \frac{\frac{1}{2}(2\pi)^4\delta^4(k_1' + k_2' - \tilde{P})}{E'_{CM} - E_{CM} - i0}\cdot\frac{g^2}{\mu^2 + (\vec{k}_f - \vec{k}_2')^2} \quad . \tag{3.40}$$

In the δ^4-function in the right-hand side of the Eq. (3.40) it is assumed that $k_{j0}' = m + \frac{\vec{k}_j'^2}{2m}$. The invariant total energy squared $s = P^2$ and $s' = \tilde{P}^2$ are related with E_{CM} and E'_{CM}:

$$s = 4m^2 + 4mE_{CM} \quad , \quad s' = 4m^2 + 4mE'_{CM} \quad . \tag{3.41}$$

Thus in Eq. (3.41) E_{CM} and E'_{CM} may be replaced by the invariants s and s'. At last, the momenta transferred and the phase space elements can also be replaced by their invariant equivalents (see Appendix B); the resulting expression for $A^{(2)}$ is valid in an arbitrary reference frame:

$$A^{(2)} = \frac{1}{2}\int_{4m^2}^{\infty} \frac{ds'}{\pi}\cdot\frac{1}{s'-s-i0}\int d\Phi_2(\tilde{P};\, k_1',\, k_2')$$

$$\times \frac{g^4}{[\mu^2-(k_i-k_2')^2][\mu^2-(k_2'-k_f)^2]} \quad ,$$

$$d\Phi_2(\tilde{P};\, k_1',\, k_2') = \frac{d^3k_1'}{2k_{10}'(2\pi)^3}\cdot\frac{d^3k_2'}{2k_{20}'(2\pi)^3}\,(2\pi)^4\delta^4(k_1'+k_2'-\tilde{P}) \quad .$$

(3.42)

Equation (3.42) is the dispersion relation for the diagram 3.1b. This procedure gives a similar dispersion integral expressions for the more complicated diagrams $A^{(n)}$. In particular, Eq. (3.31) for the elastic scattering amplitude may be written as

$$A(s,\, t) = V(t) + \int_{4m^2}^{\infty}\frac{ds'}{2\pi}\cdot\frac{1}{s-s'-i0}\int d\Phi_2(\tilde{P};\, k_1',\, k_2')V(t_f')A(s,\, t_i') \tag{3.43}$$

where

$$t_f' = (k_f-k_2')^2 \quad , \quad t_i' = (k_i-k_2')^2 \quad .$$

A comparison of the Eqs. (3.31) and (3.43) should be carried out bearing in mind that these expressions do not coincide identically. They are equivalent with a rather good accuracy only in the region of s and t values where the non-relativistic approximation is valid, i.e. at $s \sim 4m^2$, $|t| \ll m^2$. When s (or t) is far from this region the relativizated expressions for $A^{(n)}$ or A may differ significantly from their non-relativistic prototypes (like (3.31) or (3.12)). In the case when we consider only the region $s \sim 4m^2$ such a discrepancy is of course not confusing.

Now let us turn to the scattering of the composite particle by an external field in order to get the Lorentz-invariant expression for the amplitude of the diagram in Fig. 3.4b. The non-relativistic expression (3.25) may be rewritten as

$$A(\vec{q}^2) = 4me\int \frac{d^3k}{(2\pi)^3}\cdot \frac{\frac{1}{16m^3}G[(\vec{k}-\frac{1}{2}\vec{P})^2]G[(\vec{k}-\frac{1}{2}\vec{P}')^2]}{\left(-E_0+\frac{\vec{k}^2}{2m}-\frac{(\vec{P}-\vec{k})^2}{2m}\right)\left(-E_0'+\frac{\vec{k}^2}{2m}+\frac{(\vec{P}'-\vec{k})}{2m}\right)} \tag{3.44}$$

$$= 2me\int \frac{d^3k_1}{2m(2\pi)^3}\cdot\frac{d^3k_2}{2m(2\pi)^3}\cdot\frac{d^3k_1'}{2m(2\pi)^3}$$

$$\times\frac{G[\frac{1}{2}(\vec{k}_1-\vec{k}_2)^2]G[\frac{1}{4}(\vec{k}_1'-\vec{k}_2)^2](2\pi)^3\delta^3(\vec{k}_1+\vec{k}_2-\vec{P})(2\pi)^3\delta^3(\vec{k}_1'+\vec{k}_2-\vec{P}')}{\left(-E_0+\frac{\vec{k}_1^2}{2m}+\frac{\vec{k}_2^2}{2m}\right)\left(-E_0'+\frac{\vec{k}_1'^2}{2m}+\frac{\vec{k}_2^2}{2m}\right)} . \tag{3.45}$$

The following procedure is just that which provided for the transition from the Eq. (3.38) to the Eq. (3.43): the kinetic energies $E = \frac{\vec{k}_1^2}{2m}+\frac{\vec{k}_2^2}{2m}$ and $E' = \frac{\vec{k}_1'^2}{2m}+\frac{\vec{k}_2'^2}{2m}$ and the corresponding 4-vectors $\tilde{P} = (2m+E, \vec{P})$ and $\tilde{P}' = (2m+E', \vec{P}')$ are introduced, and the integrand is expressed in terms of the c.m. kinetic energies $E_{CM} = E - \frac{\vec{P}^2}{4m}$, $E_{CM}' = E' - \frac{\vec{P}'^2}{4m}$. The latter are related by Eq. (3.42) with the squares of the 4-vectors $\tilde{P}^2 = s$ and $\tilde{P}'^2 = s'$. The integrations over E and E', which we introduce together with the δ-functions $\delta(k_{10}+k_{20}-\tilde{P}_0)$ and $\delta(k_{10}'+k_{20}'-P_0')$ may be replaced by the invariant integrations over s and s' and finally we obtain the following expression for the two-particle composite system form factor (Fig. 3.4b):

$$F(\vec{q}^2) = \frac{1}{8}\int_{4m^2}^{\infty}\frac{ds}{\pi}\int_{4m^2}^{\infty}\frac{ds'}{\pi}\frac{G_D(s)G_D(s')}{(s-M^2)(s'-M^2)}\Delta(s, s', q^2) ,$$

$$\Delta(s, s', q^2) = \int d\Phi_2(\tilde{P}; k_1, k_2)d\Phi_2(\tilde{P}'; k_1', k_2')(2\pi)^3 2k_{20}'\delta^3(\vec{k}_2-\vec{k}_2') . \tag{3.46}$$

Here we introduced for simplicity a shortened notation $G_D(s) = G(\frac{s}{4} - m^2)$.

The expression $\frac{1}{2}\, emG_D(s)\Delta(s,\, s',\, q^2)G_D(s')$ is $disc_s$ $disc_s$ $A(s,\, s',\, q^2)$; it is related with the real processes: the two-particle state with the mass $\sqrt{s}$ decays into constituents, one of them is scattered by the external field and then the constituents are bound into a two-particle state with the mass $\sqrt{s}'$.

Similarly the amplitude of the process in Fig. 3.4c may be written in the form of a dispersion integral, which in the non-relativistic limit coincides with Eq. (3.32):

$$A_{wf} = \frac{1}{2}\int_{4m^2}^{\infty}\frac{ds}{\pi}\cdot\frac{G_D(s)}{s - M^2}\int d\Phi_2(\tilde{P};\, k_1',\, k_2')\cdot(2\pi)^3 2k_{20}\delta^3(\vec{k}_2 - \vec{k}_2') \quad . \tag{3.47}$$

The results obtained may be easily generalized for a many-body composite system. For example, the form factor of the N-body system (Fig. 3.6) is:

$$F(\vec{q}^{\,2}) = \frac{1}{4N}\int_{(Nm)^2}^{\infty}\frac{ds}{\pi}\int_{(Nm)^2}^{\infty}\frac{ds'}{\pi}\,\frac{G_D(s_{12},\, s_{23},\ldots)G_D(s_{12}',\, s_{23}',\ldots)}{(s - M^2)(s' - M^2)}$$

$$\times\int d\Phi_N(\tilde{P};\, k_1,\, k_2,\ldots,k_N)d\Phi_N(\tilde{P}';\, k_1',\, k_2',\ldots,k_N')$$

$$\times\prod_{j=2}^{N}(2\pi)^3 2k_{j0}\delta^3(\vec{k}_j - \vec{k}_j') \quad . \tag{3.48}$$

The developed technique of the dispersion integrals and the expressions (3.46)-(3.48) provide the desired possibility to express the amplitude of the composite system interaction in any reference frame in terms of the vertex function G_D. The latter is connected unambiguously with the wave function in the non-relativistic limit.

Let us emphasize once more that similarly to the Eq. (3.43), Eqs.(3.46)-(3.48) coincide with their non-relativistic prototypes only when the non-relativistic approximation is valid. As a result there is a certain ambiguity in the Eqs. (3.46)-(3.48). For example, the integrands in (3.46)-(3.48) may be multiplied by an arbitrary function of s_{ik}, which equals to unity at $(s_{ik} - 4m^2)/(4m^2) \ll 1$, has no singularities in the integration region and does not change the convergence interval of the integrals. We shall consider this point in more detail in the next section.

High energy collision processes are often considered in the infinite momentum frame, so it is useful to rewrite the obtained expressions in this frame. Let us start with the two-body system form factor (3.46).

We assume that the z-component of the initial momentum is very large, $P_z \to \infty$; the transferred momentum $\vec{q}$ lies in the xy plane. Then

$$\tilde{P} = (P_z + \frac{s + P_T^2}{2P_z}, P_z, \vec{P}_T) \quad ,$$

$$\tilde{P}' = (P_z + \frac{s' + P_T^2}{2P_z}, P_z, \vec{P}_T') \quad . \tag{3.49}$$

The variables x_i and x_i' determine the P_z-momentum fraction carried away by the constituents:

$$x_i = \frac{k_{iz}}{P_z} \quad , \quad x_i' = \frac{k_{iz}'}{P_z} \quad . \tag{3.50}$$

The δ-functions of the energy conservation law may be rewritten in terms of

$$\delta(k_{10} + k_{20} - \tilde{P}_0)\delta(k_{10}' + k_{20}' - \tilde{P}_0')$$

$$= 4P_z^2 \delta(\frac{m_{1T}^2}{x_1} + \frac{m_{2T}^2}{x_2} - s - \vec{P}_T^2)\delta(\frac{m_{1T}'^2}{x_1'} + \frac{m_{2T}'^2}{x_2'} - s' - \vec{P}_T'^2) \quad . \tag{3.51}$$

Here $m_{iT}^2 = m^2 + \vec{k}_{iT}^2$, $m_{iT}'^2 = m^2 + \vec{k}_{iT}'^2$. *These δ-functions allow one to* remove the integrations over s and s':

$$F(\vec{q}^2)$$

$$= \frac{1}{\pi} \int_0^1 \frac{dx_1}{x_1} \cdot \frac{dx_2}{x_2} \delta(x_1 + x_2 - 1) \int \frac{d^2k_{1T} d^2k'_{1T} d^2k_{2T}}{(2\pi)^2} \delta^2(\vec{k}_{1T} + \vec{k}_{2T} - \vec{P}_T)$$

$$\times \delta^2(\vec{k}'_{1T} + \vec{k}_{2T} - \vec{P}'_T) \frac{G_D\left(\frac{m_{1T}^2}{x_1} + \frac{m_{2T}^2}{x_2} - \vec{P}_T^2\right) G_D\left(\frac{m_{1T}'^2}{x_1} + \frac{m_{2T}'^2}{x_2} - \vec{P}_T'^2\right)}{\left(\frac{m_{1T}^2}{x_1} + \frac{m_{2T}^2}{x_2} - \vec{P}_T^2 - M^2\right)\left(\frac{m_{1T}'^2}{x_1} + \frac{m_{2T}'^2}{x_2} - \vec{P}_T'^2 - M^2\right)} \quad . \tag{3.52}$$

Let us obtain the momentum distribution of one of the constituents (for definiteness - that of the constituent 2) in the infinite momentum frame. This distribution may be easily calculated with the help of the amplitude of the process of Fig. 3.4c, which was written in the form of the dispersion integral in the Eq. (3.47). The inclusive cross section of the process of Fig. 3.4c may be obtained directly from the Eq. (3.47) or, instead of this, the discontinuity of the process in Fig. 3.8 may be calculated. In the infinite momentum frame we get:

$$\frac{d\sigma(\text{composite system} \to \text{constituent 2})}{\frac{dx_2}{x_2} \cdot \frac{d^2k_{2T}}{(2\pi)^2}}$$

$$= n(x_2, \vec{k}_{2T})$$

$$= \frac{1}{\pi} \left| \int \frac{dx_1}{x_1} d^2k_{1T} \delta(x_1 + x_2 - 1) \delta^2(\vec{k}_{1T} + \vec{k}_{2T} - \vec{P}_T) \frac{G_D\left(\frac{m_{1T}^2}{x_1} + \frac{m_{2T}^2}{x_2} - \vec{P}_T^2\right)}{\frac{m_{1T}^2}{x_1} + \frac{m_{2T}^2}{x_2} - \vec{P}_T^2 - M^2} \right|^2 \tag{3.53}$$

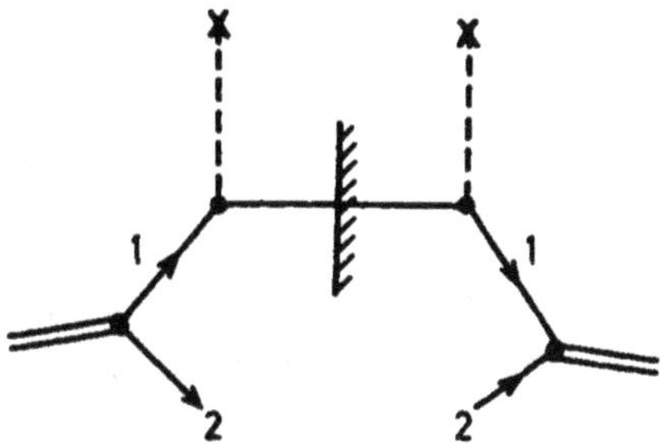

Fig. 3.8 Diagram for the inclusive production of the constituent 2.

$n(x_2, \vec{k}_{2T})$ is just the desired distribution. It is normalized by the condition $\int_0^1 \frac{dx_2}{x_2} \int \frac{d^2k_{T2}}{(2\pi)^2} n(x_2, \vec{k}_{T2}) = 1$ which is equivalent to the equality $F(0) = 1$ in the Eq. (3.52). The distributions of this type will be used in Chapter 7.

3.3 The Inclusion of Relativistic Effects in Composite Systems

The quark model, in which hadrons are treated as non-relativistic composite systems of the constituent (dressed) quarks, is a relatively rough approximation. The relativistic effects in the quark model lead to the observable contributions of the order of 20% (see e.g. Ref. 1), which is just the level of accuracy of the model predictions. Thus the correct treatment of these effects is a significant problem in view of the further development of the quark model on a higher level of predictive power. Since a great number of papers, devoted to this problem, have appeared recently, we shall briefly discuss it here. However, in the remaining part of the book our considerations as a rule will not concern truly relativistic effects.

The non-relativistic approximation is valid in the neighbourhood of the threshold $\frac{s - 4m^2}{4m^2} \ll 1$. In order to consider relativistic effects it is necessary to obtain the description of the amplitude in a broader region of s values, which are not bound by this condition. The dispersion-integral technique is quite suitable for this purpose. Indeed, the dispersion-relation method was developed exactly in order to describe the collision processes in a broad interval of the energy variables (see e.g. Ref. 2). The advantage of this particular method is its ability to clarify the problems and difficult points of other approaches to the description of relativistic effects.

Here we will restrict our considerations to the form factor of the two-body system.

First of all let us return to the Lorentz-invariant expressions (3.46). It was derived assuming $(k_{1\mu} + k'_{1\mu})\varepsilon_\mu \simeq 2m$, and we always remained in the framework of this approximation. This is very significant for non-relativistic systems: if the vertex function of the composite system interaction with some field does not vanish for zero momentum transfer, the scattering of this system is determined by the universal form factor. For example, the same form factor enters the amplitude of the composite system interactions with the electromagnetic field and with the Pomeron. This is not true for the relativistic systems, which "feel" the spin structure of the field. The form factors of such systems are different for the interactions with different fields Strictly speaking, Eq. (3.46) is the form factor of the composite system interacting with the scalar field.

Let us obtain the form factor of the interaction with the vector electromagnetic field. The vector $k_{1\mu} + k'_{1\mu}$ from the vertex coupling may be expanded over $\tilde{P}$ and $\tilde{P}'$:

$$\begin{aligned} k_{1\mu} + k'_{1\mu} &= \frac{1}{2}\alpha(s, s', q^2)(\tilde{P}_\mu + \tilde{P}'_\mu) + \beta(s, s', q^2)(\tilde{P}_\mu - \tilde{P}'_\mu) + (k_{1\mu} + k'_{1\mu})_T \\ &= \frac{1}{2}\alpha(s, s', q^2)\left[\tilde{P}_\mu + \tilde{P}'_\mu - \frac{s - s'}{q^2}(\tilde{P}_\mu - \tilde{P}'_\mu)\right] + (k_{1\mu} + k'_{1\mu})_T \quad . \end{aligned} \tag{3.54}$$

Here $(k_{1\mu}+k'_{1\mu})_T\tilde{P}_\mu = (k_{1\mu}+k'_{1\mu})_T\tilde{P}'_\mu = 0$. The factors α and β may be calculated using the relation $\tilde{P}'-\tilde{P}=q$ and taking into account that particle momenta in the Eq. (3.46) are considered on the mass shell: $k_1^2 = k_1'^2 = k_2^2 = m^2$. Then $\beta = \frac{s-s'}{-2q^2}\alpha$, while

$$\alpha(s,\ s',\ q^2) = 2\,\frac{q^4 - q^2(s+s')}{q^4 + (s-s')^2 - 2q^2(s+s')} \quad . \tag{3.55}$$

As $q^2 \to 0$, $s-s' = 0(\sqrt{q^2})$. The $q^2\to 0$ limit of (3.55) is:

$$\alpha(s,\ s',\ q^2\to 0) \equiv \alpha(s) = \frac{1}{2i\sqrt{\frac{s-4m^2}{s}}}\,\ln\frac{1+i\sqrt{(s-4m^2)/s}}{1-i\sqrt{(s-4m^2)/s}} \quad . \tag{3.56}$$

As $s \geq 4m^2$ this function is real, $\alpha(s) = \sqrt{s/(s-4m^2)}\ \mathrm{arctg}\ \sqrt{(s-4m^2)/s}$ and has no singularity on the physical sheet at $s=4m^2$; it has a square root branch point at $s=0$. The discontinuity of the triangle diagram in s and s' channels is:

$$\mathrm{disc}_s\,\mathrm{disc}_{s'}\,A(s,\ s',\ q^2) = e\varepsilon_\mu\left[\tilde{P}_\mu + \tilde{P}'_\mu - \frac{s-s'}{q^2}(\tilde{P}_\mu - \tilde{P}'_\mu)\right] \times \frac{1}{8}G_D(s)G_D(s')\Delta_V(s,\ s',\ q^2) \tag{3.57}$$

where

$$\Delta_\nu(s,\ s',\ q^2) = \int d\Phi_2(\tilde{P};\ k_1k_2)d\Phi_2(\tilde{P};\ k'_1k'_2)\alpha(s,\ s',\ q^2) \times (2\pi)^3 2k'_{20}\delta^3(\vec{k}_2 - \vec{k}'_2) \quad . \tag{3.58}$$

The limit of Δ_V as $q^2\to 0$ is

$$\Delta_V(s,\ s',\ 0) = \frac{1}{2}\sqrt{\frac{s-4m^2}{s}}\,\alpha(s)\delta(s-s') \quad . \tag{3.59}$$

The amplitude of the real process is determined by Δ_V:

$$A(-q^2) = \varepsilon_\mu (P_\mu + P'_\mu) F_V(-q^2)$$

$$F_V(-q^2) = \frac{1}{8}\int_{4m^2}^{\infty} \frac{ds}{\pi}\cdot\frac{ds'}{\pi}\,\frac{G_D(s)G_D(s')}{(s-M^2)(s'-M^2)}\,\Delta_V(s, s', q^2) \quad . \qquad (3.60)$$

Vector current conservation requires $F_V(0) = 1$. Thus, using (3.59), we get

$$\frac{1}{16\pi^2}\int_{4m^2}^{\infty} ds \sqrt{\frac{s-4m^2}{s}}\cdot\frac{\alpha(s)G_D^2(s)}{(s-M^2)^2} = 1 \quad . \qquad (3.61)$$

The expression for the vector form factor (3.60) differs from (3.46) only by the extra factor $\alpha(s, s', q^2)$ in the integrand. In the non-relativistic limit $\alpha(s, s', q^2) = 1$; it leads to the universality of the q^2-dependence of the non-relativistic system interaction.

Now let us consider the dispersion-relation technique for the low energy scattering amplitudes. An advanced method for this energy region is the well known method of N/D equations[3]. A detailed discussion of N/D method may be found e.g. in Refs. 2-5.

As before, we shall consider only the S-wave partial amplitude. It has a left-hand cut, caused by the "forces" (i.e. by the t-channel exchanges) and a right-hand cut, connected with the rescatterings (see Fig. 3.3). We shall also assume that an S-wave bound state with mass M exists, i.e. $A_0(s)$ has a pole at $s = M^2$.

Let us write $A_0(s)$ in the form

$$A_0(s) = \frac{N(s)}{D(s)} \qquad (3.62)$$

where $N(s)$ has only left-hand singularities, while $1/D(s)$ has only right-hand singularities and a bound state pole. $D(s)$ is normalized by the condition $\lim_{s\to\infty} D(s) = 1$. Assuming that $A_0(s)$ vanishes at infinity, we conclude that $N(s) \to 0$ as $s \to \infty$.

In the physical region (i.e. on the upper lip of the right-hand cut at $s > 4m^2$) $A_0(s)$ obeys the unitarity condition:

$$\mathrm{Im}A_0(s) = \rho(s)|A_0(s)|^2 \quad , \quad \text{where} \quad \rho(s) = \frac{1}{16\pi}\sqrt{\frac{s-4m^2}{s}} \tag{3.63}$$

Since $N(s)$ is real for $s > 4m^2$, the unitarity condition (3.63) gives $\mathrm{Im}D(s) = -\rho(s)N(s)$ and we get:

$$D(s) = 1 - \frac{1}{\pi}\int_{4m^2}^{\infty} ds' \frac{\rho(s')N(s')}{s'-s} \tag{3.64}$$

In fact the general solution contains besides (3.64) a sum over the Castillejo-Dalitz-Dyson poles[6] $\sum_i \gamma_i/(s-s_i)$. We shall consider here only the case $\gamma_i = 0$; amplitude $A_0(s)$ is assumed to have no zeroes in the physical region (i.e. $N(s)$ does not change sign when $s \geq 4m^2$).

The relation (3.64) gives the solution of the problem under consideration: as soon as the function $N(s)$, defined by the interaction forces, is given, one can calculate $D(s)$ and the whole amplitude $A_0(s)$.

The function $N(s)$ may be interpreted in terms of a separate potential of the constituent interaction. Indeed, using (3.64) Eq. (3.62) may be presented in the form:

$$A_0(s) = N(s) + \frac{1}{\pi}\int_{4m^2}^{\infty} ds' \frac{\rho(s')N(s')}{s'-s} A_0(s') \quad . \tag{3.65}$$

Let us compare (3.65) with the expression of the type (3.43) for the S-wave partial amplitude:

$$A_0(s) = V_0(s, s') + \frac{1}{\pi}\int_{4m^2}^{\infty} \frac{ds'}{s'-s} \rho(s')V_0(s, s')A_0(s, s') \quad . \tag{3.66}$$

Here V_0 is the S-wave interaction potential. Both the potential V_0 and the amplitude A_0 in the integrand in (3.66) depend on two variables s and s'. The reason is that in Eq. (3.43) the total energy of the intermediate state differs from that of the initial (or final) state in the c.m. frame $|\vec{k}_i| = |\vec{k}_f| = \sqrt{\frac{s}{4} - m^2}$, while $|\vec{k}'| = \sqrt{\frac{s'}{4} - m^2}$.

The expressions (3.65) and (3.66) coincide if $V_0(s, s')$ is separable, i.e. in the case when it may be presented as a product of two terms, each depending only on s or s': $V_0(s, s') = v(s) \cdot v(s')$. In this case $A_0(s, s')/A_0(s) = v(s')/v(s)$ and (3.66) turns into (3.65) when

$$V_0(s, s') = \sqrt{N(s)N(s')} \quad . \tag{3.67}$$

One must, however, emphasize here once more the difference between the Born term of the amplitude (Fig. 3.1a), which defines the interaction force, and the function N(s): the Born amplitude has only one left-hand singularity at $s = 4m^2 - \mu^2$ (see Fig. 3.3), while N(s) has all left-hand singularities.

Possible ways to connect the Born term with N(s) are discussed in Refs. 2-5.

The bound state occurs at the zero of D(s):

$$1 - \frac{1}{\pi}\int_{4m^2}^{\infty} ds' \frac{\rho(s')N(s')}{s' - M^2} = 0 \quad . \tag{3.68}$$

Near the point $s = M^2$ the amplitude $A_0(s)$ may be approximated by the one-pole expression:

$$A_0 \simeq \frac{-G^2(M^2)}{s - M^2} \tag{3.69}$$

where

$$G^2(M^2) = \frac{N(M^2)}{\frac{1}{\pi}\int_{4m^2}^{\infty} ds' \frac{\rho(s')N(s')}{(s' - M^2)^2}} \quad . \tag{3.70}$$

Let us compare Eqs. (3.61) and (3.70). They coincide when

$$G_D^2(s) = G^2(M^2)\,\frac{N(s)}{\alpha(s)N(M^2)} \quad . \tag{3.71}$$

Thus the vertex function $G_D(s)$ may be determined through the function $N(s)$. From (3.71) it is evident that $G_D^2(s)$ has only left-hand singularities.

Equation (3.68) specifies the conditions for the bound state in terms of the dispersion integrals. Thus it is similar to the Eq. (3.28) for the vertex function. This can be shown explicitly, when (3.28) is presented in Lorentz-invariant form:

$$G_D(s) = \frac{1}{\pi}\int_{4m^2}^{\infty} ds'\,\frac{\rho(s')}{s' - M^2}\,\gamma(s,\, s')V_0(s,\, s')G_D(s') \quad . \tag{3.72}$$

We have already noticed that the transformation of the non-relativistic expressions into Lorentz-invariant form is not an unambiguous procedure. Thus in the integrand in the Eq. (3.72) an a priori unknown function $\gamma(s,\, s')$ is introduced; in the non-relativistic limit it is just unity. If $V_0(s,\, s')$ is a separable potential (3.67), the Eq. (3.72) coincides with (3.68) at

$$\gamma(s,\, s') = \sqrt{\frac{\alpha(s')}{\alpha(s)}} \quad . \tag{3.73}$$

When the vertex function $G_D(s)$ is known, one may consider arbitrary processes of the composite system interaction in the framework of the dispersion-integral technique.

We have considered the case of spinless contituents. The generalization for the spin-1/2 constituents is evident.

The description of composite systems taking into account the small distance structure is, nowadays, quite relevant (see e.g. Refs. 7-16). The dispersion relation approach to this problem also has its own history[17-19]. The other approaches may be divided into two classes.

In one of these the relativistic propagator of the constituent is replaced by its residue in the pole (it is similar to the non-relativistic replacement (3.9)). For the ladder diagrams of the Fig. 3.1 this procedure leads to the Bethe-Salpeter equation. The composite system form factor was considered in the framework of this approach in Ref. 8.

This procedure has two confusing points: 1) some additional states in the composite systems appear; the fraction and the content of these additional states depend on the process considered. 2) Anticausal singularities appear in the amplitudes; it is necessary to check that these singularities are far from the region under consideration.

For these reasons recently another approach was developed, in which the composite system is considered in the infinite momentum frame (the detailed description may be found, e.g., in Ref. 16). The starting point for this approach is the parton structure of hadrons. Here one is confronted with other specific problems and difficulties. For example, the description of particles with nonzero spin requires additional efforts, since some spin states are inaccessible in the framework of the standard technique. The method of the "wave function" calculation are also not yet developed in this approach. In this case the "wave function" is in fact a Fock column[8]; its elements are determined from the experimental data. Thus, as yet, there is no adequate alternative to the N/D method, which allows us to calculate the vertex function G_D with the help of the dispersion-relation technique.

The description of the composite systems in the dispersion-relation-technique approach and in the infinite momentum frame complement each other to a certain extent. In the dispersion-relations approach we move from large to small distances, i.e. from the non-relativistic approximation to the consideration of large relative momenta. In the infinite momentum frame approach the starting point is the parton model, i.e. the hadron structure at extremely small distances.

It is necessary to emphasize that the dispersion approach developed here may also appear to be insufficient in certain cases. The reason is that in this method the dispersion relations are written only over the masses of the composite systems, while the analytic properties of the amplitude in other variables are not tested. An example of a variable for the scattering amplitude is the momentum transfer squared. However, the analytic properties in such variables were intensively investigated in the fifties and sixties, and thus one may hope that these problems will not be confusing.

References

1. I.G. Aznauryan and N.L. Ter-Isaakyan, Yad. Fiz. 31, 1680 (1980); Sov. J. Nucl. Phys. 31, 871 (1980).
2. G.F. Chew, The Analytic S-Matrix, Benjamin, New York (1966).
3. G.F. Chew and S. Mandelstam, Phys. Rev. 119, 467 (1960).
4. B.M. Udgaonkar, in High Energy Physics and Elementary Particles, Vienna (1965).
5. F. Zachariasen, Lectures given at the Pacific Int. Summer School in Physics, Honolulu, Hawaii (1965).
6. L. Castillejo, R.H. Dalitz and F.J. Dyson, Phys. Rev. 101, 453 (1956).
7. E. Salpeter and H. Bethe, Phys. Rev. 84, 1232 (1951).
8. F. Gross, Phys. Rev. 140B, 410 (1965).
9. J. Melosh, Phys. Rev. D6, 1095 (1973).
10. S.J. Brodsky and G.R. Farrar, Phys. Rev. D11, 1309 (1975).
11. M.V. Terentiev, Yad. Fiz. 24, 207 (1976); Sov. J. Nucl. Phys. 24, 106 (1976).
12. V.A. Karmanov, ZETF 71, 399 (1976); Sov. Phys. JETP 44, 210 (1976).
13. R.D. Amado and R.M. Woloshyn, Phys. Rev. Lett. 36, 1435 (1976).
14. P.V. Landshoff and J.C. Polkinghorne, Phys. Rev. 18, 153 (1978).
15. A. Donnachie, R.R. Horgan and P.V. Landshoff, Zeit. f. Phys. C10, 71 (1981).
16. L.L. Frankfurt and M.I. Strikman, Physics Reports 76C, 215 (1981).

17. S. Mandelstam, Phys. Rev. Lett. 4, 84 (1960).

18. R. Blankenbeckler and L.F. Cook, Phys. Rev. 119, 1745 (1960).

19. V.V. Anisovich, ZETF, 41, 1907 (1961); Sov. Phys. JETP 14, 1355 (1962).

20. V.A. Fock, Z. fur Phys. 75, 622 (1932); Sov. Phys. 6, 425 (1934).

Chapter 4

THE CALCULATION OF HIGH ENERGY SCREENING

A significant role in the collision processes of composite systems with a particle (or another composite system) is played by the so-called screening or shadow effects. Here we shall present the technique for the calculation of such effects for high energy collisions.

In an elastic scattering process at moderate energies the main screening effect is connected with the elastic scattering of the projectile and the constituents; it is usually called Glauber screening. As the energy increases, the inelastic screening also becomes significant. The reason is that inelastic processes at moderate energies require large values of the momentum transferred to the constituent, and this leads to the decay of the composite system. For high energies new particles may be produced with small momentum transferred.

The calculation technique will be presented here in a form suitable for the description of both the elastic and the inelastic effects.

4.1 Scattering on a Two-Body System — Glauber Approximation

The most convenient and simple example, which allows one to understand nearly all main features of the calculation technique, is the scattering of an elementary particle on a two-body composite system. Let us consider this example in detail.

We shall assume that the binding energy of the system is small in comparison with the constituent masses. Thus, if the whole system does not move too fast, the non-relativistic approximation may be applied for its description. For the sake of simplicity we shall assume all the colliding particles to be spinless.

In the laboratory frame the composite target is initially at rest and the projectile particle has high energy. If the average distances between the constituents inside the composite system are sufficiently large, the main process will be the scattering of the projectile on one of the constituents (Fig. 4.1a). Since we consider only elastic scattering, the constituents must join each other to form a composite system in the final state. In the other possible process the projectile, scattered by the first constituent, strikes the second one and scatters repeatedly; after the double scattering the constituents are joined into a composite system again (Fig. 4.1b). This process is less

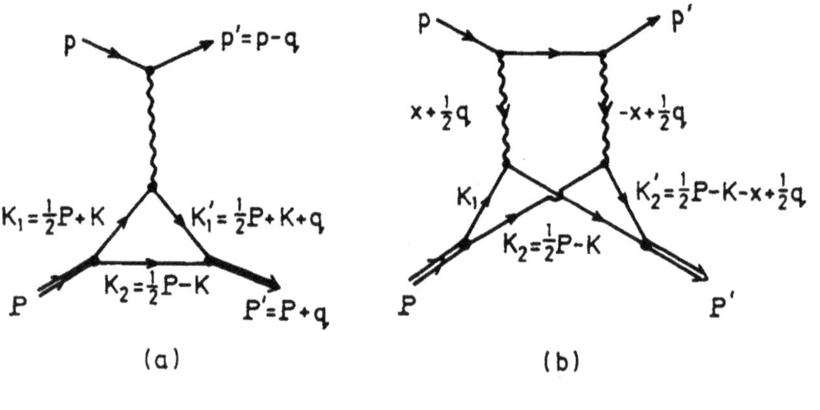

Fig. 4.1 The processes of the single (a) and double (b) scattering of the projectile by the constituents of the composite system.

probable than the first one if the size of the particle-constituent interaction region r_{int} is significantly less than the size of the composite system R: the probability suppression is of the order of r^2_{int}/R^2. The processes of multiple rescatterings have an extra factor 1/S, and for high initial energies we may neglect them. Thus, the total scattering amplitude of the projectile and a two-body system is a sum of two items: single scattering amplitude $A_1(s,q^2)$ and double scattering amplitude $A_2(s,q^2)$:

$$A(s,q^2) = A_1(s,q^2) + A_2(s,q^2) \quad . \tag{4.1}$$

The technique for amplitude calculation in the non-relativistic approximation was considered in the preceding chapter in detail. Using it for the calculation of the projectile interaction with the first constituent we obtain:

$$A_1(1) = \int \frac{d^3kdE}{i(2\pi)^4} \frac{G(\frac{1}{4}(\vec{k}_1 - \vec{k}_2)^2)a_1(s_1,q^2)G(\frac{1}{4}(\vec{k}_1' - \vec{k}_2)^2)}{(-2mE_1 + \vec{k}_1^2 - i0)(-2mE_2 + \vec{k}_2^2 - i0)(-2mE_1' + \vec{k}_1'^2 - i0)} \tag{4.2}$$

Momentum notations are clear from Fig. 4.1a. The vertex functions G of the transitions "composite system → constituents" depend only on the relative constituent momenta; the masses of the constituents are assumed to be equal $m_1 = m_2 = m$. The amplitude of the projectile scattering by the first constituent a_1 depends on the c.m. energy squared $s_1 = (p + k_1)^2 = (p' + k_1')^2$ and the momentum transferred squared $q^2 = (k_1' - k_1)^2$. Strictly speaking, a_1 depends also on the momenta squared k_1^2 and $k_1'^2$, but in (4.2) the on-shell amplitude is taken with $k_1^2 = m^2$, $k_1'^2 = m^2$. Off-shell corrections overdraw the accuracy of the non-relativistic approximation (this point was also considered in Chap. 3).

In the laboratory frame we have

$$P = (2m - \epsilon, 0, 0, 0)$$

$$p = (p_0, 0, 0, p) \simeq (p + \frac{\mu^2}{2p}, 0,0,p) \quad . \tag{4.3}$$

Here ϵ is the binding energy of the composite system; the z-axis is directed along the projectile momentum, μ is the projectile mass $\mu \ll p$.

Particle momenta in the final state are

$$P' \simeq (2m - \epsilon + \frac{\vec{q}_T^{\,2}}{4m} , \vec{q}_T, q_z)$$

$$p' \simeq (p - q_z + \frac{\mu^2 + \vec{q}_T^{\,2}}{2p} , -\vec{q}_T, p - q_z) \qquad (4.4)$$

where $q_z = \frac{\vec{q}_T^{\,2}}{4m} + \frac{\vec{q}_T^{\,2}}{2p}$; $q_z << |\vec{q}_T|$, $q_0 = q_z - \frac{\vec{q}_T^{\,2}}{2p} = \frac{\vec{q}_T^{\,2}}{4m} << |\vec{q}_T|$. Thus $q^2 = q_0^2 - \vec{q}^{\,2} \approx -\vec{q}^{\,2} \approx -\vec{q}_T^{\,2}$.

Enclosing the pole singularity $(-2mE_2 + \vec{k}_2^{\,2} - i0)^{-1}$ by the integration contour in (4.2), we obtain:

$$A_1(1) = \frac{1}{8m} \int \frac{d^3k}{(2\pi)^3} a_1(s_1, q^2) \frac{G(\vec{k}^2)G((\vec{k} + \frac{1}{2}\vec{q})^2)}{(m\epsilon + \vec{k}^2)(m\epsilon + (\vec{k} + \frac{1}{2}\vec{q})^2)} \quad . \qquad (4.5)$$

The wave function of the composite system is $\psi(\vec{k}) = \frac{1}{4\sqrt{m}} G(\vec{k}^2) \Big/ (m\epsilon + \vec{k}^2)$ (see Sec. 3.1), thus (4.5) may be presented in the form:

$$A_1(1) = 2 \int \frac{d^3k}{(2\pi)^3} \psi(\vec{k}) a_1(s_1, q^2) \psi(\vec{k} + \frac{1}{2}\vec{q}) \quad . \qquad (4.6)$$

If the elastic scattering characteristics depend weakly on the initial energy (this is just the case of high-energy collisions), the scattering amplitude may be removed in front of the integral sign in the r.h.s. of Eq. (4.6) taking $s_1 = s/2$, where $s = (p+q)^2 \simeq 4pm$. In this case the amplitude $A_1(1)$ is expressed in terms of the composite system form factor

$$A_1(1) = 2a_1(\tfrac{s}{2}, q^2) F(\vec{q}^{\,2}) \quad . \qquad (4.7)$$

The amplitude of the particle scattering by the second constituent is calculated in the same way and the final expression for $A_1(s,q^2)$ is

$$A_1(s,q^2) = A_1(1) + A_1(2) = 2\left\{a_1(\tfrac{s}{2}, q^2) + a_2(\tfrac{s}{2}, q^2)\right\} F(\vec{q}^{\,2}) \quad . \qquad (4.8)$$

Since the composite system is loose, i.e. the average distances between the constituents are large enough, at small $|\vec{q}^2|$ - values double scattering effects are negligible and a good approximation is $A(s,q^2) \approx A_1(s,q^2)$. In this approximation the total cross section for particle-composite system interaction is a sum of the cross sections of the particle-constituent interactions:

$$\sigma_{tot} \simeq \sigma_{tot}(1) + \sigma_{tot}(2) \quad . \tag{4.9}$$

Here we used the unitarity condition for the amplitude of scattering by the composite system $\mathrm{Im}\, A(s,0) = s\sigma_{tot}$ and that for the scatterings by the constituents $\mathrm{Im}\, a_i(\frac{s}{2},0) = \frac{s}{2}\cdot\sigma_{tot}(i)$ (see Appendix B).

Let us consider double scattering processes, i.e. the amplitude $A_2(s,q^2)$. One of the processes, which contribute to A_2, is shown in Fig. 4.1b: the projectile collides with the first constituent and then with the second one. We will denote the amplitude of this process as $A_2^{e\ell}(12)$. The other possible double scattering process is obtained from the first one via the permutation $1 \rightleftarrows 2$; its amplitude will be denoted as $A_2^{e\ell}(21)$.

The amplitude in Fig. 4.1b is:

$$A_2^{e\ell}(12) = \int \frac{d^3k_1 dE_1}{i(2\pi)^4} \cdot \frac{d^3k_2' dE_2'}{i(2\pi)^4} \cdot \frac{G(\frac{1}{4}(\vec{k}_1 - \vec{k}_2)^2)}{(-2mE_1 + \vec{k}_1^2 - i0)}$$

$$\times \frac{\dfrac{a_1(s_1,\, t_1,\, M^2)a_2(s_2,\, t_2,\, M^2)}{\mu^2 - M^2} G(\frac{1}{4}(\vec{k}_1' - \vec{k}_2')^2)}{(-2mE_2 + \vec{k}_2^2 - i0)(-2mE_1' + \vec{k}_1'^2 - i0)(-2mE_2' + \vec{k}_2'^2 - i0)} \, . \tag{4.10}$$

The momentum notations are shown in Fig. 4.1b. The amplitudes a_1 and a_2 depend on the squares of the invariant energies and momenta transferred $s_1 = (p + k_1)^2$, $s_2 = (p' + k_2')^2$, $t_1 = (k_1 - k_1')^2$, $t_2 = (k_2 - k_2')^2$. They are taken on the mass shell of the constituents $k_1^2 = k_1'^2 = k_2^2 = k_2'^2 = m^2$ (as in Eq. (4.2)), but off-shell for

$M^2 = (p - x - \frac{1}{2}q)^2 \neq \mu^2$. This is essential for our further calculations.

The values of kinetic energies E_i and E_i' which are significant in the integral (4.10), are much less than the space components of the momenta $|\vec{k}_i|$ and $|\vec{k}_i'|$. Hence the amplitudes a_i are approximately independent of E_i and E_i'. Indeed:

$$
\begin{aligned}
s_1 &\simeq 2pm - 2pk_z \\
s_2 &\simeq 2pm + 2p(k_z + x_z) \\
t_1 &\simeq -(\vec{x} + \tfrac{1}{2}\vec{q}_T)^2 = -(\vec{x}_T + \tfrac{1}{2}\vec{q}_T)^2 - x_z^2 \\
t_2 &\simeq -(-\vec{x} + \tfrac{1}{2}\vec{q}_T)^2 = -(-\vec{x}_T + \tfrac{1}{2}\vec{q}_T)^2 - x_z^2 \\
M^2 &= (p - x - \tfrac{1}{2}q)^2 \simeq 2px_z + \mu^2 \,.
\end{aligned}
\tag{4.11}
$$

Thus the integrations over E_1 and E_2' in (4.10) may be carried out explicitly. Enclosing the pole singularities $(-2mE_2 + \vec{k}_2^2 - i0)^{-1}$ and $(-2mE_1' + \vec{k}_1'^2 - i0)^{-1}$ by the corresponding integration contours, we obtain:

$$
A_2^{e\ell}(12) = \frac{1}{m}\int \frac{d^3k}{(2\pi)^3}\frac{d^3x}{(2\pi)^3}\frac{a_1(s_1, t_1, M^2)a_2(s_2, t_2, M^2)}{\mu^2 - M^2 - i0}
\times \frac{1}{16m}\frac{G(\vec{k}^2)G((\vec{x} + \vec{k})^2)}{(me + \vec{k}^2)(me + (\vec{x} + \vec{k})^2)} \,.
\tag{4.12}
$$

The second factor in the integrand in the r.h.s. of Eq. (4.12) is a product of the wave functions $\psi(\vec{k})\psi(\vec{x}+\vec{k})$. The first factor corresponds to the process in Fig. 4.2a. As in the case of single scattering, a good approximation for a_i is here $s_1 \simeq s_2 \simeq s/2$. In this case the first factor does not depend on k at all, while the second one gives the form factor of the composite system: $\int \frac{d^3k}{(2\pi)^3}\psi(\vec{k})\psi(\vec{x} + \vec{k}) = F(4\vec{x}^2)$. Thus

$$A_2^{e\ell}(12) = \frac{1}{m}\int \frac{d^3x}{(2\pi)^3}\,\frac{a_1(\frac{s}{2},\, t_1,\, M^2)\,a_2(\frac{s}{2},\, t_2,\, M^2)}{\mu^2 - M^2 - i0}\,F(4\vec{x}^2) \quad . \qquad (4.13)$$

The region of integration over $\vec{x}$ is practically determined by the form factor $F(4\vec{x}^2)$; the values of $|\vec{x}|$ which contribute into (4.13) significantly are of the order of $|\vec{x}| \sim 1/R$, where R is the radius of the composite system $(R \sim 1/\sqrt{m\epsilon})$. Hence the value of $M^2 \approx 2px_z$ is of the order of $M^2 \sim p/R$, i.e. it is large enough at high initial energies. This means that the region of integration in (4.13) includes the interval of M^2 values where virtual transitions of one particle into few ones (e.g. into three in Fig. 4.2b) are possible. It is a

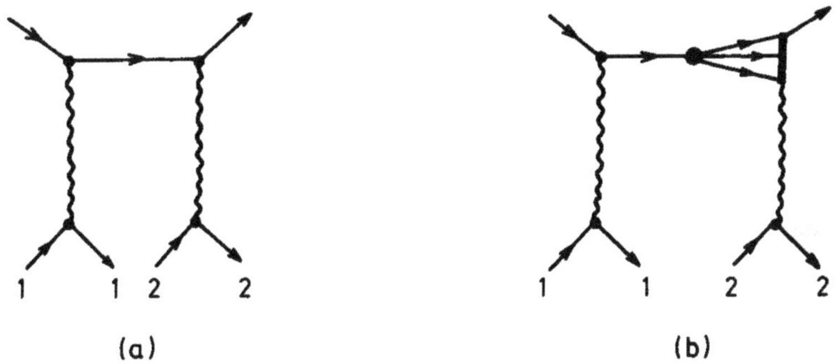

Fig. 4.2 Elastic (a) and inelastic (b) screening for hadron-deuteron collisions

significant point for the consideration of the general case of shadow effects, which may be caused by the inelastic rescatterings.

When only elastic scattering is taken into account, the Glauber approximation[1,2] corresponds to the replacement of the pole factor $(\mu^2 - M^2)^{-1}$ in the r.h.s. of Eq. (4.13) by its semiresidue:

$$(\mu^2 - M^2)^{-1} \to i\pi\delta(M^2 - \mu^2) \quad . \qquad (4.14)$$

We shall consider the status of this assumption a little further, when inelastic shadow corrections will be calculated. The replacement (4.14) gives:

$$A_2^{e\ell}(12) = \frac{i}{s}\int\frac{d^2x_T}{(2\pi)^2}\, a_1(\tfrac{s}{2},\, t_+)a_2(\tfrac{s}{2},\, t_-)F(4\vec{x}_T^2)$$

$$t_\pm = -(\pm\vec{x}_T + \tfrac{1}{2}\vec{q}_T)^2 \quad . \tag{4.15}$$

Here the amplitudes a_1 and a_2 are on-shell for all the external particles, i.e. they are physical amplitudes.

As it was mentioned, there exists a second elastic screening process, which differs from Fig. 4.1b by the permutation $1 \rightleftarrows 2$. The amplitudes of both processes are equal to each other:

$$A_2^{e\ell}(12) = A_2^{e\ell}(21) \tag{4.16}$$

and the total elastic shadow correction is the sum of these amplitudes

$$A_2^{e\ell}(s,q^2) = A_2^{e\ell}(12) + A_2^{e\ell}(21) \quad . \tag{4.17}$$

So, the total amplitude $A(s,q^2)$ with both single and double scatterings on the constituents included is

$$A(s,q^2) = 2[a_1(\tfrac{s}{2},\, q^2) + a_2(\tfrac{s}{2},\, q^2)]F(\vec{q}^2)$$
$$+ \frac{2i}{s}\int\frac{d^2x_T}{(2\pi)^2}\, a_1(\tfrac{s}{2},\, t_+)a_2(\tfrac{s}{2},\, t_-)F(4\vec{x}_T^2) \quad . \tag{4.18}$$

In order to illustrate the role of the shadow corrections, let us consider a simple example, in which the momentum transferred dependence of the scattering amplitudes and of the form factor is approximated by a simple exponent:

$$\frac{2}{s}\, a_1(\tfrac{s}{2},\, q^2) = i\sigma_{tot}(1)(1 - i\rho_1)\exp(-\frac{b_1}{2}\,\vec{q}^2)$$

$$\frac{2}{s}\, a_2(\tfrac{s}{2},\, q^2) = i\sigma_{tot}(2)(1 - i\rho_2)\exp(-\frac{b_2}{2}\,\vec{q}^2) \quad . \tag{4.19}$$

Here ρ is the ratio of the real part of the amplitude to the imaginary part: $\rho_i = \mathrm{Re}\, a_i/\mathrm{Im}\, a_i$. For high energies $\rho \ll 1$.

Similarly the form factor may be approximated by a one-parameter dependence through the radius of the composite system R:

$$F(\vec{q}^{\,2}) = \exp(-\frac{R^2}{6}\vec{q}^{\,2}) \quad . \tag{4.20}$$

A normal relation for the composite system is $R^2 \gg b_i\ R^2 \gg \sigma_{tot}(i)$ (i.e. the size of the composite system is appreciably larger than that of the constituents). In this approximation

$$\frac{1}{s} A(s,\vec{q}^{\,2}) = i \Big\{ [\sigma_{tot}(1) + \sigma_{tot}(2)] e^{-\frac{R^2}{6}\vec{q}^{\,2}} - \frac{\sigma_{tot}(1)\sigma_{tot}(2)}{\frac{16}{3}\pi R^2} e^{-\frac{q^2}{8}(b_1+b_2)} \Big\} + (\rho_1\sigma_{tot}(1) + \rho_2\sigma_{tot}(2)) e^{-\frac{R^2}{6}\vec{q}^{\,2}} \quad . \tag{4.21}$$

All specific qualitative features of the behaviour of the differential cross section on the composite system are clear from Eq. (4.21). The scattering amplitude is determined mainly by its imaginary part — the real part is comparatively small. At small $\vec{q}^{\,2}$ the main contribution comes from the first item in the braces (single scattering). This contribution decreases rapidly with the growth of $\vec{q}^{\,2}$. The contribution of the shadow correction (the second item in the braces) is small at small $\vec{q}^{\,2}$, but it decreases slowly with the growth of $\vec{q}^{\,2}$. At some $\vec{q}^{\,2} = \vec{q}^{\,2}_{min}$ the contributions of the single and double scatterings to the imaginary part of $A(s,q^2)$ cancel each other, and $\mathrm{Im}\, A(s,\vec{q}^{\,2}_{min}) = 0$. For $\vec{q}^{\,2} > \vec{q}^{\,2}_{min}$ the main contribution to the amplitude comes from the double scattering process. At $\vec{q}^{\,2} \simeq \vec{q}^{\,2}_{min}$ a dip in the differential cross section is to be observed. The value of the cross section in this region is determined by the small value of the real part of the amplitude. The differential cross section, which corresponds to the amplitude (4.21) is shown in Fig. 4.3.

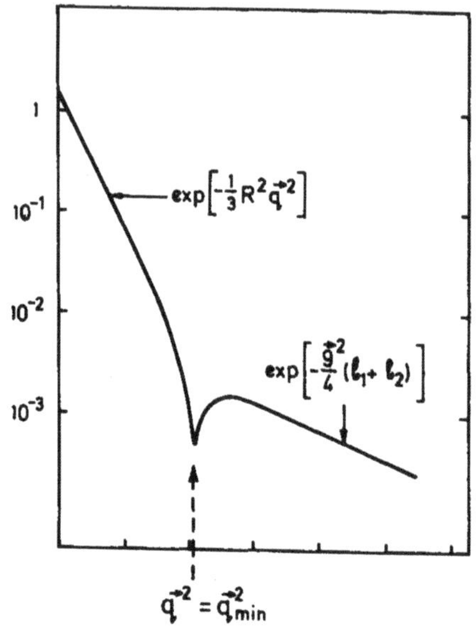

Fig. 4.3 Differential cross section of the elastic scattering by the composite system as a function of the squared momentum transfer $\vec{q}^{\,2}$.

The total cross section of the scattering on the composite system in the approximation of Eq. (4.21) is:

$$\sigma_{tot} = \sigma_{tot}(1) + \sigma_{tot}(2) - \frac{\sigma_{tot}(1)\sigma_{tot}(2)}{\frac{16}{3}\pi R^2} \quad . \tag{4.22}$$

The last item, the elastic shadow correction $\delta\sigma_{e\ell}$ diminishes the value of the cross section. Traditionally the elastic shadow correction is written as $\delta\sigma_{e\ell} = \sigma_{tot}(1)\sigma_{tot}(2)/4\pi < R^2 >$. Such a form of the expression for $\delta\sigma_{e\ell}$ is prompted by the analogy of the considered problem with the scattering of light on a system of two black spheres with the radii $r_i = \sqrt{\frac{\sigma_{ine\ell}(i)}{2\pi}}$. Though the values $<R^2>$ and R^2 are not expected to be exactly equal to each other their difference is not large.

4.2 Scattering on a Two-Body System — Inelastic Screening

In order to account for inelastic rescattering it is necessary to replace the amplitude of Fig. 4.2a in Eq. (4.13) by the amplitude which corresponds to the sum of diagrams in Fig. 4.4; we shall denote it by $A_{3\to 3}(12)$. The total shadow correction includes also the process with the permutation $1 \rightleftarrows 2$. Let us define the amplitude

$$A_{3\to 3}(s_1,\ t_1,\ M^2,\ s_2,\ t_2) = A_{3\to 3}(12) + A_{3\to 3}(21) \tag{4.23}$$

where the invariant variables are

$$\begin{aligned} s_1 &= (p + k_1)^2 \qquad & s_2 &= (p' + k_2')^2 \\ t_1 &= (k_1 - k_1')^2 \qquad & t_2 &= (k_2 - k_2')^2 \\ M^2 &= (p + k_1 - k_1')^2 \ . \end{aligned} \tag{4.24}$$

The momentum notations are shown in Fig. 4.4.

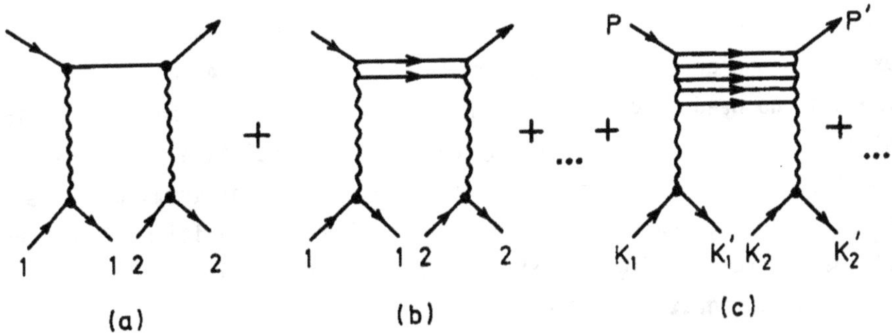

Fig. 4.4 Sum of the diagrams for the amplitude $A_{3\to 3}(12)$.

The total shadow correction is determined from Eq. (4.13) with the replacement

$$\frac{a_1(\frac{s}{2},\ t_1,\ M^2)a_2(\frac{s}{2},\ t_2,\ M^2)}{\mu^2 - M^2 - i0} \to A_{3\to 3}(\tfrac{s}{2},\ t_1,\ M^2,\ \tfrac{s}{2},\ t_2) \ . \tag{4.25}$$

Then

$$A_2(s,q^2) = \frac{1}{m}\int \frac{d^3x}{(2\pi)^3} A_{3\to3}(\tfrac{s}{2}, t_1, M^2, \tfrac{s}{2}, t_2)F(4\vec{x}^2) \quad . \tag{4.26}$$

Here t_1 and t_2 are related to $\vec{x}$ through Eq. (4.11), while x_z may be expressed via M^2:

$$x_z \simeq \frac{M^2 - \mu^2}{2p} \quad . \tag{4.27}$$

So in Eq. (4.26) we may replace the integration over x_z by that over M^2 along the real axis from $-\infty$ to $+\infty$.

Let us consider the singularities of the amplitude $A_{3\to3}$ in the complex plane of M^2. The diagram of Fig. 4.4a has a pole singularity at $M^2 = \mu^2$; the integration contour in Eq. (4.26) lies over this singularity. The diagram of Fig. 4.4b has a threshold singularity at $M^2 = M^2_{thr}$. In the case when two particles of the same kind as the projectile are produced in the intermediate state, as it is shown in Fig. 4.4b, $M^2_{thr} = 4\mu^2$. The integration contour lies over this singularity and over the associated cut. (This can be proved easily by the explicit consideration of any example of the diagram 4.4b. However, there is no need to carry out calculations; instead of that, the analogy with the diagram of Fig. 4.4a may be used: the relative positions of the two-particle singularity and the integration contour must be the same as for a one-particle singularity.) The singularities of the rest of the diagrams in Fig. 4.4 are also on the real axis at $M^2 > M^2_{thr}$ under the integration contour (see Fig. 4.5).

The other set of singularities comes from the diagrams of Fig. 4.4 with the permutation $1 \rightleftarrows 2$. Let us define for them the value M'^2, similar to M^2:

$$M'^2 = (p + k_2 - k_2')^2 = 2\mu^2 - 2\vec{x}^2 + \frac{1}{2}\vec{q}^2 - M^2 \quad . \tag{4.28}$$

The singularities are situated in the points $M'^2 = \mu^2,\ 4\mu^2,$ etc. From Eq. (4.28) one may see that they lie on the left real semiaxis M^2 over the integration contour (see Fig. 4.5).

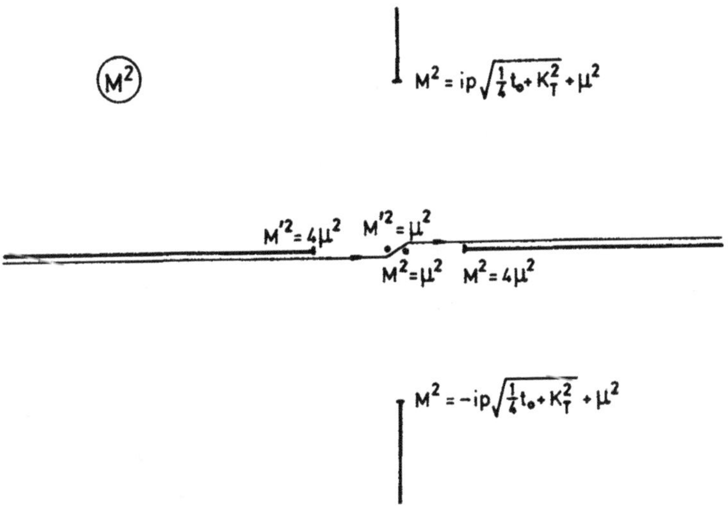

Fig. 4.5 Positions of the singularities of the integrand in Eq. (4.26) in the complex M^2 plane.

The singularities of the form factor also contribute to the integrand in Eq. (4.26). The form factor $F(-t)$ has singularities for positive t values. Let us define t_0 as a singularity of $F(-t)$ which is the nearest one to the physical region $t < 0$ (for example, in the considered case of the two-body composite system with the equal masses of the constituents $t_0 = 16m\varepsilon$). Then the singularities of the integrand are in the point $-4\vec{x}^2 = t_0$, i.e. at

$$M^2 = \pm\ ip\sqrt{\frac{t_0}{4} + \vec{x}_T^2}\ . \qquad (4.29)$$

The other singularities of the form factor are also near the imaginary M^2 axis, but farther from the small M^2 region; that is why the cuts from the singularities (4.29) are directed in Fig. 4.5 alongside the imaginary axis.

As the projectile momentum p increases, the form factor singularities (4.29) move away from the region of small M^2. However, the convergence of the integral (4.26) depends on the decrease of $F(4\vec{x}^2)$

with the growth of its argument, and the mentioned movement of the singularities is accompanied by the growth of M^2 values, which are significant in the integral (4.26). Thus the singularities of the form factor are always to be considered, but only during the integration over large $|M^2|$ values (which still obey the inequality $|M^2|/s \ll 1$, which follows from Eq. (4.27) and the condition $|x_z|/m \ll 1$). As a result, the calculation of (4.26) may be handled in the following way.

Let us introduce the asymptotic value of the amplitude $A_{3\to 3}$, which at $s \gg M^2$ and $M^2 > M_0^2$ (where M_0^2 is some large value) obeys the condition:

$$A_{3\to 3}(\frac{s}{2},\, t_1,\, M^2,\, \frac{s}{2},\, t_2) - A^{(asym)}_{3\to 3}(\frac{s}{2},\, t_1,\, M^2,\, \frac{s}{2},\, t_2) \simeq 0 \quad . \qquad (4.30)$$

The region of high M^2 values (but still $M^2 \ll s$) is the region of three-Reggeon behaviour for the amplitude $A_{3\to 3}$ (see Sec. 2.3). Here $A^{(asym)}_{3\to 3}$ may be expanded into a series over three-Reggeon amplitudes $A(R_iR_kR_1)$:

$$A^{(asym)}_{3\to 3} = \sum_{ik\ell} C_{ik\ell}\, A(R_iR_kR_\ell) \quad . \qquad (4.31)$$

The discontinuity of the amplitude $A^{(asym)}_{3\to 3}$ determines the cross section of the diffraction-dissociation processes at large M^2. There are many papers devoted to the phenomenological description of the diffraction-dissociation processes in terms of the expansion (3.31), i.e. to the determination of the coefficients $C_{ik\ell}$ from the fit of the experimental data (see Sec. 2.3). We will assume further that the value of $A^{(asym)}_{3\to 3}$ is known from these phenomenological investigations.

Let us present the amplitude (4.26) in the form:

$$A_2(s,q^2) = \frac{1}{pm}\int \frac{d^2x_T}{(2\pi)^2}\int_{-\infty}^{\infty}\frac{dM^2}{2\pi}\left[A_{3\to 3} - A^{(asym)}_{3\to 3}\right] F\left(4\vec{x}_T^{\,2} + \frac{(M^2-\mu^2)^2}{p^2}\right)$$

$$+ \frac{1}{pm}\int \frac{d^2x_T}{(2\pi)^2}\int_{-\infty}^{\infty}\frac{dM^2}{2\pi}\, A^{(asym)}_{3\to 3}\, F\left(4\vec{x}_T^{\,2} + \frac{(M^2-\mu^2)^2}{p^2}\right) . \qquad (4.32)$$

The convergence of the first integral in the r.h.s. of Eq. (4.32) is guaranteed by the rapid decrease of the difference $[A_{3\to3} - A_{3\to3}^{(asym)}]$, thus the singularities of the form factor may be neglected in this integral and the integration contour may be deformed as shown in Fig. 4.6 (from the position I to the position II). The resulting integration interval of M^2 values is $(\mu^2, +\infty)$ and the integrand is to be replaced by its discontinuity:

$$[A_{3\to3} - A_{3\to3}^{(asym)}] \to 2i \text{ disc } [A_{3\to3} - A_{3\to3}^{(asym)}] \tag{4.33}$$

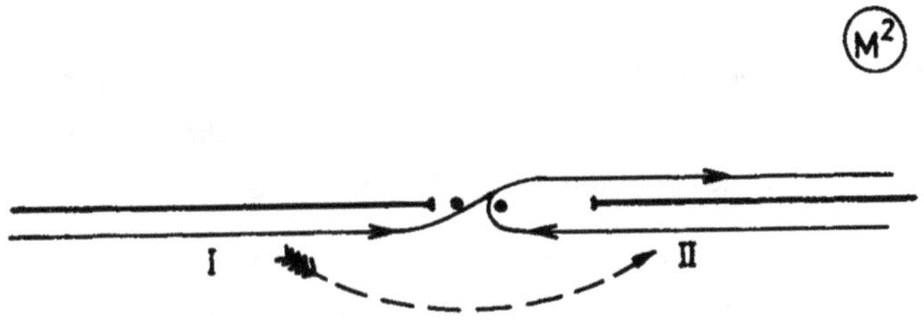

Fig. 4.6 Deformation of the integration contour (dashed arrow) in Eq. (4.32) from position I to position II.

Near the pole singularity of the diagram Fig. 4.4a the discontinuity of $A_{3\to3}$ is

$$\text{disc } A_{3\to3}(M^2 \sim \mu^2) = \pi\delta(M^2 - \mu^2) a_1(\tfrac{s}{2}, t_1, \mu^2) a_2(\tfrac{s}{2}, t_2, \mu^2) \ . \tag{4.34}$$

With the increase of M^2 the non-zero discontinuity of $A_{3\to3}$ appears at $M^2 > 4\mu^2$, where it may be expressed in terms of the production amplitudes of two, three, etc. particles (see Appendix B). The resulting expression is

$$A_2(S,q^2) = \frac{i}{2pm}\int \frac{d^2x_T}{(2\pi)^2}\, a_1(\tfrac{s}{2}\,,\, t_+)a_2(\tfrac{s}{2},\, t_-)F(4\vec{x}_T^2)$$

$$+ \frac{i}{2pm}\int \frac{d^2x_T}{(2\pi)^2}\int_{4\mu^2}^{\infty} \frac{dM^2}{\pi}\ \mathrm{disc}\ A_{3\to 3}(\tfrac{s}{2},\, t_1,\, M^2,\, \tfrac{s}{2},\, t_2)F\left(4\vec{x}_T^2 + \frac{(M^2-\mu^2)}{p^2}\right)$$

$$+ \frac{1}{2pm}\int \frac{d^2x_T}{(2\pi)^2}\int_0^{\infty} \frac{dM^2}{\pi}\left\{A_{3\to 3}^{(asym)}(\tfrac{s}{2},\, t_1,\, M^2,\, \tfrac{s}{2},\, t_2)\right.$$

$$\left. - i\ \mathrm{disc}\ A_{3\to 3}^{(asym)}(\tfrac{s}{2},\, t_1,\, M^2,\, \tfrac{s}{2},\, t_2)\right\} \times F\left(4\vec{x}_T^2 + \frac{(M^2-\mu^2)^2}{p^2}\right) \ . \qquad (4.35)$$

The first term in the r.h.s. of Eq. (4.35) is already known — it is the elastic shadow correction (Eqs. (4.15)-(4.17)). The other two terms give the inelastic shadow correction. All the values in (4.35) can in principle be determined from the experimental data.

In order to understand the structure of the Eq. (4.35) better, let us consider it in a simple case when the constituents are identical and $q^2 = 0$. Besides that, we shall neglect the real part of $A_2(s,0)$. Then $t_+ = t_- = -\vec{x}_T^2$ and $t_1 = t_2 = -\vec{x}_T^2 - \frac{(M^2-\mu^2)}{4p^2} \equiv t$, and the amplitudes of the first two items in the r.h.s. of Eq. (4.35) may be expressed in terms of the elastic differential cross section $d\sigma/d\vec{x}_T^2$ and the inclusive cross section in the region of diffraction dissociation $d^2\sigma/d\vec{x}_T^2 dM^2$:

$$-\frac{1}{s}\mathrm{Im}\ A_2(s,0) = 2\int_0^{\infty} d\vec{x}_T^2\, \frac{d\sigma}{d\vec{x}_T^2}\,(\tfrac{s}{2},\, \vec{x}_T^2)F(4\vec{x}_T^2)$$

$$+ 2\int_0^{\infty} d\vec{x}_T^2\int_{4\mu^2}^{\infty} dM^2\, \frac{d^2\sigma}{d\vec{x}_T^2 dM^2}\, F\left(4\vec{x}_T^2 + \frac{(M^2-\mu^2)^2}{p^2}\right)$$

$$+ \frac{2}{(8\pi pm)^2}\int_0^{\infty} d\vec{x}_T^2\int_0^{\infty} dM^2\left\{\mathrm{Im}\ A_{3\to 3}^{(asym)}(\tfrac{s}{2},\, t,\, M^2,\, \tfrac{s}{2},\, t)\right.$$

$$\left. - \mathrm{disc}\ A_{3\to 3}^{(asym)}(\tfrac{s}{2},\, t,\, M^2,\, \tfrac{s}{2},\, t)\right\}F\left(4\vec{x}_T^2 + \frac{(M^2-\mu^2)^2}{p^2}\right) \ . \qquad (4.36)$$

The notations in the r.h.s. of Eq. (4.36) are illustrated by Fig. 4.7.

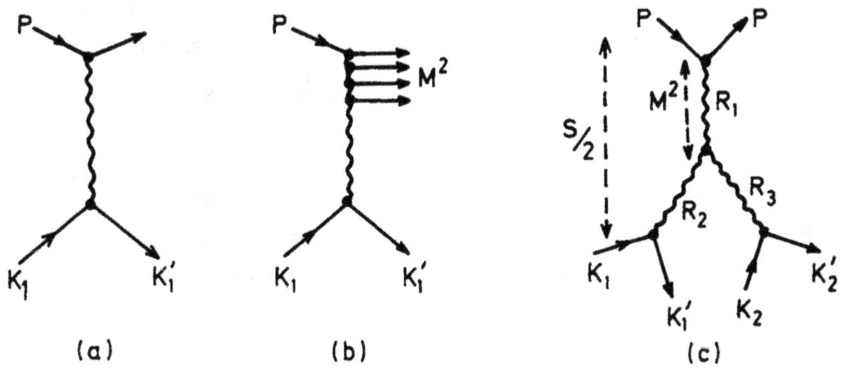

Fig. 4.7 (a) Elastic scattering process: $(k_1-k_1')^2 = -\vec{x}_T^2$, $s/2 = (p+k_1)^2$.

(b) Process of the particle-cluster production with the mass M, $M^2 = (p+k_1-k_1')^2$. The squared momentum transfer to the recoil particle is $(k_1-k_1')^2 \simeq -x_T^2 - (M^2-\mu^2)^2/4p^2$.

(c) Triple-reggeon amplitude $A\,(R_1R_2R_3)$: the sum of such amplitudes determines $A_{3\to 3}$ (see Eq. (4.31)). $s/2 = (p+k_1)^2$, $M^2 = (p+k_1-k_1')^2$, $t = (k_1-k_1')^2 = -\vec{x}_T^2 - (M^2-\mu^2)^2/4p^2$.

The first term in the r.h.s. of Eq. (4.36) is the Glauber shadow correction to the total cross section (i.e. the last term of the Eq. (4.22) in the case $\sigma_{tot}(1) = \sigma_{tot}(2)$). The second one is the inelastic shadow correction[3,4]. It has the same sign, as the elastic one and also diminishes the cross section value. The last item is negative and gives antishadow (or antiscreening) effect[5]. In order to understand it let us see which Reggeons give contributions to the last term in the r.h.s. of the Eq. (4.36).

In the region of relatively small M^2 (about $M^2 \sim 5\text{-}30\ \text{GeV}^2$) the main contribution to the inclusive cross section (and to $A_{3\to 3}^{(asym)}$ as well) comes from the three-Reggeon diagrams with $R_2 = R_3 = P$ (see Fig. 4.7c). For these diagrams $\text{Im}\, A_{3\to 3}^{(asym)} = \text{disc}\, A_{3\to 3}^{(asym)}$. This property is true also for the three-Pomeron diagram, which gives the main contribution for large s in the region $M^2 \sim \sqrt{m^2 s} >> 5\ \text{GeV}^2$. When $m^2 \frac{s}{M^2} \sim (5\text{-}30)\ \text{GeV}^2$ the main contribution comes from the diagrams with $R_1 = P,\ R_2, R_3 \neq P$. For these diagrams $\text{Im}\, A_{3\to 3}^{(asym)} < \text{disc}\, A_{3\to 3}^{(asym)}$. The cancellation of the integrand in the last item of the r.h.s. of the Eq. (4.36) may be shown explicitly by the change of the integration limits $\int_0^\infty dM^2 \to \int_{M_0^2}^\infty dM^2$ where $M_0^2 \sim (5\text{-}30)\ \text{GeV}^2$. Finally we get:

$$
\begin{aligned}
-\frac{1}{s}\text{Im}\, A_2(s,0) &= 2\int_0^\infty d\vec{x}_T^{\,2} \frac{d\sigma}{d\vec{x}_T^{\,2}} \left(\frac{s}{2}, \vec{x}_T^{\,2}\right) F(4\vec{x}_T^{\,2}) + 2\int_{4\mu^2}^{M_0^2} dM^2 \int_0^\infty d\vec{x}_T^{\,2} \\
&\times \frac{d^2\sigma}{d\vec{x}_T^{\,2} dM^2}\left(\frac{s}{2}, \vec{x}_T^{\,2}, M^2\right) F\left(4\vec{x}_T^{\,2} + \frac{(M^2-\mu^2)^2}{p^2}\right) + \frac{2}{(8\pi pm)^2}\int_0^\infty d\vec{x}_T^{\,2} \int_{M_0^2}^\infty dM^2 \\
&\times \text{Im}\, A_{3\to 3}^{(asym)}\left(\frac{s}{2}, t, M^2, \frac{s}{2}, t\right) F\left(4\vec{x}_T^{\,2} + \frac{(M^2-\mu^2)^2}{p^2}\right) \quad . \qquad (4.37)
\end{aligned}
$$

In deriving (4.37) we used the fact that at $M^2 > M_0^2$, $\frac{1}{(8\pi pm)^2}\, \text{disc}\, A_{3\to 3}^{(asym)} = d^2\sigma/d\vec{x}_T^{\,2}\, dM^2$.

The only composite system for which a detailed analysis of the elastic and inelastic shadow corrections has been made, is the deuteron. In the Table 4.1 the values of different contributions to the total shadow correction for pd scattering at $P_{lab} = 10^3\ \text{GeV/c}$[6)] are presented as an illustration of the Eq. (4.37). One can see, that the inelastic screening is about 30% of the elastic one. The numbers in the Table 4.1 characterize the possible accuracy of the calculations at the available energies in the case when inelastic screening is neglected.

The inelastic correction grows with the energy increase as $\ell n\, p$. According to the calculations of Ref. 6, when p increases from

Table 4.1 The values of the elastic screening ($\delta\sigma_{e\ell}$ — the first term in the r.h.s. of the Eq. (4.37)); of the resonance screening ($\delta\sigma_{res}$ — the second term at $M_0^2 = 7\ \text{GeV}^2$) and of the three-Pomeron screening ($\delta\sigma_{PPP}$ — one of the contributions to the third item) for pd scattering at $p_{lab} = 10^3$ GeV/c. The sum of the remaining three-Reggeon contributions $\sum_R \delta\sigma_{PRR}$ is negligible.

$\delta\sigma_{e\ell}$(mb)	$\delta\sigma_{ine\ell} = \delta\sigma_{res} + \delta\sigma_{PPP} + \sum_R \delta\sigma_{PRR}$		
	$\delta\sigma_{res}$(mb)	$\delta\sigma_{PPP}$(mb)	$\sum_R \delta\sigma_{PRR}$(mb)
3.62	0.52	0.55	0.06

10^2 GeV/c to 10^3 GeV/c, the value of the inelastic correction increases by about 0.45 mb. It is difficult to estimate the role of inelastic screening at superhigh energies: it depends on the relation between the growth of the elastic cross section and that of the diffraction dissociation.

If, in the shadow correction calculations, the antiscreening item is neglected, i.e. $\text{Im}\, A_{3\to3}^{(asym)}$ is replaced by $\text{disc}\, A_{3\to3}^{(asym)}$ in the r.h.s. of the Eq. (4.37) (see Ref. 7), the contribution of the large M^2 region will increase by about 0.5 mb (due to the replacement $\sum_R \delta\sigma_{PRR} = 0.06\ \text{mb} \to 0.6\ \text{mb}$). In fact, the antishadow item cancels the contribution of the large M^2 region. This allows us to apply a phenomenological trick in the Eq. (4.37): the third item is neglected, while M_0^2 in the second one is considered as a free parameter.

4.3 Scattering on N-Body Systems. General Formulas

Our aim is to obtain here the formulae for the scattering of a fast particle on a composite system of N constituents. We will consider the same case as in Sec. 3.1, namely: the constituents are different particles with equal masses.

The amplitude of n repeated elastic collisions on a slow N-body composite system (see Fig. 4.8) is:

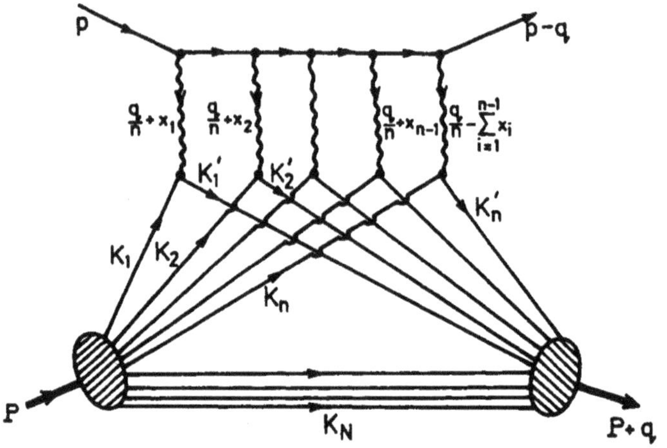

Fig. 4.8 n-fold scattering off the N-particle composite system.

$$\int \frac{G(k_{12}^2,\ldots)}{-2mE_1 + \vec{k}_1^2 - i0} \prod_{j=2}^{N} \left(\frac{-i(2\pi)^{-4} dE_j d^3k_j}{-2mE_j + \vec{k}_j^2 - i0} \right) \prod_{\ell=1}^{n-1} \left(\frac{-i(2\pi)^{-4} dE'_\ell d^3k'_\ell}{-2mE'_\ell + \vec{k}'^2_\ell - i0} \right)$$

$$\times \frac{G(k'^2_{12},\ldots)}{-2mE'_n + k'^2_n - i0} \cdot \frac{a_1(s_1, t_1)\ldots a_n(s_n, t_n)}{(\mu^2 - M_{12}^2 - i0)\ldots(\mu^2 - M_{n-1,n}^2 - i0)} \quad . \tag{4.38}$$

The momentum labels are illustrated by Fig. 4.8. In Eq. (4.38) the following conditions are fulfilled:

$$\vec{k}'_\ell - \vec{k}_\ell = \frac{\vec{q}}{n} + \vec{x}_\ell \ , \quad \vec{P} = \sum_{i=1}^{N} \vec{k}_i \ , \quad \vec{P}' = \sum_{i=1}^{n} \vec{k}'_i + \sum_{i=n+1}^{N} \vec{k}_i \ ,$$

$$E_0 = -e + \frac{\vec{P}^2}{2Nm} = \sum_{i=1}^{N} E_i \ , \quad E'_0 = -e + \frac{\vec{P}'^2}{2Nm} = \sum_{i=1}^{n} E'_i + \sum_{i=n+1}^{N} E_i \ . \tag{4.39}$$

The toal energies squared s_ℓ and the momenta transferred squared t_ℓ, which are the arguments of the amplitudes a_ℓ, are for large p values equal to:

$$s_\ell \simeq 2mp, \quad t_\ell \simeq -\left(\frac{\vec{q}}{n} + \vec{x}_\ell\right)^2 = -\left(\frac{\vec{q}}{n} + \vec{x}_{\ell T}\right)^2 - x_{\ell z}^2 \quad . \tag{4.40}$$

In the same approximation

$$\mu^2 - M_{\ell-1,\ell}^2 = \mu^2 - \left(p - \sum_{i=1}^{\ell-1}\left(\frac{q}{n} + x_i\right)\right)^2 \simeq -2p \sum_{i=1}^{\ell-1} x_{iz} \quad . \tag{4.41}$$

In order to obtain the amplitude in the Glauber approximation it is necessary, according to Eqs. (4.33) and (4.34), to replace in Eq. (4.38) $(\mu^2 - M_{\ell-1,\ell}^2)^{-1}$ by $2\pi i\delta(M_{\ell-1,\ell}^2 - \mu^2)$. Carrying out the integrations over $E_2, \ldots E_N$ and $E_1', \ldots E_{n-1}'$ we obtain for the amplitude of the projectile interaction with the constituents 1,2,...,n:

$$A_{1,2,\ldots n}^{e\ell}(s,q^2) = \int \prod_{j=1}^{N} \frac{d^3k_j}{2m(2\pi)^3} \prod_{\ell=1}^{n} \frac{d^3k_\ell'}{2m(2\pi)^3}$$

$$\cdot \frac{G(k_{12,\ldots}^2)G(k_{12,\ldots}'^2)}{\left(E_0 - \sum_{j=1}^{N} \frac{\vec{k}_j^2}{2m}\right)\left(E_0' - \sum_{j=1}^{n} \frac{\vec{k}_j'^2}{2m} - \sum_{j=n+1}^{N} \frac{\vec{k}_j^2}{2m}\right)}$$

$$\times (2\pi)^3\delta^3\left(\vec{P} - \sum_{j=1}^{N} \vec{k}_j\right)(2\pi)^3\delta^3\left(\vec{P}' - \sum_{j=1}^{n} \vec{k}_j' - \sum_{j=n+1}^{N} \vec{k}_j\right) a_1(s_1, t_1)\ldots a_n(s_n, t_n)$$

$$\times (2\pi i)\delta(M_{12}^2 - \mu^2)\ldots(2\pi i)\delta(M_{n-1,n}^2 - \mu^2) \quad . \tag{4.42}$$

The total number of the amplitudes of the projectile interaction with n constituents is $N!/n!(N-n)!$. These amplitudes differ from (4.42) by the set of amplitudes $a_{\ell_1}, a_{\ell_2} \ldots a_{\ell_n}$ and by the corresponding replacements of the momentum subscripts. The total amplitude with n interactions is

$$A_n^{e\ell}(s,q^2) = \sum_{\ell_1,\ldots\ell_n} A_{\ell_1,\ell_2,\ldots\ell_n}^{e\ell}(s,q^2) \quad . \tag{4.43}$$

The sum in (4.43) runs over all possible combinations of the subscripts

$\ell_1,\ldots,\ell_n$. The total scattering amplitude is the sum of all $A_n^{e\ell}$ amplitudes:

$$A^{e\ell}(s,q^2) = \sum_{n=1}^{N} A_n^{e\ell}(s,q^2) \quad . \tag{4.44}$$

The expression (4.42) is convenient for the transformation into a Lorentz-invariant form. The detailed consideration of this transformation was carried out in the Sec. 3.2. The same recipe is valid here: it is necessary just to introduce the Lorentz-invariant integration over the constituent momenta

$$\frac{d^3k_j}{2m} \to \frac{d^3k_j}{2k_{j0}} , \quad \frac{d^3k'_\ell}{2m} \to \frac{d^3k'_\ell}{2k'_{\ell 0}} \quad . \tag{4.45}$$

and to introduce the integrations over the total energies squared (s and s'), i.e. to replace

$$\frac{1}{\left(E_0 - \sum\limits_{j=1}^{N} \frac{\vec{k}_j^2}{2m}\right)\left(E'_0 - \sum\limits_{j=1}^{n} \frac{k'^2_j}{2m} - \sum\limits_{j=n+1}^{N} \frac{\vec{k}_j^2}{2m}\right)}$$

$$\to \int_{(Nm)^2}^{\infty} ds \int_{(Nm)^2}^{\infty} ds' \frac{\delta\left(\tilde{P}_0 - \sum\limits_{j=1}^{N} k_{j0}\right)\delta\left(\tilde{P}'_0 - \sum\limits_{j=1}^{n} k'_{j0} - \sum\limits_{j=n+1}^{N} k_{j0}\right)}{(s - M_N^2)(s' - M_N^2)} \quad . \tag{4.46}$$

Here M_N is the mass of the composite system; the zeroth components of the four-momenta $\tilde{P}$ and $\tilde{P}'$ are $\tilde{P}_0 = \sqrt{s - \vec{P}^2}$, $\tilde{P}'_0 = \sqrt{s' - \vec{P}'^2}$. Thus the Lorentz-invariant expression for $A_{1,2,\ldots,n}^{e\ell}$ is:

$$A_{1,2,\ldots n}^{e\ell}(s,q^2) = \frac{1}{(2\pi)^2}\int_{(Nm)^2}^{\infty} ds \int_{(Nm)^2}^{\infty} ds' d\phi_N(\tilde{P};k_1,\ldots k_N) d\phi_N(\tilde{P}';k'_1,\ldots k'_N) \times$$

$$\times \prod_{j=R+1}^{N} 2k_{j0}(2\pi)^3\delta^3(\vec{k}_j - \vec{k}'_j) \frac{G_D(s_{12,\ldots})G_D(s'_{12,\ldots})}{(s - M_N^2)(s' - M_N^2)} \prod_{\ell=1}^{n-1} (2\pi i)\delta(M^2_{\ell,\ell+1} - \mu^2) \times$$

$$\times \prod_{j=1}^{N} a_j(s_j, t_j) \quad . \tag{4.47}$$

Here similarly to the Chap. 3 we redefined the vertex function

$G(\frac{s_{12}}{4} - M^2,\dots) \to G_D(s_{12,\dots})$.

As in the case of the two-body composite system, the inelastic screening may be significant in the general case, considered here, too. The amplitude of the projectile interaction with n constituents (including inelastic collisions) is defined by the expression (4.42) (or (4.47)) with the replacement

$$\prod_{\ell-1}^{n-1} \delta(M^2_{\ell,\ell+1} - \mu^2) \prod_{i=1}^{n} a_i(s_i, t_i) \to A_{n+1 \to n+1}(s_1, t_1,\dots; M^2_{12},\dots) \tag{4.48}$$

where $A_{n+1 \to n+1}$ is the amplitude given by the sum of the diagrams shown in Fig. 4.9.

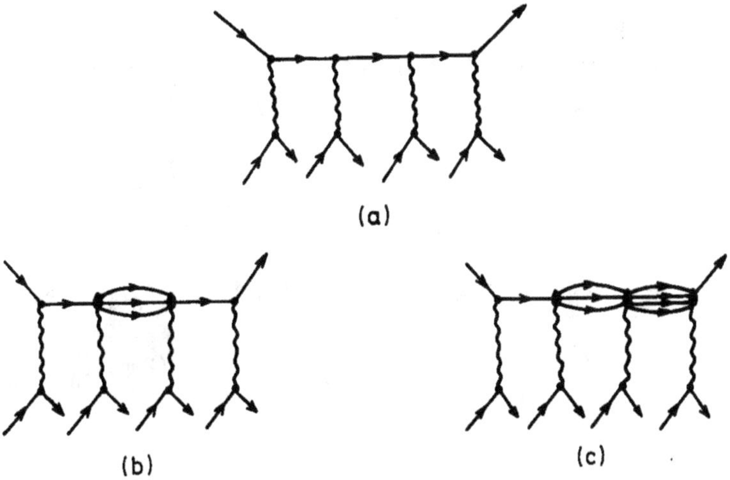

Fig. 4.9 Projectile interactions with N-particle composite system: (a) elastic rescatterings of the Glauber-shadow corrections; (b) production and absorption of the shower of particles; (c) rescattering of the shower of particles.

Inelastic screening in the process of Fig. 4.9b has the same structure as that for a two-body system: a shower of particles is produced by one of the constituents and absorbed by the other one. This type of screening may be handled similarly to the case of the two-body system. However, there is another possible type of screening process: the shower of particles produced may be rescattered or produces another shower before the absorption takes place (see Fig. 4.9c). These processes are experimentally unknown, and the corresponding screening may be calculated only in the framework of some model assumptions about the structure of these processes. Recent methods of the approximate calculations of the inelastic corrections will be considered later in this chapter.

Let us return to the calculation of the elastic screening. In many cases it is convenient to express (4.42) in terms of the wave functions in the coordinate representation. In order to rewrite (4.42) through the N-particle wave functions (see Sec. 3.2) it is necessary to replace the integrations over $\vec{k}'_\ell$ by the integrations over $\vec{x}_\ell$ (see Fig. (4.8) and to remove the integrations over $x_{\ell z}$ using the δ-functions $\Pi\,\delta(M^2_{\ell,\ell+1} - \mu^2)$ (see Eq. (4.41)). If so,

$$A^{e\ell}_{1,\ldots n}(2Nmp,\vec{q}^{\,2}) = N\left(\frac{i}{4mp}\right)^{n-1}\int\prod_{j=1}^{N}\frac{d^3k_j}{(2\pi)^3}\prod_{\ell=1}^{n}\frac{d^2x_{\ell T}}{(2\pi)^2}$$
$$\times\,\psi(\vec{k}_1,\ldots\vec{k}_N)\psi(\vec{k}_1+\frac{\vec{q}}{n}+\vec{x}_{1T},\ldots\vec{k}_n+\frac{\vec{q}}{n}+\vec{x}_{nT},\vec{k}_{n+1},\ldots\vec{k}_N)$$
$$\times\,(2\pi)^3\delta^3(\vec{P}-\sum_{j=1}^{N}\vec{k}_j)(2\pi)^2\delta^2(\sum_{\ell=1}^{n}\vec{x}_{\ell T})\cdot\prod_{\ell=1}^{n}a_\ell(2mp,(\frac{\vec{q}}{n}+\vec{x}_{\ell T})^2)\;. \tag{4.49}$$

The wave function in the coordinate representation is connected with $\psi(\vec{k}_1,\ldots\vec{k}_N)$ by the relation

$$\psi(\vec{k}_1,\ldots\vec{k}_N) = \int\prod_{j=1}^{N}d^3r_j\;\varphi(\vec{r}_1,\ldots\vec{r}_N)\delta(\sum_{j=1}^{N}\vec{r}_j - N\vec{R}_0)\exp[i\sum_{j=1}^{N}\vec{r}_j\cdot\vec{k}_j] \tag{4.50}$$

where $\vec{R}_0$ is the coordinate of the centre of mass for the composite system.

Let us introduce the n-particle density

$$\rho(\vec{r}_1,\ldots\vec{r}_n) = \int \prod_{j=n+1}^{N} d^3r_j\, \varphi^2(\vec{r}_1,\ldots\vec{r}_N)\delta(\sum_{j=1}^{N}\vec{r}_j - N\vec{R}_0) \qquad (4.51)$$

with the normalization condition

$$\int \prod_{j=1}^{n} d^3r_j \rho(\vec{r}_1,\ldots\vec{r}_n) = 1 \qquad (4.52)$$

Substituting (4.50) into (4.49) and carrying out the integrations over $\vec{k}_j$ and $\vec{r}_{n+1},\ldots\vec{r}_N$ we obtain the following expression for $A^{e\ell}_{1,2,\ldots,n}$ in terms of $\rho(\vec{r}_1,\ldots\vec{r}_n)$

$$A^{e\ell}_{1,2,\ldots n}(2Nmp, \vec{q}^{\,2}) = N\left(\frac{i}{4mp}\right)^{n-1}\int \prod_{j=1}^{n} d^3r_j\rho(\vec{r}_1,\ldots\vec{r}_n)(2\pi)^2\delta^2(\sum_{j=1}^{n}\vec{x}_{jT})$$

$$\times \prod_{\ell=1}^{n}\frac{d^2x_{\ell T}}{(2\pi)^2}\, a_\ell(2mp, (\frac{\vec{q}}{n} + \vec{x}_{\ell T})^2)\exp[i\vec{r}_\ell\cdot(\frac{\vec{q}}{n} + \vec{x}_{\ell T})] \,. \qquad (4.53)$$

The typical situation of the scattering by the composite system is the case when the characteristic distances in the scattering amplitudes a_ℓ are small compared to the composite system size. In this case for small $|\vec{q}|$ a good approximation for Eq. (4.53) is

$$a_\ell(2mp, (\frac{\vec{q}}{n} + \vec{x}_{\ell T})^2) \simeq a_\ell(2mp, 0) \quad , \qquad (4.54)$$

and the scattering of the projectile on every constituent proceeds at the same value of the impact parameter. Indeed, substituting into (4.53) $\int d^2\vec{r}_T\exp(i\Sigma\vec{x}_{jT}\cdot\vec{r}_T)$ instead of $(2\pi)^2\delta^2(\Sigma\vec{x}_{jT})$ and carrying out integrations over $\vec{x}_{jT}$ we shall obtain:

$$A^{e\ell}_{1,2,\ldots n}(2Nmp, \vec{q}^{\,2}) = N\left(\frac{i}{4mp}\right)^{n-1}\int d^2r_T e^{i\vec{q}\cdot\vec{r}_T}\rho_{1\ldots n}(\vec{r}_T)\prod_{\ell=1}^{n} a_\ell(2mp, 0) \qquad (4.55)$$

where

$$\rho_{1\ldots n}(\vec{r}_T) = \int \prod_{\ell=1}^{n} d^3r_\ell\,\delta^2(\vec{r}_{\ell T} - \vec{r}_T)\rho_{1\ldots n}(\vec{r}_1,\ldots\vec{r}_n) \qquad (4.56)$$

is the probability for all the constituents 1,2,...,n to have the same impact parameter $\vec{r}_T$. This probability determines the amplitude $A^{e\ell}_{1,\ldots,n}$ at small $|\vec{q}|$.

4.4 Nuclear Target Scattering (A $\gtrsim$ 10)

When a composite system consists of a great number of constituents, there is an approximation, which simplifies the calculation of the scattering amplitude. The examples of such composite systems are nuclei heavier than beryllium (i.e. with the nucleon number $A \gtrsim 10$). Here we shall consider the scattering on such nuclear targets; as it is adopted in nuclear physics, we shall use the notation A instead of N for the number of constituents.

If a fast particle is scattered by the nucleus with $A \gtrsim 10$ on a small angle, the number of interactions of this particle with the nucleons (n) is small with respect to A:

$$n << A \quad . \tag{4.57}$$

Indeed, as we have already considered, the scattering proceeds at the same value of the impact parameter. The number of nucleons with a fixed value of the impact parameter is $n \lesssim A^{1/3} << A$. A good approximation for the density distribution of n nucleons in this case is a factorized expression

$$\rho(\vec{r}_1,\ldots\vec{r}_n) = \rho(\vec{r}_1)\ldots\rho(\vec{r}_n) \tag{4.58}$$

Let us emphasize, that in the case when n is of the order of A, the facotrized expression (4.58) becomes invalid, for in this case an exact account of the motion of the centre of mass is required. The wave function depends on the relative coordinates of the constituents, and only a comparatively small number of particles may be considered as moving in the potential created by the constituents. The reverse example is the deuteron; its wave function depends only on $|\vec{r}_1 - \vec{r}_2|$ (the centre mass motion removed) and there is, of course, no factorization of the proton and neutron distributions.

For high initial energies the proton and neutron target scattering amplitudes are approximately equal:

$$a_\ell(2mp,\ 0) = a(0) \qquad \ell = 1,\ldots n \ . \tag{4.59}$$

Thus using Eqs. (4.55) and (4.56) and taking $A!/n!(A-n)! \approx \frac{A^n}{n!}$ we may write for the elastic scattering amplitude the expression (see Eqs. (4.43), (4.44)):

$$A^{e\ell}(2Amp,q^2) = \frac{i}{4Amp}\int d^2r_T\, e^{i\vec{q}\cdot\vec{r}_T} \sum_{n=1}^{\infty} \frac{1}{n!}\left(i\,\frac{a(0)T(\vec{r}_T)}{4mp}\right)^n$$

$$= \frac{1}{4Amp}\int d^2r_T\ e^{i\vec{q}\cdot\vec{r}_T}\left[1 - e^{-\frac{1}{2}T(\vec{r}_T)\frac{a(0)}{2mpi}}\right] \ . \tag{4.60}$$

Here $T(\vec{r}_T)$ is a profile function, which describes the nucleon distribution in the impact parameter space:

$$T(\vec{r}_T) = A\int dr_z \rho(\vec{r}) \ . \tag{4.61}$$

If proton- and neutron-target scattering amplitudes are different (a_p and a_n), then $aT(\vec{r}_T)$ in Eq. (4.60) should be replaced by $a_nT_n(\vec{r}_T) + a_pT_p(\vec{r}_T)$ where T_n and T_p are the profile functions of the neutrons and protons, correspondingly.

At high energy with a good accuracy $a(0) = 2mpi\sigma_{tot}$ where σ_{tot} is a particle-nucleon total cross section. Since, due to the unitarity condition, $\text{Im}\ A^{e\ell}(2Amp,\ 0) = 2Amp\sigma_{tot}(A)$ (see Appendix B), we obtain for the total particle-nucleus cross section:

$$\sigma_{tot}(A) = 2\int d^2r_T\left(1 - e^{-\frac{1}{2}\sigma_{tot}T(\vec{r}_T)}\right) \ . \tag{4.62}$$

The differential elastic cross section for the nuclear target scattering is

$$\frac{d\sigma_{e\ell}(A)}{d\vec{q}^2} = \frac{1}{4\pi}\left[\int d^2r_T\ e^{i\vec{q}\cdot\vec{r}_T}\left(1 - e^{-\frac{1}{2}\sigma_{tot}T(\vec{r}_T)}\right)\right]^2 \tag{4.63}$$

while the total elastic cross section is:

$$\sigma_{e\ell}(A) = \int d^2r_T\left(1 - e^{-\frac{1}{2}\sigma_{tot}T(\vec{r}_T)}\right)^2 \ . \tag{4.64}$$

Inelastic particle-nucleus cross section is

$$\sigma_{\text{ine}\ell}(A) = \sigma_{tot}(A) - \sigma_{e\ell}(A) = \int d^2r_T(1 - e^{-\sigma_{tot}T(\vec{r}_T)}) \quad . \tag{4.65}$$

Up to now we used the condition (4.54). In fact all the previous expressions may be derived in a more general case, if there is only a factorization of the distributions of the interacting nucleons (4.58). Let us rewrite the Eq. (4.49), using the factorization assumption in the momentum representation:

$$\psi(\vec{k}_1,\ldots\vec{k}_n, \vec{k}_{n+1},\ldots\vec{k}_A) = \psi(\vec{k}_1)\ldots\psi(\vec{k}_n)\psi(\vec{k}_{n+1},\ldots\vec{k}_A) \quad . \tag{4.66}$$

Then instead of (4.49) we immediately obtain:

$$A^{e\ell}_{1\ldots n}(2Amp,\vec{q}^{\,2}) = A\left(\frac{i}{4mp}\right)^{n-1} \int \prod_{\ell=1}^{n} \frac{d^2x_{\ell T}}{(2\pi)^2}$$

$$\times\, S(\vec{x}^{\,2}_{\ell T})a(\vec{x}^{\,2}_{\ell T})(2\pi)^2\delta^2\left(\sum_{j=1}^{n}\vec{x}_{jT} - \vec{q}\right) \quad . \tag{4.67}$$

Here we redefined $\frac{\vec{q}}{n} + \vec{x}_{\ell T} \to \vec{x}_{\ell T}$ and introduced the form factor of the separate nucleon inside the nucleus:

$$S(\vec{x}^{\,2}) = \int \frac{d^3k}{(2\pi)^3}\, \psi(\vec{k})\psi(\vec{k} + \vec{x}) \quad . \tag{4.68}$$

The same calculation that turns (4.53) into (4.55) gives for Eq. (4.67):

$$A^{e\ell}_{1\ldots n}(2Amp,\vec{q}^{\,2}) = -i\,\frac{4mp}{A^{n-1}} \int d^2r_T\, e^{i\vec{q}\cdot\vec{r}_T}(i\chi(\vec{r}_T))^n \tag{4.69}$$

where

$$\chi(\vec{r}_T) = \frac{A}{4mp}\int \frac{d^2x_T}{(2\pi)^2}\, e^{-i\vec{x}_T\cdot\vec{r}_T} a(\vec{x}^{\,2}_T)S(\vec{x}^{\,2}_T) \quad . \tag{4.70}$$

For the total scattering amplitude a sum like (4.60) over n should be taken:

$$\frac{-i}{4Amp}\, A^{e\ell}(2Amp,\vec{q}^{\,2}) = \int d^2r_T e^{i\vec{q}\cdot\vec{r}_T}(1 - e^{i\chi(\vec{r}_T)}) \quad . \tag{4.71}$$

Equation (4.71) is the generalization of the Eq. (4.60), which was obtained without the assumption (4.54). At $\vec{r}_T$ of the order of several nucleon radii (i.e. at $|\vec{r}_T| \gtrsim (7\text{-}10)\ \text{Gev}^{-1}$) the amplitude $a(2mp, \vec{x}^2)$ depends only slightly on $\vec{x}^2$; taking $\vec{x}^2 = 0$ in the Eq. (4.71) we just obtain Eq. (4.60). The function $\chi(\vec{r}_T, 2mp)$ is then, up to a factor, the nucleon distribution over $\vec{r}_T$. For small $\vec{r}_T$ such an interpretation of $\chi(\vec{r}_T, 2mp)$ is not valid. Equation (4.71) may be used for the scattering process at comparatively large momentum transfers.

4.5 Diffraction Excitation of the Nuclei and Noncoherent Scattering

Now let us consider another type of scattering on nuclear targets: a fast particle with momentum p_{lab} undergoes elastic scattering on the nucleon (new particles are not produced), while the nucleus transits into some excited state — e.g. the excited nucleus A^* — or disintegrates partially or completely. Examples of such processes are shown in Fig. 4.10. We shall look for the sum of such noncoherent processes. In other words, we shall consider the sum of all possible final states (f) with fixed momentum transfer $\vec{q}_T$. The effective masses of the nucleons in different final states M_f are close to each other and to the nucleus mass M_A, but nevertheless they are slightly different. The noncoherent cross section is a cross section summed up over these masses, i.e. integrated over a small q_z region (since $|q_z| = \frac{1}{2p}(M_f^2 + \vec{q}_T^{\,2} - M_A^2)$).

Let us consider for the beginning the transition of the nucleus A to some definite excited state f. The amplitude with n interactions for this process is described by Eq. (4.49) with the replaced wave function of the final state:

$$A^f_{1\ldots n} = A\left(\frac{i}{4mp}\right)^{n-1} \int \prod_{j=1}^{A} \frac{d^3k_j}{(2\pi)^3} \prod_{\ell=1}^{n} \frac{d^2x_{\ell T}}{(2\pi)^2} (2\pi)^3\delta^3(\vec{P} - \sum_{j=1}^{A} \vec{k}_j)$$
$$\times (2\pi)^2\delta^2(\sum_{\ell=1}^{n} \vec{x}_{\ell T} - \vec{q})\, a(\vec{x}_1^{\,2})\ldots a(\vec{x}_n^{\,2})\psi(\vec{k}_1\ldots\vec{k}_A)\psi_f^*(\vec{k}_1+\vec{x}_1,\ldots\vec{k}_n+\vec{x}_n,\vec{k}_{n+1},\ldots\vec{k}_A). \tag{4.72}$$

The sum of the elastic cross section and the noncoherent cross section, which we shall denote as $d\sigma_{scat}/d\vec{q}_T^{\,2}$, is

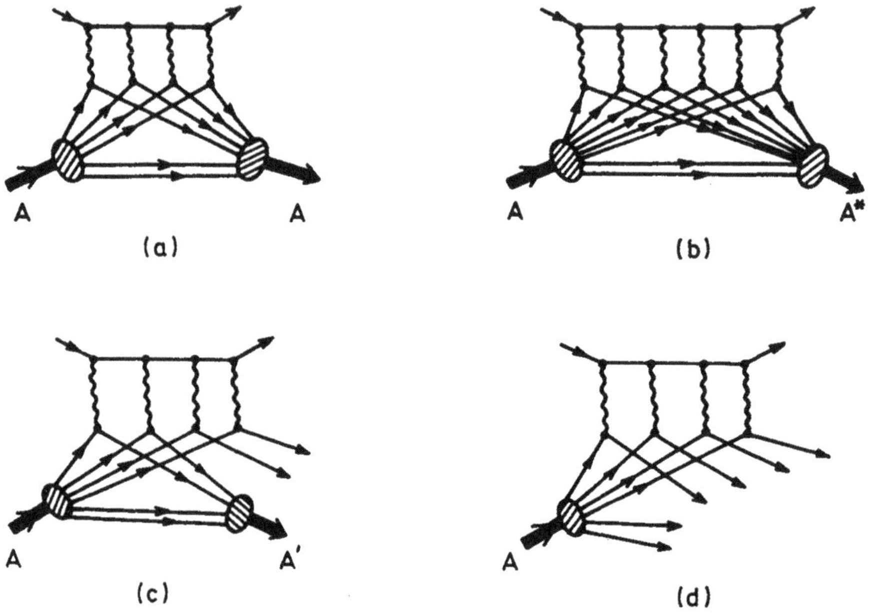

Fig. 4.10 (a) Coherent elastic scattering off the composite system; (b-d) different incoherent processes, caused by the elastic scattering of the projectile by constituents: target excitation (b), partial (c) or total (d) target disintegration.

$$\frac{d\sigma_{scat}}{d\vec{q}_T^2} = \frac{1}{4\pi(4mpA)^2} \sum_f |A^f(2Amp,q^2)|^2 \quad , \tag{4.73}$$

where A^f is the amplitude of the transition $A \to f$. Let us express $d\sigma_{scat}/d\vec{q}_T^2$ in terms of the amplitudes $A^f_{1\ldots n}$. Expanding A^f into the amplitudes with different interaction multiplicities, we obtain

$$\sum_f |A^f|^2 = \sum_f \sum_{n,n'} \sum_{\ell_1\ldots\ell_n} \sum_{\ell_1'\ldots\ell_{n'}'} A^f_{\ell_1\ldots\ell_n} A^{f*}_{\ell_1'\ldots\ell_{n'}'} \quad . \tag{4.74}$$

When the expressions of the type of Eq. (4.72) are substituted into Eq. (4.74), the sum over f may be calculated using the completeness condition:

$$\sum_f \psi_f(\vec{k}_1,\vec{k}_2\ldots)\psi_f^*(\vec{k}_1',\vec{k}_2'\ldots) = (2\pi)^3\delta^3(\vec{k}_1-\vec{k}_1')(2\pi)^3\delta^3(\vec{k}_2-\vec{k}_2')\ldots \tag{4.75}$$

Then items of the series (4.74) are the discontinuities of the diagrams of Fig. 4.11 type. These discontinuities may be easily calculated using

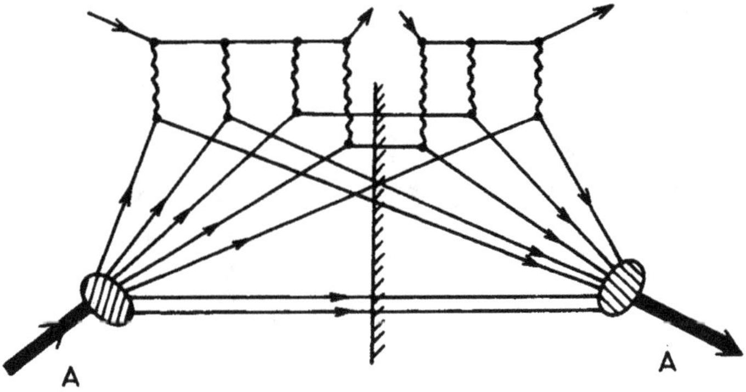

Fig. 4.11 The discontinuity of the diagram, which determines Eq. (4.76).

the factorization assumption (4.66) in much the same way as the expression (4.67):

$$(4mpA)^2 \int \prod_{j=n-\nu+1}^{n} \left[\frac{d^2x_{jT}}{(2\pi)^2} \frac{ia(\vec{x}_{jT}^2)}{4mp} S(\vec{x}_{jT}^2)\right] \prod_{k=n'-\nu+1}^{n'} \left[\frac{d^2x_{kT}}{(2\pi)^2} \frac{(-i)a^*(\vec{x}'^2_{kT})}{4mp} S(\vec{x}'^2_{kT})\right]$$

$$\times \prod_{\ell=1}^{\nu} \left[\frac{d^2x_{\ell T}d^2x'_{\ell T}}{(2\pi)^4} S((\vec{x}_{\ell T} - \vec{x}'_{\ell T})^2) \frac{a(\vec{x}^2_{\ell T})a^*(\vec{x}'^2_{\ell T})}{(4mp)^2}\right]$$

$$\times (2\pi)^4 \delta^2(\sum \vec{x}_{jT} - \vec{q})\delta^2(\sum \vec{x}'_{\ell T} - \vec{q}) \quad . \qquad (4.76)$$

The number of contributions of the type (4.76) for fixed n, n' and ν taking into account all possible permutations of the nucleons is

$$\frac{A!}{\nu!(n-\nu)!(n'-\nu)!(A-n-n')!} \simeq \frac{A^{n+n'}}{\nu!(n-\nu)!(n'-\nu)!} \quad . \qquad (4.77)$$

Summation over n, n' and ν is easily carried out in the impact parameter space with the help of the same transformations which we used in the derivations of the Eqs. (4.55) and (4.69). Finally we obtain:

$$\frac{d\sigma_{scat}}{d\vec{q}_T^{\,2}} = \frac{1}{4\pi}\int d^2r_T d^2r_T'\, e^{i\vec{q}_T\cdot(\vec{r}_T-\vec{r}_T')}\Big[1-e^{-i\chi(\vec{r}_T)} - e^{-i\chi^*(\vec{r}_T')}$$

$$+ e^{i(\chi(\vec{r}_T)-\chi^*(\vec{r}_T'))+\eta(\vec{r}_T,\vec{r}_T')}\Big] \quad . \tag{4.78}$$

The function $\chi(\vec{r}_T)$ is defined by the Eq. (4.70); $\eta(\vec{r}_T, \vec{r}_T')$ is

$$\eta(\vec{r}_T, \vec{r}_T) = \frac{A}{(4mp)^2}\int \frac{d^2x_T}{(2\pi)^2}\frac{d^2x_T'}{(2\pi)^2}\, e^{-i\vec{x}_T\cdot\vec{r}_T+i\vec{x}_T'\cdot\vec{r}_T'} S((\vec{x}_T-\vec{x}_T')^2)\, a(\vec{x}_T^{\,2})\, a^*(\vec{x}_T'^{\,2}) \ . \tag{4.79}$$

The total cross section $\sigma_{scat}(hA) = \int \frac{d\sigma_{scat}}{d\vec{q}_T^{\,2}}\frac{d^2q_T}{\pi}$ has a simple form

$$\sigma_{scat}(hA) = \int d^2r_T\left[1 - 2e^{-\frac{1}{2}\sigma_{tot}T(\vec{r}_T)} + e^{-\sigma_{ine\ell}T(\vec{r}_T)}\right] \tag{4.80}$$

since the dependence of $a(\vec{x}_T^{\,2})$ on $\vec{x}_T^{\,2}$ is appreciably weaker than that of $S(\vec{x}_T^{\,2})$. Thus the total inelastic cross section $\sigma_{ine\ell}(hA)$ (Eq. (4.65)) may be divided into two parts: the cross section of all inelastic processes, accompanied by the excitation or disintegration of the nucleus without new particle production (i.e. noncoherent scattering cross section)

$$\sigma_{dis}(hA) = \sigma_{scat}(hA) - \sigma_{e\ell}(hA) = \int d^2r_T\, e^{-\sigma_{ine\ell}T(\vec{r}_T)}(1- e^{-\sigma_{e\ell}T(\vec{r}_T)}) \tag{4.81}$$

and the cross section of the processes with the production of at least one secondary hadron:

$$\sigma_{prod}(hA) = \sigma_{tot}(hA) - \sigma_{scat}(hA) = \int d^2r_T(1 - e^{-\sigma_{ine\ell}T(\vec{r}_T)}) \quad . \tag{4.82}$$

Let us note that Eq. (4.82) may be derived from simple macroscopic considerations: in the r.h.s. of Eq. (4.82) the probability of at least one inelastic interaction of the projectile with the nucleon is integrated over the impact parameter.

So we have seen that within the framework of the presented diagram technique it is possible to obtain all main results[8-10] of the Glauber approximation for the nuclear target scattering in a comparatively simple way.

4.6 Inelastic Screening in Nuclear Target Scattering and the Method of Eigenstates

A correct calculation of the inelastic screening effects in the nuclear target scattering is not a simple problem. The main difficulty is related to the contribution of the processes like Fig. 4.9c: they include the transitions of the inelastic states into other states. The amplitudes of such transitions cannot be expressed in terms of the experimentally observable values, and thus the calculation of the inelastic screening effects is model dependent. We will consider here one of the recently famous models of the nuclear inelastic screening calculation — the so-called eigenstate method[11-13].

When considering deuteron target scattering, we have established that the inelastic screening contributions may be divided into two types. The first one is the shadow effect of the resonance region with comparatively low values of the effective mass of the produced particle shower. The second one is the contribution of the three-Reggeon diagrams (high effective masses).

Let us consider resonance screening. Suppose that in the scattering process of the projectile hadron (h) on the nucleon target there are k+1 different channels of the diffraction transitions (including elastic ones):

$$h \to h, \quad h \to h_\ell \qquad \ell = 1,2,\ldots k \quad .$$

The states h_ℓ may be e.g. resonances. The set of the amplitudes of all possible transitions between these states is a $k \times k$ matrix. It may be diagonalized by the eigenstates $|\alpha\rangle$. In other words, the eigenstates $|\alpha\rangle$ may be only scattered elastically or absorbed by the nucleons. Thus the interactions of each of the states $|\alpha\rangle$ with the nucleus are described by the formulae of the previous sections. Let us expand hadronic states h_ℓ over the eigenstates $|\alpha>$:

$$|h_\ell> = \sum C_{\ell\alpha}|\alpha> \tag{4.83}$$

and introduce the elastic scattering amplitudes of the eigenstates (and the corresponding total cross sections):

$$<\alpha|a(q^2)|\alpha> = a_\alpha(q^2)$$

$$a_\alpha(0) = \frac{\sigma_\alpha}{2mpi} \quad . \tag{4.84}$$

Then the matrix element of any power of the amplitude a^n between the states h_ℓ and h_j is

$$<h_j|a^n|h_\ell> = \sum_\alpha C^*_{j\alpha} C_{\ell\alpha} a^n_\alpha(q^2) \quad . \tag{4.85}$$

With the help of Eq. (4.85) it is possible to generalize the formulae of the preceding section for the considered multichannel problem. For example, the generalizations of Eqs. (4.62) and (4.64) for $\sigma_{tot}(hA)$ and $\sigma_{e\ell}(hA)$ are

$$\sigma_{tot}(hA) = 2\int d^2r_T(1 - \sum_\alpha |C_{h\alpha}|^2 e^{-\frac{1}{2}\sigma_\alpha T(\vec{r}_T)})$$

$$\sigma_{e\ell}(hA) = \int d^2r_T(1 - \sum_\alpha |C_{hA}|^2 e^{-\frac{1}{2}\sigma_\alpha T(\vec{r}_T)})^2 \quad . \tag{4.86}$$

Explicit calculations in the eigenstate method require the knowledge of all the transition amplitudes $h_\ell \to h_j$. In practice only part of these amplitudes may be determined from the experiment. Thus it is necessary to define them according to some chosen model. Usually the equality of all diagonal transitions $<h_j|a|h_j> = a_0$ as well as of all non-diagonal ones $<h_j|a|h_\ell> = a_1 (j \neq \ell)$ is assumed.

Strictly speaking, the eigenstate method is valid only for the screening effects from the resonance region (region of small effective masses of the produced particles). However, one may hope that the variation of the parameters a_α will allow one to describe the total inelastic screening effectively.

4.7 Calculation of the Noncoherent Process Cross Sections through Diagram-Cutting Rules

In this section we shall consider the high energy collisions of two composite systems. The generalization of the considered methods for these processes does not lead to any new problems if only the transferred momenta are sufficiently small (as it was considered in the Sec. 3.2, the transferred energies are to be negligible with respect to the space components of the momenta transferred). However, the straightforward calculations seem to be very bulky. Here we would like to present the technique for the noncoherent cross section calculation for the composite systems, which is based on the diagram-cutting rules. It is closely related to the AGK rules[14] (see also the Sec. 2.3).

Let us start with a simple example of deuteron disintegration by a high energy projectile h. The cross section of the reaction $h + d \rightarrow h + p + n$ with the nucleon final state interaction included may be graphically written as

$$\sigma_{dis}(hd) = \frac{1}{J} \int d\phi_3 \left| \begin{array}{c} \end{array} \right|^2$$

$$= \frac{1}{S} \left\{ \begin{array}{c} \end{array} \right\}$$

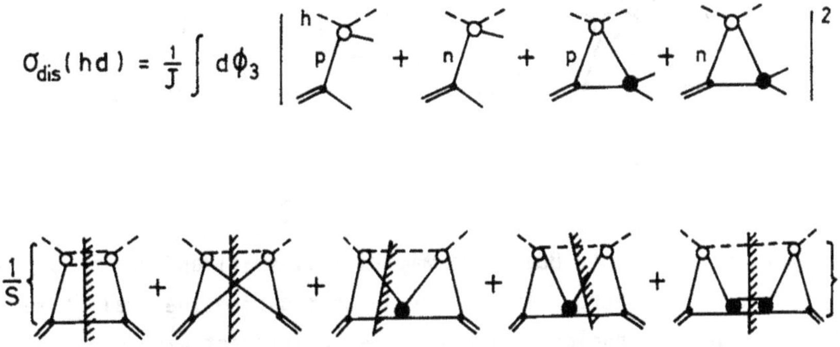

$$(4.87)$$

The diagrams in Eq. (4.87) represent the corresponding amplitudes. The dashed lines stand for the projectile h, the solid ones, for the nucleons. The open blobs are hN high energy amplitudes, the full blobs, low energy proton-neutron amplitudes. The amplitude modulus squared in the l.h.s. of the Eq. (4.87) is integrated over the three-particle (h + p + n) phase space, S is the hd c.m. energy squared, J, invariant flux factor (see Appendix B). The r.h.s. of the Eq. (4.87) is the sum of the discontinuities of the elastic hd amplitude.

The sum squared of the first two diagrams in the l.h.s. of the Eq. (4.87) is equal to the sum of the first and the second items in the r.h.s. of (4.87). The first one is the sum of the elastic cross sections $\sigma_{e\ell}(hp) + \sigma_{e\ell}(hn)$, while the second one is the shadow correction, which we shall neglect. In order to calculate the sum of the other three items in the r.h.s. of Eq. (4.87), let us use the equality

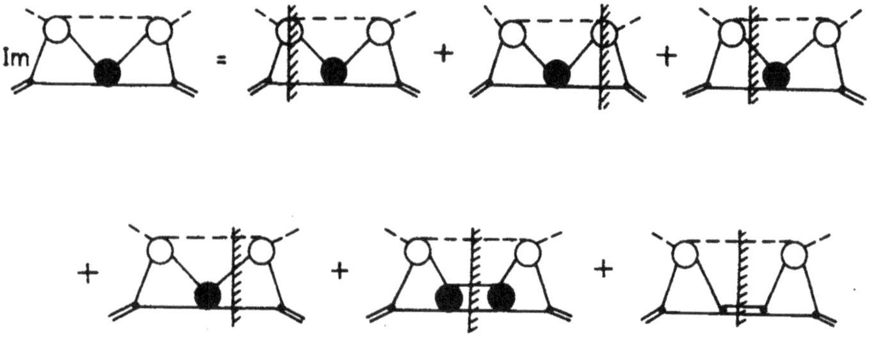

(4.88)

The l.h.s. of the Eq. (4.88) is parametrically small and may be approximately taken to be zero. The reason is that during the small time of particle flight the nucleons have no time to interact with each other. Similarly, the first two items in the r.h.s. of (4.88) are also negligible. Thus the sum of the last four items in the r.h.s. of (4.88) is equal to zero. The last one is $S\sigma_{e\ell}(hd)$, while the remaining three items are just the desired contributions to the r.h.s. of the Eq. (4.87). As a result up to the shadow corrections we obtain

$$\sigma_{dis}(hd) = \sigma_{e\ell}(hp) + \sigma_{e\ell}(hn) - \sigma_{e\ell}(hd) \quad . \tag{4.89}$$

Let us now apply this technique to the collision process of two composite systems. We shall treat[15)] the cross section of the diquark-spectator production in the proton-nucleon collision, which is required in Chapter 7. The "diquark" is assumed to be a pair of dressed quarks with small relative momentum (e.g. with the same momentum distribution as in the initial proton). For the sake of definiteness the production cross section of the diquark d_{uu}, consisting of two u-quarks will be calculated. To the first order in the number of interactions it equals to:

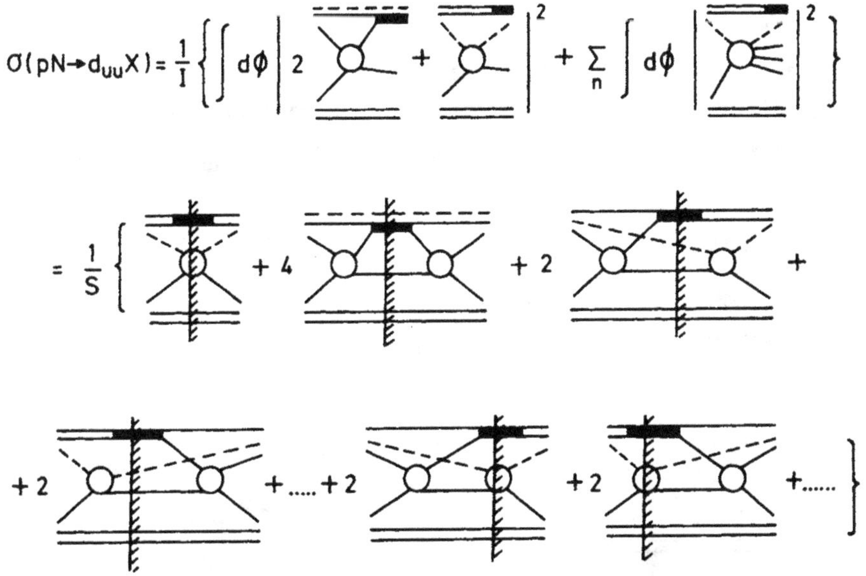

(4.90)

The quarks of the projectile proton are in the upper part of each diagram, while the quarks of the target nucleon, in the lower parts. The d-quark of the proton is shown by the dashed line, the u-quarks by the solid lines, the d_{uu} diquark is blackened. In the l.h.s. of the Eq. (4.90) the contributions with and without the production of new quarks are separated. The factor 2 at the first item in the l.h.s. of (4.90) expresses the possibility for both u-quarks to interact with the target. For brevity in the r.h.s. of (4.90) those contributions, in which different quarks of the target interact with the projectile to the left and to the right of the cut line, are omitted.

Let us consider the items of the r.h.s. of the Eq. (4.90) one by one. The contribution of the first one equals the total cross section of d-quark interaction with the nucleon

$$\sigma_{qN} \quad . \tag{4.91}$$

The second item corresponds to the process with d-quark-spectator. The factor 4 accounts for the possibilities of different u-quarks of the diquark to interact with the target; this interaction does not lead to the disintegration of the diquark. All possible states of the target quarks are summed up, including the processes with and without target-nucleon disintegration. Thus the second item contribution is the sum of the elastic $d_{uu}N$ scattering cross section and $d_{uu}N$ diffraction dissociation

$$\sigma_{e\ell}(d_{uu}N) + \sigma_{DD}(d_{uu}N \rightarrow d_{uu}X) \quad . \tag{4.92}$$

The next item may be expressed through the imaginary part of the same diagram with the help of the AGK rules:

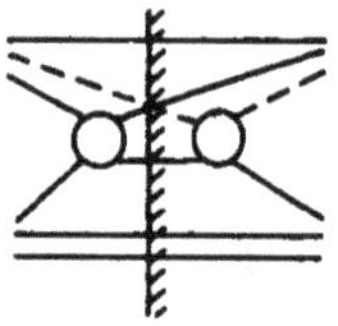 $= -\mathrm{Im}$ 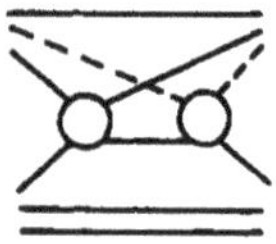 $= \Delta\sigma$ (4.93)

Thus the sum of the third and the fourth items in the r.h.s. of the Eq. (4.90) equals

$$4\Delta\sigma \quad . \tag{4.94}$$

These four items exhaust the contributions to $\sigma(pN \rightarrow d_{uu}X)$ in the impulse approximation. There are some other contributions of the same order in the interaction number; they are also shown in the r.h.s. of (4.90) and may be expressed through $\Delta\sigma$ according to the AGK rules:

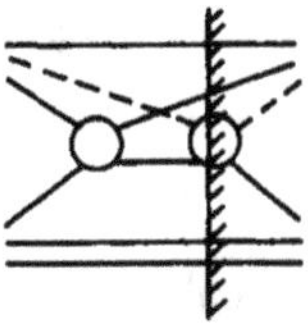 $= 2\,\mathrm{Im}$ 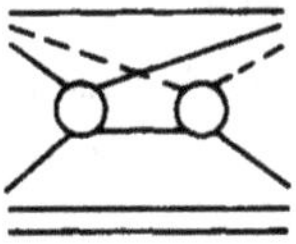 $= -2\Delta\sigma$ (4.95)

The sum of the last two items in the r.h.s. of (4.90) is then equal to

$$-8\Delta\sigma \quad . \tag{4.96}$$

Let us show now that the sum of the contributions (4.91), (4.94) and (4.96) may be written as $\sigma_{tot}(pN) - \sigma_{tot}(d_{uu}N)$. Considering only the diagrams of the same type as that in Eq. (4.90), we may write:

$\sigma_{tot}(pN) = \mathrm{Im}$  $+ 2\mathrm{Im}$  $+ \mathrm{Im}$ 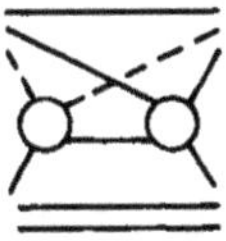

Im 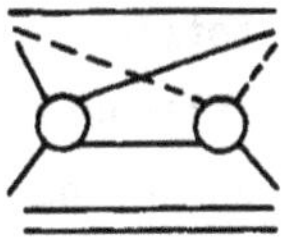 $+ 4\mathrm{Im}$  $= 3\sigma_{qN} - 6\Delta\sigma$

$\sigma_{tot}(d_{uu}N) = 2\mathrm{Im}$  $- 2\mathrm{Im}$ 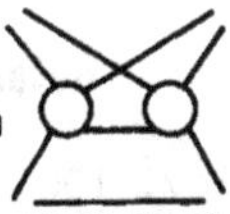 $= 2\sigma_{qN} - 2\Delta\sigma$

(4.97)

Using (4.97), we obtain for the sum of the contributions considered in Eq. (4.90):

$$\sigma(pN \to d_{uu}X) = \sigma_{tot}(pN) - \sigma_{tot}(d_{uu}N) + \sigma_{e\ell}(d_{uu}N) + \sigma_{DD}(d_{uu}N \to d_{uu}X) \quad . \tag{4.98}$$

It is easy to show that Eq. (4.98) is valid also when the contributions not treated in (4.90) are accounted for.

A useful notation for (4.98) is

$$\sigma_{react}(d_{uu}N) = \sigma_{tot}(d_{uu}N) - \sigma_{e\ell}(d_{uu}N) - \sigma_{DD}(d_{uu}N \to d_{uu}X) \quad . \tag{4.99}$$

In other words, σ_{react} is the sum of the cross sections of all inelastic processes without the disintegration of the projectile subtracted by the diffraction dissociation of the target. Then the r.h.s. of Eq. (4.98) may be rewritten as $\sigma_{tot}(pN) - \sigma_{react}(d_{uu}N)$.

The expression (4.98) presents the sum of all noncoherent processes with the production of the diquark-spectator d_{uu}. It includes the processes in which this diquark remains in the initial proton, which does not disintegrate, namely, the processes of the elastic scattering $pN \to pN$ and target-diffraction dissociation $pN \to pX$. In some cases these processes are to be subtracted from (4.98):

$$\sigma_{react}(pN \to d_{uu}X) = \sigma(pN \to d_{uu}X) - \sigma_{e\ell}(pN) - \sigma_{DD}(pN \to pX)$$

$$= \sigma_{react}(pN) - \sigma_{react}(d_{uu}N) \quad . \qquad (4.100)$$

We have considered the noncoherent production of only one chosen diquark d_{uu}. The production cross sections for other diquarks are also determined by the expressions (4.98) or (4.100). The total cross section of diquark production summed up over all possible flavours is determined by the r.h.s. of (4.98) or (4.100) tripled according to the number of different diquarks in the proton.

References

1. R.J. Glauber, Phys. Rev. 100, 242 (1955).
2. A.G. Sitenko, Ukr. Fiz. Zhur. 4, 152 (1959), (in Russian).
3. V.N. Gribov, ZhETF 56, 892 (1969); Sov. Phys. JETP 29, 483 (1969).
4. J. Pumplin and M. Ross, Phys. Rev. Lett. 21, 1778 (1968).
5. V.V. Anisovich, L.G. Dakhno and P.E. Volkovitski, Phys. Lett. 42B, 224 (1972).
6. L.G. Dakhno, Yad. Fiz. 37, 993 (1981); Sov. J. Nucl. Phys. 37, 590 (1981).
7. S.A. Gurvitz and M.S. Marinov, Phys. Lett. 32B, 551 (1970).
8. R.J. Glauber, High Energy Physics and Nuclear Structure, Amsterdam, 1967.

9. V. Franco and R.J. Glauber, Phys. Rev. 142, 1195 (1966).

10. R.H. Bassel and C. Wilkin, Phys. Rev. 174, 1179 (1968).

11. B.Z. Kopeliovich and L.I. Lapidus, Pis'ma v ZhETF 28, 664 (1978); JETP Lett. 28, 614 (1978).

12. H.I. Miettinen and J. Pumplin, Phys. Rev. D18, 1969 (1978).

13. N.N. Nikolaev, ZhETF 81, 814 (1981); Sov. Phys. JETP 54, 434 (1981).

14. V.A. Abramovsky, V.N. Gribov and O.V. Kancheli, Yad. Fiz. 18, 595 (1973); Sov. J. Nucl. Phys. 18, 308 (1974).

15. V.V. Anisovich, V.M. Braun and Yu. M. Shabelski, Yad. Fiz. 39, 932 (1984); Sov. J. Nucl. Phys. 39 (to be published).

Chapter 5

QUARK MODEL AND STATIC FEATURES OF HADRONS

The quark structure of hadrons is the starting point for the classification of hadrons and the description of their static features as well as for high energy physics. Hence, even if one intends to investigate only high energy interactions, it is necessary to have a clear picture of the status of the quark model in the low energy region.

As it will be seen from our further considerations, in the quark model there exists a direct link between the physics of high and low energies: the description of the multiparticle production processes is based to a large extent on the quark systematics of hadrons. Having in mind such an application of the quark model, we present in this chapter the hadron zoology and the rules of construction for the quark wave functions. We will also demonstrate here the possibilities of potential approaches to the quark model which were developed rather intensely in the last ten years. Indeed, progress in the description of multiparticle production processes seems to be impossible without the consideration of higher resonance production. The spectroscopy of higher resonances is so far not a very clear field, and potential models may serve as a guide to it.

In the present chapter the problems mentioned are considered only very briefly. A more detailed description can be found, e.g. in Refs. 1-4.

5.1 Hadron Zoology

Experiments show, that low energy interactions of light quarks almost do not depend on their spin and flavour. This can be taken into account using the language of spin-flavour symmetry $SU(6)_{sf}$. Such a symmetry is, of course, essentially non-relativistic, and it can be used only in the non-relativistic model.

The wave functions of hadrons consisting of m quarks and n antiquarks are written here as superpositions of terms each of which is a product of separated spin-flavour and colour parts:

$$\Psi = \sum_{\ell} A_{\ell} C\big(\alpha(1),\ldots,\alpha(m);\bar{\alpha}(1),\ldots,\bar{\alpha}(n)\big) h_{\ell}\big(q(1),\ldots,q(m);\bar{q}(1),\ldots,\bar{q}(n)\big)\cdot$$

$$\cdot\ \Phi_{\ell}(\vec{r}_1,\ldots,\vec{r}_m;\vec{r}\,'_1,\ldots,\vec{r}\,'_n) \tag{5.1}$$

The colour part $C(\alpha(1),\ldots,\alpha(m);\bar{\alpha}(1),\ldots,\bar{\alpha}(n))$ of the wave function (where $\alpha(i)$ and $\bar{\alpha}(j)$ represent the colour of the i-th quark and the j-th antiquark, respectively) is an $SU(3)_c$ singlet. It depends only on the number of quarks and antiquarks which are contained in the hadron, and is the same, e.g., for all mesons consisting of a quark and an antiquark or for all baryons, consisting of three quarks.

The spin-flavour part $h_{\ell}(q(1),\ldots,q(m),\bar{q}(1),\ldots,\bar{q}(n))$ of the wave function realizes a definite $SU(6)_{sf}$ representation for a system of m quarks and n antiquarks. In the non-relativistic potential model $SU(6)_{sf}$ multiplets correspond to excitation levels (N) of the system with a definite value of the total orbital momentum (L) of all particles. The usual notation for such a multiplet is $[M,L^P]_N$, where M is the total number of states in the multiplet (i.e. the dimension of the representation), while P is the parity which is the same for all states in the multiplet.

The coordinate part Φ_{ℓ} of the wave function is the same for all states belonging to one $SU(6)_{sf}$ multiplet. $SU(6)_{sf}$ is an approximate symmetry; it is broken by both the peculiarities of the strange quark (which, e.g. has a mass different from that of the u and d-quarks) and by the spin-spin and spin-orbital interactions of the quarks.

This circumstance makes it convenient to choose the basis for the $SU(6)_{sf}$ symmetric wave function in a form for which the main breaking interactions are diagonal.

The observable hadrons might be, generally speaking, mixtures of states which belong to different $SU(6)_{sf}$ multiplets. The ground state with $N = 0$ can be mixed up with excited states $N = 2$, etc. Therefore one writes the wave functions for real hadrons in the form (5.1), where in this case the index ℓ counts the different $SU(6)_{sf}$ states. However, if such a phenomenon exists at all, these admixtures are rather small, and we will neglect them.

In the following we will denote the spin-flavour wave function as a ket-vector with the notation of the considered hadron and with the spin projection index J_z. For instance $|\rho\rangle_1$ is the spin-flavour wave function of a ρ-meson with the spin projection +1, or $|p\rangle_{1/2}$ that of a proton with +1/2 spin projection. Analogously, the quark states will also be denoted by ket vectors, in which the symbols of quarks (antiquarks) from left to right correspond to the increasing number of the arguments in the coordinate part of the wave function. Besides, for the S_z indices 1/2 and -1/2 corresponding to the spin projections of quarks (and antiquarks) we will write $\uparrow$ and $\downarrow$. For example, the spin-flavour part of the wave function of a three-quark system, in which the quark with the coordinate $\vec{r}_1$ is a u-quark with $s_z = +1/2$, to $\vec{r}_2$ corresponds a u-quark with $s_z = -1/2$ and to $\vec{r}_3$ a d-quark with $s_z = +1/2$, will be written as $|u^\uparrow u^\downarrow d^\uparrow\rangle$, etc.

Let us consider now concrete examples.

5.1.1 *Mesons*

The simplest quark systems are mesons, consisting of a quark and an antiquark: $q\bar{q}$. The colour wave function is the same for all these mesons and equals

$$c(\alpha;\bar{\alpha}) = \frac{1}{\sqrt{3}}\,\alpha_i\bar{\alpha}_i \tag{5.2}$$

where α_i (i=1,2,3) describe the colour of the quark, $\bar{\alpha}_i$ that of the antiquark.

The coordinate part $\Phi(\vec{r}_1;\vec{r}_1')$ of the wave function of a meson with a definite L momentum state of the quark-antiquark pair can be expanded in terms of the standard spherical functions $Y_{LL_z}(\frac{\vec{r}}{r})$, where $\vec{r} = \vec{r}_1 - \vec{r}_1'$, $r = |\vec{r}|$. The non-trivial coordinate part of the wave function depends only on the modulus of the distance between the quark and the antiquark: $\Phi(r)$. The lowest $SU(6)_{sf}$ meson multiplets correspond to $L = 0,1$ states, in which radial excitations are absent.

In the $L = 0$ state there are two SU(6) multiplets: the 35-plet $[35,0^+]$ and the singlet $[1,0^+]$. The corresponding mesons are listed in Table 5.1. The spin-flavour part of the singlet wave function $|\eta^1\rangle$ (here the index 1 shows that a singlet state is considered) is symmetric in all flavour indices and antisymmetric in the spin indices, which corresponds to the total quark spin $S_{q\bar{q}} = 0$ of the $q\bar{q}$ system

$$|\eta^1\rangle = \frac{1}{\sqrt{6}}(|u^\uparrow\bar{u}^\downarrow\rangle - |u^\downarrow\bar{u}^\uparrow\rangle + |d^\uparrow\bar{d}^\downarrow\rangle - |d^\downarrow\bar{d}^\uparrow\rangle + |s^\uparrow\bar{s}^\downarrow\rangle - |s^\downarrow\bar{s}^\uparrow\rangle) \tag{5.3}$$

For the pseudoscalar meson octet

$$h_o = \pi^+,\pi^o,\pi^-,\eta,\eta',K^+,K^o,\bar{K}^o,K^- \tag{5.4}$$

the spin-flavour parts of the wave functions are

$$|\pi^+\rangle = \frac{1}{\sqrt{2}}(|u^\uparrow\bar{d}^\downarrow\rangle - |u^\downarrow\bar{d}^\uparrow\rangle)$$

$$|\pi^o\rangle = \frac{1}{2}(|u^\uparrow\bar{u}^\downarrow\rangle - |d^\uparrow\bar{d}^\downarrow\rangle - |u^\downarrow\bar{u}^\uparrow\rangle + |d^\downarrow\bar{d}^\uparrow\rangle)$$

$$|\eta^8\rangle = \frac{1}{2\sqrt{3}}(|u^\uparrow\bar{u}^\downarrow\rangle + |d^\uparrow\bar{d}^\downarrow\rangle - 2|s^\uparrow\bar{s}^\downarrow\rangle - |u^\downarrow\bar{u}^\uparrow\rangle - |d^\downarrow\bar{d}^\uparrow\rangle + 2|s^\downarrow\bar{s}^\uparrow\rangle). \tag{5.5}$$

All the other wave functions of pseudoscalar mesons can be obtained from (5.5) by replacing the flavour indices. For instance, $|K^+\rangle = \frac{1}{\sqrt{2}}(|u^\uparrow\bar{s}^\downarrow\rangle - |u^\downarrow\bar{s}^\uparrow\rangle)$, i.e. the wave function $|K^+\rangle$ differs from

Table 5.1 Mesons of the lowest S-wave $1 \oplus 35$ - plet.

$SU(6)_{sf}$		singlet	35 - plet									
$SU(3)_f$		singlet	octet					singlet + octet				
	J^P	0^-	0^-					1^-				
S	I \ I_z		1	1/2	0	-1/2	-1	1	1/2	0	-1/2	-1
+1	1/2			K^+		K^0			K^{*+}		K^{*0}	
0	0	η'			η					$\omega\ \Phi$		
0	1		π^+		π^0		π^-	ρ^+		ρ^0		ρ^-
-1	1/2			$\bar{K}^0$		K^-			$\bar{K}^{*0}$		K^{*-}	

$|\pi^+>$ by the replacement $\bar{d}\to\bar{s}$; in other words, one can write $|K^+> = |\pi^+>(\bar{d}\to\bar{s})$. Similarly $|K^0> = |K^+>(u\to d)$, $|\pi^-> = |\pi^+>(u\rightleftarrows d)$. The wave functions $|K^->$ and $|\bar{K}^0>$ are obtained by charge conjugation of $|K^+>$ and $|K^0>$.

In (5.5) for the wave function of the η-meson the index 8 means, that here we consider a pure (flavour) octet state.

The wave functions of vector mesons $|h_1>_{J_z}$ (where $h_1 = \rho^+, \rho^0, \rho^-, \omega, K^{*+}, K^{*0}, \bar{K}^{*0}, K^{*-}, \Phi$) can be written as

$$|\rho^+>_1 = |u^\uparrow\bar{d}^\uparrow>$$

$$|\rho^+>_0 = \frac{1}{\sqrt{2}}(|u^\uparrow\bar{d}^\downarrow> + |u^\downarrow\bar{d}^\uparrow>)$$

$$|\rho^0>_1 = \frac{1}{\sqrt{2}}(|u^\uparrow\bar{u}^\uparrow> - |d^\uparrow\bar{d}^\uparrow>)$$

$$|\rho^0>_0 = \frac{1}{2}(|u^\uparrow\bar{u}^\downarrow> - |d^\uparrow\bar{d}^\downarrow> + |u^\downarrow\bar{u}^\uparrow> - |d^\downarrow\bar{d}^\uparrow>)$$

$$|\omega>_1 = \frac{1}{\sqrt{2}}(|u^\uparrow\bar{u}^\uparrow> + |d^\uparrow\bar{d}^\uparrow>)$$

$$|\omega>_0 = \frac{1}{2}(|u^\uparrow\bar{u}^\downarrow> + |d^\uparrow\bar{d}^\downarrow> + |u^\downarrow\bar{u}^\uparrow> + |d^\downarrow\bar{d}^\uparrow>)$$

$$|\Phi>_1 = |s^\uparrow\bar{s}^\uparrow>$$

$$|\Phi>_0 = \frac{1}{\sqrt{2}}(|s^\uparrow\bar{s}^\downarrow> + |s^\downarrow\bar{s}^\uparrow>) \quad . \tag{5.6}$$

The wave functions with $J_z = -1$ can be obtained from the functions with $J_z = 1$ by replacing the quark spins to the opposite: $\uparrow\rightleftarrows\downarrow$. For the wave functions of the K^* and ρ^--mesons one have $K^{*+} = \rho^+(\bar{d}\to\bar{s})$, $K^{*0} = K^{*+}(u\to d)$, $\rho^- = \rho^+(u\rightleftarrows d)$.

The observable η and η' mesons are not pure η^1 and η^8 states, but are superpositions of them

$$|\eta> = |\eta^8>\cos\Theta - |\eta^1>\sin\Theta$$

$$|\eta'> = |\eta^8>\sin\Theta + |\eta^1>\cos\Theta \quad . \tag{5.7}$$

The mixing angle Θ is negative, and different estimates give

$|\Theta| \simeq 10°$-$25°$. However, the real situation with η and η' is more complicated[5]; there are reasons to believe, that η and η' contain to a noticeable extent a gluonium component[6].

The real ω and Φ mesons are not pure states of only strange or only non-strange quarks: the mixing angle of these is about 1°-3°.

In states with $L \neq 0$ there exist SU(6) multiplets $35 \otimes (2L+1)$ and $1 \otimes (2L+1)$. Hence, for $L = 1$ we have the $[35 \otimes 3, 1^+]$ meson multiplets with $J^P = 0^+$, 1^+ and 2^+ and the axial singlet $[1 \otimes 3, 1^+]$ with $J^{PG} = 1^{+-}$: The latter singlet state mixes with the G-odd axial isoscalar meson of the $35 \otimes 3$ plet; experimentally so far only one state H(1190) is supposed to be seen from the two states with $J^{PG} = 1^{+-}$. Therefore the mixing parameters for these SU(6) multiplets are not determined with certainty; the recent Review of Particle Properties (see e.g. Ref. 7) reflects the current level of our knowledge in this field. In Table 5.2 we list the mesons belonging to the P-wave multiplets; according to the above mentioned reason we do not consider here the SU(6) singlet.

The coordinate part of the wave functions of the P-wave mesons belonging to the $35 \otimes 3$-plet can be written in the form $\Phi_{1L_z}(\vec{r}) = Y_{1L_z}(\frac{\vec{r}}{r})\Phi_1(r)$, while the spin-flavour part is determined by the functions $|h_0\rangle$ and $|h_1\rangle_{s_z}$ cited in (5.5) and (5.6). The complete wave functions for the axial octet with $J^P = 1^+$ and $s_{q\bar{q}} = 0$ is

$$\Psi_{Jj_z} = C(\alpha;\bar{\alpha})|h_0\rangle\Phi_{1j_z}(\vec{r}) \tag{5.8}$$

where j_z is the projection of the total momentum; for the mesons with $J^P = 0^+$, $1^+(s_{q\bar{q}} = 1)$ and 2^+ we write

$$\Psi_{Jj_z} = C(\alpha;\bar{\alpha}) \sum_{s_z,L_z} C^{Jj_z}_{1s_z 1L_z} |h_1\rangle_{s_z} \Phi_{1L_z}(\vec{r}) \quad . \tag{5.9}$$

$C^{Jj_z}_{1s_z 1L_z}$ are Clebsch-Gordan coefficients.

Table 5.2 Mesons of the P-wave $[1 \oplus 35] \otimes 3$ multiplet

S	I	I_z \ J_P	0^+	$1^+(s_{q\bar{q}} = 1)$	$1^+(s_{q\bar{q}} = 0)$	2^+
0	1	1	$\delta^+(980)$	$A_1^+(1270)$	$B^+(1235)$	$A_2^+(1320)$
		0	δ^o	A_1^o	B^o	A_2^o
		-1	δ^-	A_1^-	B^-	A_2^-
1	$\frac{1}{2}$	$\frac{1}{2}$	$\kappa^+(1350)$	Q_A^+ *)	Q_B^+ *)	$K^{*+}(1430)$
		$-\frac{1}{2}$	κ^o	Q_A^o	Q_B^o	K^{*o}
-1		$\frac{1}{2}$	$\bar{\kappa}^o$	$\bar{Q}_A^o$	$\bar{Q}_B^o$	$\bar{K}^{*o}$
		$-\frac{1}{2}$	κ^-	Q_A^-	Q_A^-	K^{*-}
0	0	0	$\varepsilon(1300)$	D(1285)	H(1190)	f(1270)
			S*(975)	E(1420)	?	f'(1515)

*) The observed strange axial mesons $Q_1(1280)$ and $Q_2(1400)$ are mixtures of the states Q_A and Q_B with the mixing angle $\Phi \simeq 50°$-$60°$.

5.1.2 *Baryons*

Baryons consist of three quarks qqq; the colour part of the wave function is the same for all baryons

$$C(\alpha(1),\alpha(2),\alpha(3)) = \frac{1}{\sqrt{6}}\varepsilon_{ik\ell}\alpha_i(1)\alpha_k(2)\alpha_\ell(3) \quad . \tag{5.10}$$

Since 10 is antisymmetric in the permutation of any quark pair and quarks obey Fermi statistics, the remaining part of the wave function (i.e. the coordinate and the spin-flavour one) has to be completely symmetric.

The coordinate wave function $\Phi(\vec{r}_1, \vec{r}_2, \vec{r}_3)$ might be completely symmetric, in which case the spin-flavour part must be also symmetric: this corresponds to the 56-plet representation of the SU(6) group. If $\Phi(\vec{r}_1, \vec{r}_2, \vec{r}_3)$ is totally antisymmetric, the spin-flavour part has to be also antisymmetric (20-plet representation). Φ can be also of mixed symmetry (i.e. correspond to a mixed Young scheme ): this leads to the mixed symmetry of the spin-flavour wave function, which corresponds to the 70-plet representation. All the baryons observed up to now seem to belong to either the 56-plet or the 70-plet; so far no states belonging to the 20-plet are established with certainty.

Assembling the baryon wave functions it is convenient to write down the spin-flavour part $h(q(1), q(2), q(3))$ in the form of a direct product of the spin function $|ss_z\rangle$ (where s is the total spin of three quarks, s_z its z-projection) and the flavour function $|q_1q_2q_3\rangle$ (q_i = symbols of the u, d, s quarks). The symmetric spin-functions (spin 3/2) are

$$|3/2\ 3/2\rangle = |\uparrow\uparrow\uparrow\rangle$$

$$|3/2\ 1/2\rangle = \frac{1}{\sqrt{3}}\left(|\uparrow\uparrow\downarrow\rangle + |\uparrow\downarrow\uparrow\rangle + |\downarrow\uparrow\uparrow\rangle\right) \tag{5.11}$$

etc., for spin 1/2 (mixed symmetry) two orthogonal combinations can be written

$$|1/2\ 1/2\rangle_\alpha = \frac{1}{\sqrt{6}}\left(|\uparrow\uparrow\downarrow\rangle + |\uparrow\downarrow\uparrow\rangle - 2|\downarrow\uparrow\uparrow\rangle\right)$$

$$|1/2\ 1/2\rangle_\beta = \frac{1}{\sqrt{2}}\left(|\uparrow\uparrow\downarrow\rangle - |\uparrow\downarrow\uparrow\rangle\right) \quad . \tag{5.12}$$

Analogously, the SU(3) decuplet flavour function is symmetric

$$|10\rangle = \frac{1}{\sqrt{6}}\left(|q_1q_2q_3\rangle + |q_1q_3q_2\rangle + |q_2q_1q_3\rangle + |q_2q_3q_1\rangle \right.$$
$$\left. + |q_3q_1q_2\rangle + |q_3q_2q_1\rangle\right) \tag{5.13}$$

the singlet is antisymmetric

$$|1\rangle = \frac{1}{\sqrt{6}}\left(|q_1q_2q_3\rangle + |q_2q_3q_1\rangle + |q_3q_1q_2\rangle - |q_2q_1q_3\rangle \right.$$
$$\left. - |q_3q_2q_1\rangle - |q_1q_3q_2\rangle\right) \tag{5.14}$$

while for the octet there are two orthogonal combinations

$$|8\rangle_\alpha = \frac{1}{2\sqrt{3}}\left(|q_1q_2q_3\rangle + |q_1q_3q_2\rangle + |q_2q_1q_3\rangle + |q_2q_3q_1\rangle \right.$$
$$\left. - 2|q_3q_1q_2\rangle - 2|q_3q_2q_1\rangle\right)$$

$$|8\rangle_\beta = \frac{1}{2}\left(|q_1q_2q_3\rangle - |q_1q_3q_2\rangle - |q_2q_3q_1\rangle + |q_2q_1q_3\rangle\right) \quad . \tag{5.15}$$

If two of the flavours in (5.13)-(5.15) coincide, the normalization factor changes; all the functions (5.13)-(5.15) are normalized to unity.

The direct product of the spin and flavour functions form the spin-flavour baryon wave function, e.g.

$$|8\rangle_\beta|1/2\ 1/2\rangle_\alpha = \frac{1}{2\sqrt{6}}\left(|q_1^\uparrow q_2^\uparrow q_3^\downarrow\rangle + |q_1^\uparrow q_2^\downarrow q_3^\uparrow\rangle - 2|q_1^\downarrow q_2^\uparrow q_3^\uparrow\rangle \right.$$
$$- |q_1^\uparrow q_3^\uparrow q_2^\downarrow\rangle - |q_1^\uparrow q_3^\downarrow q_2^\uparrow\rangle + 2|q_1^\downarrow q_3^\uparrow q_2^\uparrow\rangle$$
$$- |q_2^\uparrow q_3^\uparrow q_1^\downarrow\rangle - |q_2^\uparrow q_3^\downarrow q_1^\uparrow\rangle + 2|q_2^\downarrow q_3^\uparrow q_1^\uparrow\rangle$$
$$\left. + |q_2^\uparrow q_1^\uparrow q_3^\downarrow\rangle + |q_2^\uparrow q_1^\downarrow q_3^\uparrow\rangle - 2|q_2^\downarrow q_1^\uparrow q_3^\uparrow\rangle\right) \quad . \tag{5.16}$$

The lowest baryon multiplets correspond to the values $L = 0$ and $L = 1$ without radial excitations.

The baryons of the lowest multiplet $[56,0^+]_0$ have a totally symmetric coordinate part of the wave function — the orbital momentum of any quark pair equals zero. The spin-flavour part is totally symmetric too; to a symmetric flavour part (decuplet) corresponds the spin value 3/2, to a flavour function of mixed symmetry (octet) the spin 1/2:

$$[56,0^+]_0 = {}^4 10_{3/2} + {}^2 8_{1/2} \quad . \tag{5.17}$$

(We denote the SU(3) multiplets by ${}^{2s+1}M_J$ where J is the baryon spin.) Hence

$$|{}^4 10_{\frac{3}{2}}>_{J_z} = |10>|\tfrac{3}{2}J_z>$$

$$|{}^2 8_{\frac{1}{2}}>_{J_z} = \frac{1}{\sqrt{2}}\left(|8>_\alpha|\tfrac{1}{2}J_z>_\alpha + |8>_\beta|\tfrac{1}{2}J_z>_\beta\right) \quad . \tag{5.18}$$

All the 56-plet wave functions are listed in the Appendix C. The hadron content of this multiplet is given in Table 5.3.

The coordinate part of the wave function of the multiplet with $L = 1$ is of mixed symmetry — only one quark pair is in a P-wave state. Because of that, the spin-flavour part has to be also of mixed symmetry, i.e. we have a multiplet $[70,1^-]_1$ [8]. The symmetric and antisymmetric flavour functions correspond here to the quark spin 1/2, the mixed flavour function to spin 1/2 or spin 3/2. Combining the quark spins with the angular momenta, we can obtain the SU(3) multiplets:

$$\begin{aligned}[70,1^-] = {}& {}^4 8_{5/2} + {}^4 8_{3/2} + {}^4 8_{1/2} + {}^2 8_{3/2} + {}^2 8_{1/2} + {}^2 10_{3/2} \\ & + {}^2 10_{1/2} + {}^2 1_{3/2} + {}^2 1_{1/2} \quad .\end{aligned} \tag{5.19}$$

To describe the angular dependence of the coordinate function it is convenient to expand it in the terms of an orthonormal basis. For the P-wave 70-plet it is natural to consider functions $Y_{1\ell}(\vec{n}_{23})$ (P-wave between quarks with coordinates $\vec{r}_2$ and $\vec{r}_3$) and $Y_{1\ell}(\vec{r}_{1,23})$ ($\vec{n}_{1,23} \sim 2\vec{r}_1 - \vec{r}_2 - \vec{r}_3$, P-wave between quark $\vec{r}_1$ and the S-wave pair $\vec{r}_2$, $\vec{r}_3$). The expansion over this basis (together with the corresponding spin-flavour functions) gives

Table 5.3 Baryons of the S-wave $[56,0^+]_0$ multiplet.

$SU(6)_{sf}$		$[56,0^+]_0$											
${}^{2s+1}SU(3)_{J^P}$		${}^{2}8_{1/2}{}^{+}$					${}^{4}10_{3/2}{}^{+}$						
S	I	I_z					I_z						
		1	1/2	0	-1/2	-1	3/2	1	1/2	0	-1/2	-1	-3/2
0	1/2		p		n								
	3/2						Δ^{++}		Δ^{+}		Δ^{0}		Δ^{-}
-1	0			Λ									
	1	Σ^{+}		Σ^{0}		Σ^{-}		Σ^{*+}		Σ^{*0}		Σ^{*-}	
-2	1/2		Ξ^{0}		Ξ^{-}				Ξ^{*0}		Ξ^{*-}		
-3	0									Ω^{-}			

$$|{}^4 8_J\rangle_{J_z} = \frac{1}{\sqrt{2}} \sum_{\ell,\sigma} C^{JJ_z}_{1\ell\frac{3}{2}\sigma} \left\{ |8\rangle_\alpha |\tfrac{3}{2}\sigma\rangle Y_{1\ell}(\vec{n}_{1,23}) + |8\rangle_\beta |\tfrac{3}{2}\sigma\rangle Y_{1\ell}(\vec{n}_{23}) \right\}$$

$$|{}^2 8_J\rangle_{J_z} = \frac{1}{\sqrt{2}} \sum_{\ell,\sigma} C^{JJ_z}_{1\ell\frac{1}{2}\sigma} \left\{ [-|8\rangle_\alpha |\tfrac{1}{2}\sigma\rangle_\alpha + |8\rangle_\beta |\tfrac{1}{2}\sigma\rangle_\beta] Y_{1\ell}(\vec{n}_{1,23}) \right.$$

$$\left. + [|8\rangle_\alpha |\tfrac{1}{2}\sigma\rangle_\beta + |8\rangle_\beta |\tfrac{1}{2}\sigma\rangle_\alpha] Y_{1\ell}(\vec{n}_{23})] \right\} . \qquad (5.20)$$

$$|{}^2 10_J\rangle_{J_z} = \frac{1}{\sqrt{2}} \sum_{\ell,\sigma} C^{JJ_z}_{1\ell\frac{1}{2}\sigma} \left\{ |10\rangle |\tfrac{1}{2}\sigma\rangle_\alpha Y_{1\ell}(\vec{n}_{1,23}) + |10\rangle |\tfrac{1}{2}\sigma\rangle_\beta Y_{1\ell}(\vec{n}_{23}) \right\}$$

$$|{}^2 1_J\rangle_{J_z} = \frac{1}{\sqrt{2}} \sum_{\ell,\sigma} C^{JJ_z}_{1\ell\frac{1}{2}\sigma} \left\{ -|1\rangle |\tfrac{1}{2}\sigma\rangle_\beta Y_{1\ell}(\vec{n}_{1,23}) + |1\rangle |\tfrac{1}{2}\sigma\rangle_\alpha Y_{1\ell}(\vec{n}_{23}) \right\}$$

$$(5.21)$$

All wave functions of the 70-plet are listed in the Appendix C, where the notations $Y_{1\ell}(\vec{n}_{1,23}) = |\ell\rangle_\alpha$, $Y_{1\ell}(\vec{n}_{23}) = |\ell\rangle_\beta$ is introduced. The real baryons are mixtures of states with different total quark spins. This fact causes some difficulties in the classification of the hadrons belonging to the 70-plet. The mixing coefficients are determined from the analysis of mass formulae or of decay amplitudes (see, e.g., Ref. 9, 10). They can be obtained also in the framework of potential models.

5.2 Magnetic Moments of Baryons and Radiative Decays of Vector Mesons

From the point of view of the quasi-nuclear quark structure it is a very important question to determine to what extent the features of dressed quarks depend on the hadron to which they belong. If the dressed quarks can be handled as quasiparticles, they have to be practically the same in different hadrons. It is convenient to test this by the investigation of the magnetic moments of the S-wave 56-plet baryons. (Indeed, the calculation of the magnetic moments was historically the first serious success of the quasi-nuclear quark model of hadrons[11].) The same magnetic moment values of quarks give the possibility to calculate the widths of the radiative decays of S-wave vector mesons of 35-plet into pseudoscalar mesons.[12]

Consider first the magnetic moments. In the quasi-nuclear hadron picture it is natural to assume that the masses of the dressed u and d-quarks are simply 1/3 of the proton mass $m_u = m_d = \frac{1}{3} m_p$, and that the magnetic moments of these quarks are equal to the usual Dirac value, i.e. they are defined entirely by the charges and the masses of the quarks.

The baryon magnetic moment (for definiteness, consider a proton) is determined by the interaction of the magnetic field with the quarks of the proton.

$$\frac{e}{2m_p}\mu_p = \int d\Omega \Psi^*(p_{\frac{1}{2}}) \sum_{i=1}^{3} \frac{e_q(i)\sigma_z(i)}{2m_q(i)} \Psi(p_{\frac{1}{2}}) = 3\left\langle p_{\frac{1}{2}} \left| \frac{e_q(1)\sigma_z(1)}{2m_q(1)} \right| p_{\frac{1}{2}} \right\rangle . \tag{5.22}$$

Here e and $e_q(i)$ are the charges of the proton and the i-th quark respectively; m_p and $m_q(i)$ are their masses. $\Psi(p_{\frac{1}{2}})$ and $|p_{\frac{1}{2}}>$ denote here the total wave function of the proton with $J_z = 1/2$ and its corresponding spin-flavour part. The integration is carried out over the quark coordinate variables. The magnetic moment μ_p is expressed in units $e/2m_p$ (i.e. in nuclear magnetons). The calculated this way baryon magnetic momenta are given in Table 5.4, in which the notation

$$\xi = \frac{m_s - m_u}{m_u}$$

is used. The $\xi = 1/3$ corresponds to $m_s - m_u \simeq 150$ MeV, which is a rather fundamental quantity for both the quark model and the quark theory based on QCD[15].

The agreement between calculation and experiment is quite satisfactory (and typical for the quark model): the deviations are not more than 20-25%. However, if one tries to understand these deviations literally, the results will be distressing. For example, calculating the quark masses on the basis of data on μ_{Ξ^0} and μ_{Ξ^-}, one gets $m_u > m_s$. One has to remember, that the non-relativistic quark model is

Table 5.4 Magnetic moments of baryons in nuclear magnetons

Particle	Quark model prediction (ξ=1/3)	Experiment [7]
p	3	2.79
n	-2	-1.91
Λ	$-1 + \xi = -0.67$	-0.613 ± 0.004
Σ^+	$3 - \frac{1}{3}\xi = 2.89$	2.38 ± 0.02
Σ^-	$-1 - \frac{1}{3}\xi = -1.11$	-1.10 ± 0.05
Ξ^0	$-2 + \frac{4}{3}\xi = -1.56$	-1.25 ± 0.01
Ξ^-	$-1 + \frac{4}{3}\xi = -0.56$	-0.69 ± 0.04 [13,14] -1.85 ± 0.75

a rough approach, and such discrepancies are more or less natural. Small variations of the magnetic moments (in comparison with the calculated values) can be, for instance, consequences of either relativistic corrections, or the structure of the dressed quarks themselves. Introducing, e.g. a relatively small anomalous magnetic moment for the u, d and s quarks[16)] one can get a rather good agreement with the data.

The same quark magnetic moments determine the probabilities of the radiative decays of vector mesons. In Table 5.5 the calculated values and the experimental data are given. We use here $\Gamma(V \to P + \gamma)$ since this quantity is proportional to the quark magnetic moment, and is, therefore, suitable for the comparison with the calculated magnetic moment. The predictions for the radiative widths satisfy the experimental data within the same accuracy of 20-25%. It is a rather impressive fact, that the quark magnetic moments are the same in mesons and baryons; this shows, that the dressed quarks appear in hadrons as

Table 5.5 Values of $\sqrt{\Gamma(V \to P + \gamma)}$, $KeV^{1/2}$ for vector meson decays

Decay mode	$\sqrt{\Gamma(V \to P + \gamma)}$, $(KeV)^{1/2}$ Quark model prediction	Experiment
$\omega \to \pi^o\gamma$	34.6	28.1 ± 1.6
$\rho^{\pm} \to \pi^{\pm}\gamma$	11.0	8.2 ± 0.4
$\rho^o \to \eta^o\gamma$	8.4	8.1 ± 0.9
$\Phi \to \eta^o\gamma$	10.4	8.3 ± 0.6
$K^{*\pm} \to K^{\pm}\gamma$	7.0	7.1 ± 0.4
$K^{*o} \to K^o\gamma$	13.7	8.7 ± 2.0

somewhat independent objects — quasiparticles.

5.3 Hadron Masses and Potential Models

It was understood already relatively long ago[17], that the mass splitting of light hadrons can be well described in the framework of the non-relativistic quark model by the spin-spin quark interaction. The next step was made by de Rujula, Georgi and Glashow: according to Ref. 18 the hadron-mass splitting is due only to the short-range part (the spin-spin part) of the interaction, which is connected with the gluon exchange. The obtained effective potential for the interaction of two quarks (i and j) is given by

$$V_{ij} = \pm \alpha_s\left(\frac{\vec{\lambda}(i)}{2}\frac{\vec{\lambda}(j)}{2}\right)\left(-\frac{2\pi}{3}\cdot\frac{\vec{\sigma}(i)\vec{\sigma}(j)}{m(i)m(j)}\,\delta(\vec{r}_{ij})\right) \tag{5.23}$$

where α_s is the squared gluon-quark coupling constant (the analogue of the electrodynamical $\alpha = 1/137$), $\vec{\lambda}$ are the Gell-Mann matrices, acting on the colour indices of the i-th and j-th quarks, and the signs ± stand for the interactions of two quarks or a quark and an antiquark respectively. It is assumed, that the remining part of the interaction which is due to the gluon exchange, is averaged, and gives a contribution to the potential which confines the quarks. The interaction (5.23) leads in Born approximation to the following mass splitting:

$$\Delta m_{meson} = \frac{8\pi}{9} \alpha_s |\Phi_M(0)|^2 < h_M \left| \frac{\vec{\sigma}(1)\vec{\sigma}(2)}{m_q(1)m_q(2)} \right| h_M >$$

$$\Delta m_{baryon} = \frac{4\pi}{9} \alpha_s \sum_{i \neq j} \int d^3r_k |\Phi_B(\vec{r}_{ij} = 0, \vec{r}_k)|^2 < h_B \left| \frac{\vec{\sigma}(i)\vec{\sigma}(j)}{m_q(i)m_q(j)} \right| h_B > \quad . \tag{5.24}$$

The spin-flavour part of matrix elements is calculated exactly; however, in this treatment of the question it is impossible to define the coordinate part of the wave function. Because of that, the expression $\alpha_s|\Phi_M(0)|^2$ and $\alpha_s \int d^3r_k |\Phi_B(0,\vec{r}_k)|^2$ have to be considered as phenomenological constants, which can be obtained from the comparison of masses in the meson and baryon multiplets. The result of the comparison of formulae (5.24) with experiment is demonstrated in Table 5.6. Note, that in the calculations $|\Phi_M(0)|^2 = \int d^3\vec{r} |\Phi_B(0,\vec{r})|^2$ is taken. This also shows that it is roughly equiprobable to find in a hadron on a relatively small distance two quarks or a quark-antiquark pair. This feature will be used later, when we will consider the hadron production processes in the framework of quark statistics. It is a surprising fact, that the relations (5.24) work in the case of charmed particles too (D and D* are states of $c\bar{q}$, where q = u,d, with $J^P = 1^-$ and 0^-; F* and F-states of $c\bar{s}$ with $J^P = 1^-$ and 0^-). The constant $\alpha_s|\Phi_M(0)|^2$ is the same as for light hadrons (see Table 5.6).

The de Rujula-Georgi-Glashow approach allows us to understand and concretize the mass formula on a rather elementary level of the quark model. This possibility was discussed in Ref. 19, where, for the masses of the S-wave 56-plet baryons, the formula

$$m_B = \sum_i m_q(i) + b \sum_{i>k} \frac{\vec{\sigma}(i)\vec{\sigma}(k)}{m_q(i)m_q(k)} \tag{5.25}$$

was suggested. The phenomenological parameter is found from the experiment, and there is an astonishingly good description of the baryon

Table 5.6 Mass-splitting values, calculated in the model of de Rujula-Georgi-Glashow. It is assumed, that $m_u = m_d = 360$ MeV, $m_s/m_u = 3/2$, $m_c = 1440$ MeV, $|\Phi_M(0)|^2 = \int d^3\vec{r}\,|\Phi_\beta(0,\vec{r})|^2$.

Δ_m	Calculated (MeV)	Exp. (MeV)	Δ_m	Calculated (MeV)	Exp. (MeV)
$m_\Delta - m_N$	300	295	$m_\rho - m_\pi$	600	630
$m_\Sigma - m_\Lambda$	68	77	$m_{K*} - m_K$	400	398
$m_{\Sigma*} - m_\Lambda$	267	274	$m_{D*} - m_D$	150	140
$m_{\Xi*} - m_\Xi$	200	217	$m_{F*} - m_F$	100	120

Table 5.7a Baryon masses obtained from the Glashow mass formula and the corresponding experimental values.

Particle	mass (MeV)		Particle	mass (MeV)	
	Prediction	Experiment		Prediction	Experiment
N	930	937	Σ^*	1377	1384
Δ	1230	1232	Ξ	1329	1318
Σ	1178	1193	Ξ^*	1529	1533
Λ	1110	1116	Ω	1675	1672

masses (see Table 5.7a). The discrepancies between predictions and measured data are about 5-6 MeV. However in trying to write a similar formula for mesons, one fails: the systematic deviations between calculation and experiment are of the order of 100 MeV (the calculated mass values for the ρ and π mesons are $m_\rho = 875$ MeV, $m_\pi = 275$ MeV). *The reason of this discrepancy becomes obvious when one calculates the* average quark mass in a meson and in a baryon:

$$\langle m_q \rangle_M = \frac{1}{2}\left(\frac{1}{4}m_\pi + \frac{3}{4}m_\rho\right) = 303 \text{ MeV}$$

$$\langle m_q \rangle_B = \frac{1}{3}\left(\frac{1}{2}m_N + \frac{1}{2}m_\Delta\right) = 363 \text{ MeV} \quad . \tag{5.26}$$

(In these combinations of hadron masses the contribution of the splitting interaction (5.24) cancels completely.) The quark masses in mesons are "eaten" by some additional interactions. Their nature is so far not clear, and there are only guesses about the reasons of their existence.

It was assumed, for instance, that this can be connected with the special status of the pseudoscalar mesons: an example of a model in the framework of which, additional, relatively large forces can appear in the pseudoscalar channel, is the instanton model of Shuryak[20).]

The features of baryons provide us with a suitable field for the application of simple ideas of the quark model. This is demonstrated also by the good description of their features in different versions of potential models. Let us consider here one of these models, the model of Isgur and Karl[21)]. The Isgur-Karl model and the de Rujula-Georgi-Glashow approach is based on the same ideology. It is assumed, that the mass splitting is caused by an interaction which is connected with the gluon exchange between quarks. Besides, only one part of this interaction is taken into account, namely: the spin-spin (or hyperfine) interaction. Different from the approach[18)] is that not only the contact term is considered, but also the long-range term:

$$V^{ij}_{hyp} = \frac{2\alpha_s}{3m(i)m(j)}\left\{\frac{2\pi}{3}\vec{\sigma}(i)\vec{\sigma}(j)\delta^{(3)}(\vec{r}_{ij}) + \frac{1}{4r^3_{ij}}\left[\frac{3(\vec{\sigma}(i)\vec{r}_i)(\vec{\sigma}(j)\vec{r}_j)}{r^2_{ij}} - \vec{\sigma}(i)\vec{\sigma}(j)\right]\right\} \quad . \tag{5.27}$$

This interaction coincides (except for a factor 2/3) with the spin-spin term of the Breit Hamiltonian, describing the interaction of two fermions with photon exchange (see, e.g., Ref. 22).

The supplementary factor 2/3 appears because the gluon colour is taken into account. The square of the quark-gluon interaction coupling constant, α_s, is considered as a model parameter.

For the remaining part of the Hamiltonian a simple form is used, which allows an exact solution if V^{ij}_{hyp} is handled as a perturbation:

$$H = \sum_{i=1}^{3}(m_i + \frac{p_i^2}{2m_i} + \frac{k}{2}\sum_{i>j} r^2_{ij} + \sum_{i>j} V^{ij}_{hyp}) \tag{5.28}$$

where k is a parameter, characterizing the harmonic potential of the quark interaction. The introduction of non-harmonic additions to this potential is possible — they are considered in the model as perturbations.

The Hamiltonian (5.28) depends only on few parameters and it is able to describe different baryon features very well. The Karl-Isgur model preserves all the results of the de Rujula-Georgi-Glashow approach in the description of the S-wave baryon 56-plet. This is due to the fact, that for these baryons the contribution of the second term of the hyperfine interaction is small — it becomes essential only for higher baryon states. The quark masses are, however, in the Isgur-Karl model somewhat different: $m_u \simeq m_d \simeq 330$ MeV, $m_s \simeq 550$ MeV (see Table 5.6).

The main success of the model is the description of higher baryon resonances, especially of the P-wave baryons of the 70-plet and of higher multiplets with $N = 2$. First of all, the masses of these resonances can be calculated. Other possibilities for predictions are given by the fact, that in the potential model the wave functions are given explicitly.

The wave functions allow us to calculate the resonance decay amplitudes in the emission model (see Fig. 5.1). (This model is discussed widely in the literature[23].) The meson-quark coupling constant is in this case a model parameter; still, as hundreds of decay amplitudes have been measured up to now, there are very many possibilities to test the Hamiltonian (5.28). We give here only a few examples of the predicted and the measured decay widths (see Table 5.7b).

The inclusion of the admixture of the excited $N = 2$ state in the ground state wave functions leads to a negative value of the average squared charge radius of the neutron $\langle R_n^2 \rangle$ in agreement with the

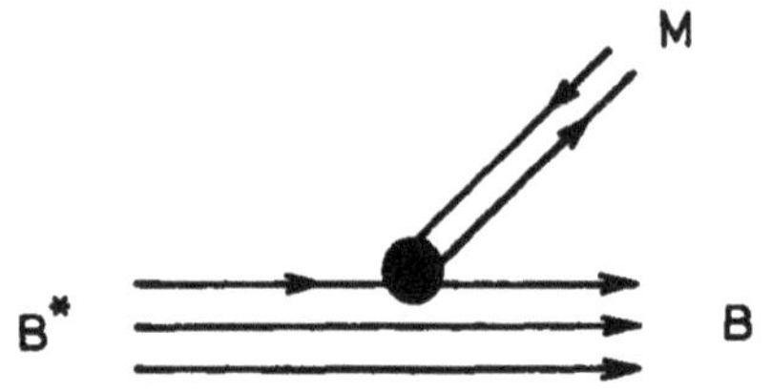

Fig. 5.1 Baryon resonance B* decay into baryon B and meson M in the quark emission model.

Table 5.7b Another example for applying the wave functions is the calculation of electromagnetic mass differences in the S-wave 56-plet. They can be calculated with the help of only one newly introduced parameter, the mass difference of the u and d quarks (Table 5.8).

Particle	J^P	Partial decay mode		
		Mode	Fraction %	
			Theor.	Experiment
N(1520)	$3/2^-$	$N\pi$	59	50-60
		$\Delta\pi$	11.5	15-25
N(1535)	$1/2^-$	$N\pi$	37.5	35-50
		$N\eta$	24	40-65
N(1650)	$1/2^-$	$N\pi$	50	55-65
		$\Delta\pi$	30	4-15
		ΛK	4	5-10
		ΣK	1.8	3-10
N(1675)	$5/2^-$	$N\pi$	37.5	30-40
		$\Delta\pi$	44.5	50-65
		$N\eta$	4.5	< 2
N(1700)	$3/2^-$	$N\pi$	12	8-12
		$\Delta\pi$	65	15-40

Table 5.8 Electromagnetic mass differences in the Karl-Isgur model[21].

Δm	prediction (MeV)	experiment (MeV)
p-n	-1.3	-1.3
$\Sigma^+-\Sigma^0$	-3.3	-3.1 ± 0.1
$\Sigma^--\Sigma^0$	4.9	4.9 ± 0.1
$\Xi^--\Xi^0$	6.8	6.4 ± 0.6
$\Delta^{++}-\Delta^0$	-3.0	-2.6 ± 0.4
$\Delta^{++}-\Delta^-$	-6.9	-5.9 ± 3.1
$\Sigma^{*+}-\Sigma^{*-}$	-5.8	-5.1 ± 0.7
$\Sigma^{*+}-\Sigma^{*0}$	3.7	5.4 ± 2.6
$\Xi^{*-}-\Xi^{*0}$	3.8	3.2 ± 0.6

experimental data. At the same time, the used wave functions fail to describe the size of proton: the calculated value of $\langle R_p^2 \rangle$ is approximately 1.5 times less than the experimental one. This seems to be a serious difficulty of the Carl-Isgur model, since the values of parameters of the Hamiltonian (5.28) are fixed by the fit of numerous mass splittings and decay rates.

5.4 Short-Range Interaction in the Quark Model

The quark model, as we see, describes the statistical features of the hadrons rather successfully. This means, presumably, that the structure of the hadrons is well guessed. The further progress of the model, i.e. the determination of more detailed features requires, however, significant efforts.

First of all the problem of relativistic effects in the quark model should be mentioned. It is obvious that the description of the pseudoscalar mesons in the non-relativistic approximation of the model is a very rough one. Thus, the inclusion of relativistic effects is a seriou point for the detailization of the quark model.

Another urgent question is the structure of the quark-quark forces. The potential models definitely indicated that the contact chromomagnetic interaction, which is caused by the gluon exchange (Eq. (5.23)) is responsible for hadron mass splittings. The conclusions of these models about other components of the quark-quark forces are much less definite. In particular, it is not clear, whether these forces are short-range or long-range ones. In view of the quasinuclear hadron structure it seems natural that long-range components should be suppressed.

We will consider here an example of the construction of low-energy amplitudes $qq \to qq$ and $q\bar{q} \to q\bar{q}$ with the inclusion of relativistic effects[24,25]. This example gives some arguments in favour of the effective short-range interaction of the dressed quarks.

In Chap. 3 it was shown that the dispersion relations provide a convenient technique for the description of relativistic composite systems. In Ref. 25 partial amplitudes in the diquark channels $qq \to qq$ and in meson channels $q\bar{q} \to q\bar{q}$ were calculated with the help of dispersion relations in the framework of the N/D method. Let us recall (see also Sec. 3.3) that the N-function characterizes the interaction forces, while the D-function provides the fulfillment of the unitarity condition.

Following the potential models we assume that the quark forces are caused mainly by vector colour-octet interactions (effective one-gluon exchange). However, we consider these gluon forces as short-range ones (because of the large effective mass of the gluon M_g — see the discussion in Sec. 1.2). If so, at low energies the one-gluon exchange diagram of Fig. 5.2a may be approximated by the contact four-fermion interaction (Fig. 5.2b):

$$g_V(\bar{\psi}\gamma_\mu\vec{\lambda}\psi)(\bar{\psi}\gamma_\mu\vec{\lambda}\psi) \tag{5.29}$$

Since the interaction (5.29) is not weak, all quark rescatterings are to be taken into account, i.e. the infinite series of the diagrams

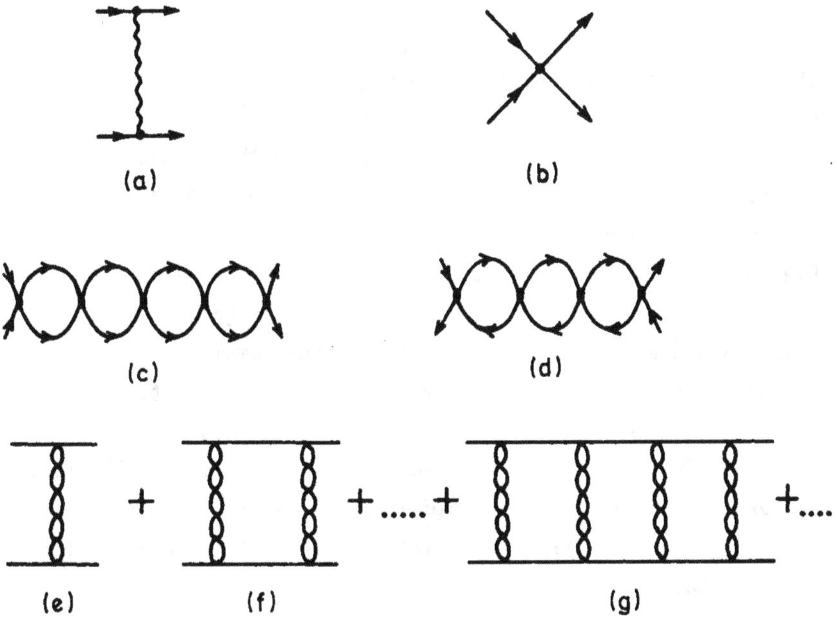

Fig. 5.2 Quark-quark interaction via massive gluon exchange (a) which at low energies may be replaced by an effective four-fermion interaction (b). Iterations of this interaction in s- and t-channels are shown in figs. (c-g).

of Figs. 5.2c, d, etc. is to be summed up. The resulting sum, however, does not give the final answer. The amplitudes in Figs. 5.2c, d, etc. considered in the crossing channels are also "interaction forces", and the total amplitude must include rescatterings of the types shown in Figs. 5.2e, f, g.

In the scale of the dressed quark size r_q (see Sec. 1.2) the effective heavy gluon exchange (5.29) is a short-range interaction; the effective radius of the interaction of Fig. 5.2e type is of the order of inverse meson mass m_ρ^{-1}, i.e. it is an effectively long-range interaction in the quasinuclear quark model. Hereafter we shall consider all the interactions with characteristic scale of *the inverse hadron mass* as long-range ones. Thus, the resulting interaction forces in the considered example include both short-range (Eq. (5.29)) and long-range (e.g. Fig. 5.2e) components.

It is convenient to construct $qq \to qq$ and $q\bar{q} \to q\bar{q}$ amplitudes in the framework of the iterative bootstrap procedure[24,25]. Equation (5.29) determines the zeroth-order N-functions of the partial amplitudes; with the help of these functions the D-functions and the partial amplitudes of the zeroth order are constructed. These amplitudes include all possible rescatterings of Figs. 5.2b, c, d types. They also determine the first-order N-functions of the crossed channels (Fig. 5.2e). These N-functions are used to construct the first-order amplitudes which include the diagrams 5.2e, f, g. Similarly the next iterations are made. The dispersion integrals are regularized using the cut-off procedure; the cut-off parameter Λ is the same for all states. The stable solution of N/D equations is obtained after 4-6 iterations if the model parameters (g_V and Λ) are renormalized at every step, so as to fix any two masses of the bound states (e.g. ρ and η).

All the calculations are made for the low-energy amplitudes; in the $q\bar{q} \to q\bar{q}$ channel the states with $J^P = 0^-, 1^-, 0^+$ are considered, while in the diquark channel $qq \to qq$ those with $J^P = 0^-, 1^+, 0^+$. The resulting masses of the bound states are presented in Table 5.9. The one-gluon exchange interaction (5.29) cannot describe successively the observed η, η' -meson mass splitting it requires an additional interaction in the colourless pseudoscalar channel. Thus a short-range term

$$g_p(\bar{\psi}\gamma_5\psi)(\bar{\psi}\gamma_5\psi) \tag{5.30}$$

is introduced into the zeroth-order bare interaction. As one may see from Table 5.9, the meson masses are calculated in the bootstrap procedure in a good agreement with the experimental data. The discrepancies for the masses of $J^P = 0^+$ states are not surprising, since higher waves with bound states in the mass region 1-1.5 GeV were not accounted for. The pion mass in these calculations is appreciably overestimated; a similar situation takes place in potential models. There must be probably a fundamental explanation for such a discrepancy. Indeed, there are some reasons to consider a pion as not only a $q\bar{q}$ bound state, but also as Goldstone-like particle; this is of course not taken into

Table 5.9 Meson masses in the bootstrap procedure (masses of the dressed quarks are taken to be equal to $m_u = m_d = 380$ MeV, $m_s = 550$ MeV).

$J^P = 0^-$			$J^P = 1^-$			$J^P = 0^+$		
meson	calculation (GeV)	exp.	meson	calculation (GeV)	exp.	meson	calculation (GeV)	exp.
π	0.38	0.14	ρ	0.77	0.77	δ	0.85	0.98
η	0.54	0.55	ω	0.77	0.78	S*	0.85	0.98
η'	1.10	0.96	Φ	1.10	1.02	ε	1.10	1.30
K	0.47	0.50	K*	0.93	0.89	κ	1.00	1.40

account in both the potential models and the bootstrap approach.

The amplitude obtained in Ref. 25 obeys the OZI rule[11,26,27]: the annihilation transitions $u\bar{u}\to d\bar{d}$, $u\bar{u}\to s\bar{s}$ and $d\bar{d}\to s\bar{s}$ are suppressed at low energies by a factor of the order of 0.1-0.2 in agreement with the experimental data.

Let us now consider the structure of quark-quark forces obtained in the bootstrap procedure. We have already mentioned, that they include both short-range and long-range components; the latter are related with the exchanges by the coloured and colourless qq and $q\bar{q}$ states. The detailed calculations show that the resulting interactions are effectively short-ranged. It is, however, not a result of the smallness of every long-range contribution. For example, in $q\bar{q}\to q\bar{q}$ colourless amplitudes the exchanges by the colour-octet $q\bar{q}$ and diquark qq states are significant (see Fig. 5.3). The contributions of these

Fig. 5.3 $q\bar{q}\to q\bar{q}$ colourless amplitudes with the exchanges by the colour-octet $q\bar{q}$ (a) and diquark qq (b) states.

exchanges interfere in the resulting force in such a manner, that it weakly depends on the momentum transferred in the low-energy region. The corresponding N-functions also weakly depend on s at $s > 4m^2$ (let us recall, that N-functions in this region determine the "interaction forces" — see Sec. 3.3).

The calculations of Ref. 25 allow one to estimate the electromagnetic radius of the dressed quark as well. This is determined by the diagrams of the type shown in Fig. 5.4a, the estimated value is

Table 5.10 Dressed quark radius for the different types of interactions.

Type of the current, c	V	P	S
r_c^2 (GeV^{-2})	2.9	20.4	-2.0

presented in Table 5.10. It is small (r_V^2 = 2.9 GeV^{-2}) as compared with the nucleon radius $R_N^2 \simeq 17$ GeV^{-2} and close to the value r_q^2, estimated in hadron-hadron collisions (see Sec. 1.2). If one tries to calculate the dressed quark radius in the vector dominance model (as it was made e.g. in Ref. 28), an appreciably larger value is obtained: $r_V^2 \simeq 12$ GeV^{-2}. The sum of Fig. 5.4a diagrams has a pole at $t = m_\rho^2$, as well as the diagram of Fig. 5.4b. However the former is a rapidly varying function of t only at $t \sim m_\rho^2$ and a smooth function near $t \sim 0$, while the latter changes rapidly both at $t \sim m_\rho^2$ and $t \sim 0$. That is why the quark electromagnetic radius in Ref. 25 appears to be small.

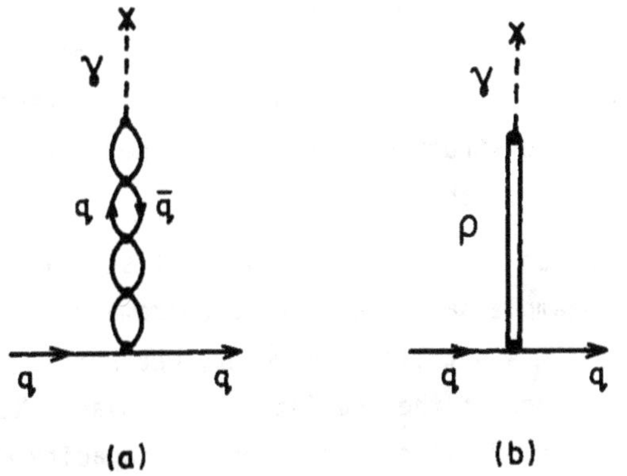

Fig. 5.4 (a) Diagram for the estimation of the electromagnetic dressed quark radius r_V in the bootstrap approach;

(b) VMD diagram which was used in Ref. 28 for the calculation of r_q^2.

The dressed quark radii for some other currents are also presented in Table 5.10. The scalar radius squared r_s^2 is also small, while the pseudoscalar one, r_p, is comparatively large. The reason is that the pion pole is close to the point $t = 0$. It is a remarkable property of the model: the dressed quark radius may be essentially different for the different current and interaction types, and thus the dressed quark may not be regarded as a quasipoint-like object with respect to some kinds of interaction. In the framework of the quasinuclear quark model it is desirable that the interactions, which this model intends to describe, should be effectively short-range ones. As we have seen in the considered example, this is quite possible in hadron-hadron interactions: quark-quark forces appeared to be of the effectively short-range type in the amplitudes $qq \to qq$ and $q\bar{q} \to q\bar{q}$.

The considered four-fermion bare interaction leads to one significant shortcoming of the model: the dispersion integrals are divergent and a regularization (cut-off) procedure is required. Generally speaking, the cut-off parameters may be different in different channels. Thus it is probably more consistent to describe quark-quark forces by some effective interaction. In other words, one may consider the possibility of a reformulation of the quark model in terms of some phenomenological Lagrangian. Such a Lagrangian may be even renormalizable, and thus one may try construct closed and self-consistent picture of the low-energy interaction of the dressed quarks.

Let us consider an illustrative example of such a phenomenological Lagrangian. This example satisfies many requirements of the quasinuclear structure; it is not clear, however, whether it may be considered as an adequate prototype of the realistic Lagrangian. Nevertheless, it suits well the purpose of illustration for the capacity of such an approach. Let us begin with the customary QCD Lagrangian:

$$
\begin{aligned}
L_{QCD} &= -\frac{1}{4} G^2_{\mu\nu} + \bar{q}(i\hat{D} - m)q \\
G^a_{\mu\nu} &= \partial_\mu A^a_\nu - \partial_\nu A^a_\mu + g f^{abc} A^b_\mu A^c_\nu \\
D_\mu &= \partial_\mu - \frac{i}{2} g \lambda^a A^a_\mu \quad .
\end{aligned}
\tag{5.31}
$$

In the quasinuclear quark model it is necessary to require the effective short-range character of the gluon forces. We will suppose, that this short-rangeness will not destroy the global colour symmetry. A simple mechanism of obtaining the global colour symmetry and massive gluons is to introduce a scalar Higgs field which is a triplet of both the local colour group $SU(3)_\ell$, and some global $SU(3)_g$. Such a mechanism for the realization of a global symmetry was first considered in Refs. 29-30 and applied to the colour group in Refs. 31-33. Let us introduce nine Higgs fields $\Phi^{k\ell}$: $SU(3)_g$ triplet (index k) and an $SU(3)_\ell$ triplet (index ℓ). The Lagrangian for the Higgs-particles is

$$L_H = -(1/2)\mathrm{Tr}((D_\mu\Phi)^+(D_\mu\Phi) + V(\Phi,\Phi^+)) \quad . \tag{5.32}$$

The potential V provides the spontaneous breaking of the local symmetry. It is enough to restrict oneself in this potential to the fourth degree of Φ:

$$V(\Phi,\Phi^+) = a\,\mathrm{Tr}(\Phi^+\Phi) + b\,\mathrm{Tr}(\Phi^+\Phi\Phi^+\Phi) + c(\mathrm{Tr}(\Phi^+\Phi))^2 + d(\det\Phi + \det\Phi^+) \ . \tag{5.33}$$

The symmetry $SU(3)_\ell \times SU(3)_g$ is broken, if $a < 0$ and $b + 3c > 0$. In the present example it is sufficient to consider such a version of the Higgs interaction, in which $b \gg c-1$, $d \simeq \sqrt{-a/b}$ (i.e. $d^2 \ll a$). In this case the average vacuum value of the Higgs field is equal to $\langle\Phi^{k\ell}\rangle = \Phi_0\delta_{k\ell}$, where $\Phi_0 \simeq \sqrt{-a/2b}$. Besides, the Lagrangian (5.32) turns out to be invariant under the global symmetry $[SU(3)_{1+g}]_{global}$. The Higgs particles form two singlets and an octet of this global (colour) symmetry. They have the following masses:

$$\mu_1^2 = -\frac{3}{\sqrt{2}}d\Phi_0 \quad , \quad M_1^2 = M_8^2 = -2a \quad . \tag{5.34}$$

The total Lagrangian

$$L = L_{QCD} + L_H \tag{5.35}$$

is also invariant under the global colour symmetry. The gluons form an octet of this symmetry and have the mass

$$M^2_{gluon} = \frac{1}{2}(g^2\Phi_o^2) \quad . \tag{5.36}$$

The quarks form a triplet of the global colour symmetry. The Lagrangian (5.35) corresponds to the interaction structure of the dressed quarks required by the quasinuclear model. Indeed, the parameters of the potential (5.33) can be chosen in such a way that the quark masses become relevantly smaller than the masses of gluons and Higgs particles, namely:

$$m \ll M_{gluon} \ , \quad \mu_1 \ll M_1, M_s \quad . \tag{5.37}$$

The large boson masses lead to short-range interactions of the dressed quarks. The one-gluon exchanges are responsible for the quark-quark interaction forces on the distances of the order of $r \sim 1/M_{gluon}$, the two-gluon exchanges and other many-gluon exchanges — for interactions at smaller distances ($r \lesssim 1/(2M_{gluon})$. This structure of quark-quark forces corresponds to the bootstrap procedure discussed above, if the main contribution to the bare interaction at low energies is determined by one-gluon processes, and the parameters of the potential V_H define the value of the cut-off in the multiple quark scattering diagrams. The Higgs sector is presumably "rich" enough to have such features of the quark-quark interaction at small distances. As in any theory with spontaneously broken symmetry, Higgs fields are some auxiliary equipment, used in order to obtain the required properties of the interaction. They may not be related with some observable particles.

The results of the above considerations definitely show that the hypothesis of short-range forces which is assumed in the quasinuclear model is quite viable. It gives an opportunity to describe, at least, the low-lying hadron states. At the same time as we shall see later from the analysis of the experimental data, the quark interactions at high energies remain effectively short-range ones. The following chapters deal with quark-quark high energy interactions.

References

1. J.J.J. Kokkedee, The Quark Model, W.A. Benjamin, New York-Amsterdam, 1969.

2. R. Feynman, Photon-Hadron Interactions, W.A. Benjamin, New York-Amsterdam, 1972.

3. F.E. Close, An Introduction to Quarks and Protons, Academic Press, London-New York-San Francisco, 1979.

4. P. O'Donnel, Rev. Mod. Phys. 53, 673 (1981).

5. S. Glashow, R. Jackiw and S.S. Shei, Phys. Rev. 187, 1917 (1969); S. Weinberg, Phys. Rev. D11, 3583 (1975). E. Witten, Nucl. Phys. B156, 269 (1979); G. Veneziano, Nucl. Phys. B159, 213 (1979).

6. M.A. Shifman, A.I. Vainstein and V.I. Zakharov, Nucl. Phys., B147, 385, 448, 519 (1979).

7. Particle Data Group, Rev. Mod. Phys. 56, No. 2, Part 2 (1984).

8. R.R. Horgan and R.H. Dalitz, Nucl. Phys. B66, 135 (1973); R.R. Horgan, Nucl. Phys. B71, 514 (1974).

9. P.J. Litchfield, R.I. Cashmore, and A.J. Hey, in Proc. Top. Conf. on Baryon Resonances, Oxford, 1976, p. 477.

10. A.J. Hey and C. Kelly, Phys. Rep., 96C, 71 (1983).

11. G. Zweig, CERN preprint 8419/TH 412 (1964).

12. V.V. Anisovich, A.A. Anselm, Ya. I. Azimov, G.S. Danilov, and I.T. Dyatlov, Phys. Lett. 16, 194 (1965); W.E. Thirring, Phys. Lett. 16, 335 (1965); S.D. Solovjev, Phys. Lett. 16, 345 (1965); C. Becchi and G. Morpurgo, Phys. Rev. 140, 687 (1965).

13. J.P. Mariner et al. Preprint FERMILAB-Conf-82/85-EXP (1982).

14. L.G. Pondrom et al. Paper presented to the V Int. Symp. on High Energy Spin Physics, Brookhaven (1982).

15. J. Gasser and H. Leutwyler, Phys. Rep. 87C, 77 (1982).

16. I.G. Aznauryan and N. Ter-Isaakyan, Yad. Fiz. 31, 1680 (1980); Sov. J. Nucl. Phys. 31, 871 (1980).

17. Ya. B. Zeldovich and A.D. Sakharov, Yad. Fiz. 4, 395 (1966); Sov. J. Nucl. Phys. 4, 283 (1967).

18. A. de Rujula, H. Georgi, and S.L. Glashow, Phys. Rev. D12, 147 (1975).

19. S.L. Glashow, "Particle Physics Far from High Energy Frontier", Harvard Preprint, HUPT-80/A089 (1980).

20. E.V. Shuryak, Nucl. Phys. B203, 93, 116, 140 (1982).

21. N. Isgur and G. Karl, Phys. Lett. B72, 109 (1977); Phys. Rev., D18, 4187 (1979); D19, 2653 (1979); D23, 155 (1981).

22. V.B. Berestetski, V.B. Lifshitz and L. Pitaevski, Relativistic Quantum Theory, Part I, p. 280, Pergamon Press (1971).

23. H.J. Lipkin, Physics Reports 8C, 173 (1973); J.L. Rosner, Physics Reports 11C, 189 (1974).

24. V.V. Anisovich, S.M. Gerasyuta and I.V. Keltuyala, Yad. Fiz. 38, 200 (1983); Sov. J. Nucl. Phys. 38, 117 (1983); On the Stability of the Bootstrap Procedure in the Quasinuclear Quark Model, Preprint LNPI-979 (1984).

25. V.V. Anisovich and S.M. Gerasyuta, "Bootstrap Procedure in the Quasinuclear Three Flavour Quark Model", Preprint LNPI-992 (1984).

26. S. Olubo, Phys. Lett. 5, 165 (1963).

27. J. Iizuka, Prog. Theor. Phys. Suppl. 32-38, 21 (1966).

28. R. Hwa, Oregon preprint, OITS-194 (1982).

29. K. Bardakci and M.B. Halpern, Phys. Rev. D6, 696 (1972).

30. B. de Witt, Nucl. Phys. B51, 237 (1973).

31. J.C. Pati and A. Salam, Phys. Rev. D8, 1240 (1973); D10, 275 (1974).

32. R.N. Mohapatra, J.C. Pati and A. Salam, Phys. Rev. D13, 1733 (1976).

33. E. Ma, Phys. Rev. D17, 623 (1978).

Chapter 6

BINARY PROCESSES IN THE QUARK MODEL

6.1 Confinement and Impulse Approximation

In the previous chapter it was demonstrated that in the framework of the quasinuclear quark model one can obtain a reasonable description of the static hadron features. The starting point has to be a plausible assumption for the character of quark interactions at low energies (e.g. an interaction via massive gluon exchange). The simple example which was considered in Sec. 5.4 shows, that there is hope to succeed in constructing an universal quark-quark interaction of the same type. However, one has to be careful not to ignore the problem of quark confinement. Indeed, the quark model treats quarks as objects with definite masses and magnetic moments. In the model one operates with quark diagrams like diagrams of real particles; dealing with hadron collision processes the notion of quark-quark cross section is introduced. All this may give the impression that the fact that quarks cannot exist as free particles outside the region of hadron interactions is not taken into account.

The potential models gave practically no information about the effect of the confinement on the structure of the quark-quark interactions. In these models it was sufficient to introduce a retaining potential *increasing unlimitedly with the increase of the* distance between the quarks. Inelastic hadron collision processes are outside the sphere of applicability of the potential models; when these processes are considered, it seems to be more probable, that quarks are confined not because of the existence of some infinite potential barrier, but mainly due to the fact that "released" quarks with a large energy of the relative motion emit new quarks and together with them turn into hadrons.

The hypothesis of quasinuclear hadron structure assumes dressed quarks which are weakly bound, almost free objects. At the same time, we know that because of the confinement of quarks these particles cannot exist as really free particles. If so, we have to find arguments supporting the physical meaning of e.g. the quark wave functions which we use frequently.

Let us return once more to the expression (3.34) for the form factor of a composite system in nonrelativistic approximation and rewrite it in the form (two-particle system with $N = 2$ is considered):

$$F(-q^2) = \frac{1}{2}\int \frac{d^4k}{i(2\pi)^4} G(k_1^2, k^2)D(k_1^2) \cdot \text{disc}\, D(k^2)G(k_1'^2, k^2)D(k_1'^2) \cdot f(k_1^2, k_1'^2, q^2) \tag{6.1}$$

where $D(k_i^2)$ are the propagators of the constituents, $f(k_1^2, k_1'^2, q^2)$ is the interaction vertex of the constituent and the external field, G is the transition vertex of the constituents into the composite system. For a usual particle which exists on the mass shell the propagator $D(k^2)$ has a pole, and its discontinuity turns into a δ-function $\delta(k^2-2mE)$ cancelling one integration in (6.1). Consequently the expression for the form factor can be written in terms of the wave functions of constituents

$$F(-q^2) = \frac{1}{2}\int \frac{d^3k}{(2\pi)^3} \psi((k_1 - k)^2)\psi^*((k_1' - k)^2) f(k_1^2, k_1'^2, q^2) , \tag{6.2}$$

where

$$\psi((k_1 - k)^2) = G(k_1^2, m^2)D(k_1^2) \cdot \frac{1}{2\sqrt{m}} \tag{6.3}$$

we call the wave function. If the vertex $f(k_1^2, k_1'^2, q^2)$ depends only weakly on k_1^2, $k_1'^2$ in the $k_1^2 \sim m^2$, $k_1'^2 \sim m^2$ region, one can ignore the fact of going off the mass shell, and obtain an expression for $F(q^2)$ in terms of constituent wave functions and form factors $f(m^2, m^2, q^2)$ on the mass shell.

In the non-relativistic quark model it is supposed from the very beginning, that the meson form factor can be described by the quark diagram of Fig. 3.4b. Since the quarks do not exist as free particles on the mass shell, the propagators $D(k_i^2)$ do not contain poles on the real axis at $k_i^2 = m_q^2$. There may be other singularities (in the complex k_i^2 plane, or singularities of non-pole type and so on; we shall discuss this later once more). The expression (6.1) for the form factor is obtained by closing the integration contour at singularities of $D(k^2)$. The discontinuity disc $D(k^2)$ is here not a δ-function any more. Still, if one assumes, that disc $D(k^2)$ is appreciably different from zero only in a relatively narrow strip

$$m_q - \delta^2 < k^2 < m_q^2 + \delta^2 \quad , \tag{6.4}$$

where $\delta << m_q$, then the value m_q can be considered as the quark mass. Supposing again, that all the vertex functions in (6.1) depend on their arguments only weakly in the region (6.4), we can conclude, that the meson form factor is determined by the quark wave functions (6.3) and the quark interaction vertex $f(m_q^2, m_q^2, q^2)$ on the "mass shell" $k_1^2 \approx k_1'^2 \approx m_q^2$.

Similarly considerations are possible for more complicated quark diagrams, corresponding to hadron interaction processes, too. The cuts of such diagrams must not lead to real processes with the formation of free quarks. This means, that the amplitudes which we express in the form of quark diagrams should not contain quark singularities, but singularities corresponding to real intermediate hadron states due to the general principles of unitarity and causality. That one is able to construct such hadron amplitudes in terms of quark diagrams is far from being obvious.

In the following we give an example which demonstrates the possibility to reconcile the model of quarks with their confinement. There is no contradiction, if the quark diagrams are considered as spectral integrals over quark masses. These spectral integrals can cancel those singularities from the quark diagrams which correspond to free quarks, and introduce singularities corresponding to intermediate states of real hadrons. For the description of analytic properties of hadron amplitudes spectral integrals were introduced relatively long ago[1-4]; in the framework of the quark model these integrals were applied to soft processes in Ref. 5.

Let us consider a simple process: the transition of a meson A into a meson B under the influence of some perturbation V which acts on one of the quarks (Fig. 6.1a). If the quarks were freely existing particles, the amplitude of this process could be written in the form

$$A(q^2, M_A^2, M_B^2) = \int \frac{d^4k}{i(2\pi)^4} \mathrm{Tr}\left\{ G_A(k_1^2, k_2^2, M_A^2) \frac{1}{-\hat{k}_1+m-i0} \cdot V(k_1^2, k_3^2, q^2) \frac{1}{-\hat{k}_3+m-i0} G_B(k_3^2, k_2^2, M_B^2) \frac{1}{\hat{k}_2+m-i0} \right\} . \tag{6.5}$$

This amplitude has, naturally, singularities corresponding to free quarks. In order to avoid them, one has to introduce an integration in (6.5) over the quark masses:

$$A(q_1^2, M_A^2, M_B^2) = \int \prod_{\ell=1}^{3} dm_\ell^2 \int \frac{d^4k}{i(2\pi)^4} \mathrm{Tr}\left\{ G_A(k_1^2, k_2^2, M_A^2; m_1^2, m_2^2) \cdot \frac{1}{-\hat{k}_1+m_1-i0} V(k_1^2, k_3^2, q^2; m_1^2, m_3^2) \frac{1}{-\hat{k}_3+m_3-i0} G_B(k_3^2, k_2^2, M_B^2; m_3^2, m_2^2) \cdot \frac{1}{\hat{k}_2+m_2-i0} \right\} . \tag{6.6}$$

The vertex functions V, G_A and G_B depend on both the quark momenta squared and their squared masses. According to this, the spectral integral (6.6) has to be interpreted as a sum of graphs with various quarks of different masses.

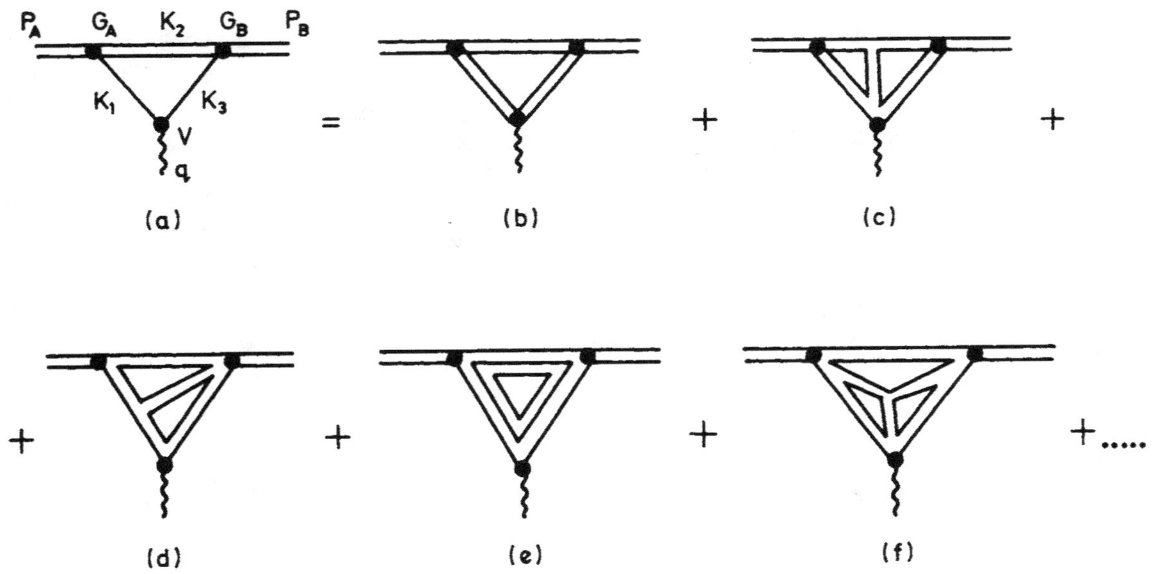

Fig. 6.1 Quark diagram (a) and its representation in the form of superposition of hadronic processes (b-f etc.): double lines stand for mesons, triple for baryons.

The integration over the quark masses cancels the singularities which are due to the poles of the quark propagators. In the following we show that the same integration may introduce in (6.6) singularities corresponding to real hadron states; there will be no contradiction with the quark confinement, if the spectral integrals will contain only these latter singularities.

In order to find out which real hadrons saturate the quark loop, Fig. 6.1a, one has to decompose the quark graph over real hadron processes. This can be done by introducing additional quark loops; in terms of the quark diagrams the condition of confinement means, that the sum of graphs in Fig. 6.1b-f, etc. has to be equal to the graph in Fig. 6.1a. Hence, the singularities of the quark amplitude in q^2 and meson masses squared M_A^2 and M_B^2 are to occur on the thresholds of real particle production: meson pair, three mesons, baryon-antibaryon pairs and so on, i.e. at the values

$$M_A^2,\ M_B^2,\ q^2 = (\mu_i + \mu_j)^2,\ (\mu_i + \mu_j + \mu_\ell)^2,\ \ldots \tag{6.7}$$

where $\mu_i, \mu_j, \ldots$ are the masses of the particles, which saturate the quark loop. In order to get a proper analytic behaviour for the amplitude (6.6), which corresponds to the diagrams Figs. 6.1 b,c,..., the integration over quark masses must start from the lowest threshold:

$$(m_1 + m_2)^2,\ (m_1 + m_3)^2,\ (m_2 + m_3)^2 \geq \min(\mu_i + \mu_j + \ldots)^2 \tag{6.8}$$

If mesons A and B are built up of nonstrange quarks, $\min(\mu_i+\mu_j+\ldots)^2 = 4\mu_\pi^2$ (μ_π is the pion mass) for G-even states and $9\mu_\pi^2$ for G-odd states. The conditions (6.8) are very "soft", for they cover even the region of negative masses. The region of admitted m_i values may be restricted, taking

$$m_i \geq \mu_\pi \ . \tag{6.9}$$

Vertex functions in (6.6) will have in this case the singularities in quark masses of the δ-function type and its derivatives. Let us consider in detail the way in which the spectral integral (6.6) reconstructs the singularities of the amplitude. For example, let us look for the discontinuity of the amplitude across the cut in the q^2 plane, going from some point $q^2 = q_s^2$:

$$\mathrm{disc}\, A(q^2, M_A^2, M_B^2) = \int_{\Omega(m_1^2, m_2^2, m_3^2)} dm_1^2 dm_3^2 \frac{1}{16\pi} \left[\frac{(q^2-(m_1+m_3)^2)(q^2-(m_1-m_3)^2}{4q^2} \right]^{1/2}$$

$$\times V(k_1^2, k_3^2, q^2;\ m_1^2, m_3^2) \int dm_2^2 A^*_{q\bar{q}\to AB} \quad . \tag{6.10}$$

Here $A^*_{q\bar{q}\to AB}$ is the transition amplitude for $q\bar{q}\to AB$ (a part of the diagram in Fig. 6.2 over the cut line). The integration domain $\Omega(m_1^2,\ m_2^2,\ m_3^2)$ is determined by the conditions:

$$m_1 \geqslant \mu_\pi, \quad m_2 \geqslant \mu_\pi, \quad (m_1 + m_3)^2 \leqslant q^2 \ . \tag{6.11}$$

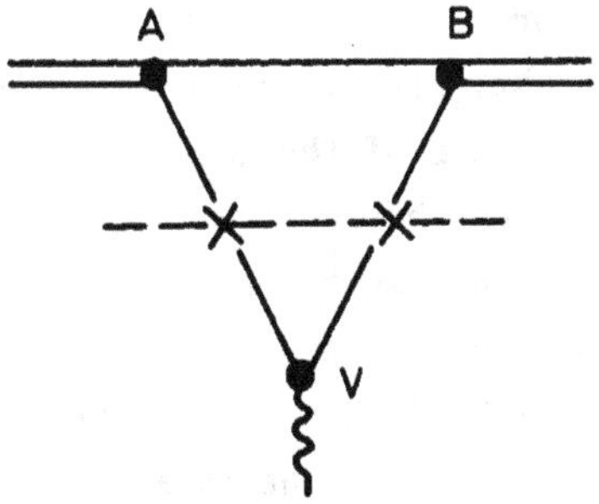

Fig. 6.2 Cut diagram which determines the discontinuity in Eq. (6.10). The upper block is the amplitude of the transition $q\bar{q}\to AB$.

It is clear from Eqs. (6.10) and (6.11) that the dependence of the vertex function V on m_1 and m_2 defines unambiguously the analytic properties of disc A in the q^2 plane. If the vertex function V contains δ-like singularities $\delta(q^2-(m_1+m_3)^2)$ or $\delta(q^2-(\mu_\pi+m_i)^2)$, $i = 1,3$ or similar θ-like singularities, then the amplitude has branching points $(q^2-q_s^2)^{N+½}$ where N is a non-negative integer value. The sign of the disc A, continued on the lower lip of the cut is determined in this case by the change of the sign of the factor $\sqrt{q_s^2-(m_1+m_3)^2}$

in (6.10). If amplitude is assumed to have a logarithmic singularity at $q^2 = q_s^2$ like $(q^2 - q_s^2)^N \ell n(q^2 - q_s^2)$, the discontinuity (6.10) has the same sign on the both lips of the cut. Then the vertex function V must contain a square root singularity factor $(q_s^2 - (m_1 + m_3)^2)^{\tilde{N}+\frac{1}{2}}$ which compensates for the change of the sign due to the factor $\sqrt{q^2 - (m_1 + m_3)^2}$ in the integrand.

Thus we see that spectral integrals over two-quark intermediate states may simulate threshold singularities of the amplitude. This integral may probably reconstruct even more complicated singularities. However, there is no reason to demand that the diagram of Fig. 6.1a should alone be responsible for all the singularities of the considered amplitude. As a matter of fact the two-particle composite system $q\bar{q}$ gives only some approximate description of the lower meson states. Obviously, there are some transitions from few-quark states to multi-quark ones like $q\bar{q} \to qq\bar{q}\bar{q}$ or $qqq \to qqqq\bar{q}$. Examples for such transitions are the decays $\rho \to \pi\pi$ or $\Delta \to N\pi$. The large values of widths of these decays may indicate the existence of the non-negligible admixture of the multiquark states into the lower hadrons (of the order of 10%). Thus the real hadrons, including the lowest ones, are some superpositions of states with different numbers of the dressed quarks. Multiquark states also correspond to some contributions of multihadron intermediate states. This means that some singularities of the total amplitude are defined not by the integral (6.6), but by similar spectral integrals for multi-quark states.

It is an open question whether the vertex function V in (6.6) (or more complicated functions, which describe the interactions between the dressed quarks) indeed has the desired analytical properties. The additive quark model cannot solve this question — in this model we simply assume that V has a proper set of singularities. Such an assumption prompts a possible way to discover the connection between QCD and the additive quark model.

Let us now consider the quark model diagrams as spectral integrals of the considered type. They can realize a quasinuclear quark model, if the main contribution to the integral (6.6) and other similar integrals comes from the mass region close to the value m_q of the dressed quark mass. Another necessary condition is that the vertex functions (V in Fig. 6.1a) and transition functions "hadron→quarks" (G_A and G_B in Fig. 6.1a) must be universal functions. That is, in the quark model the amplitude of the process in Fig. 6.3a may be obtained from the Eq. (6.6) by the replacement $V \to W$, that in Fig. 6.3b by the replacement $G_B \to G_C$. More complicated processes, like that in Fig. 6.3c are also expressed in terms of the same vertex functions (G_A, G_B, V, W).

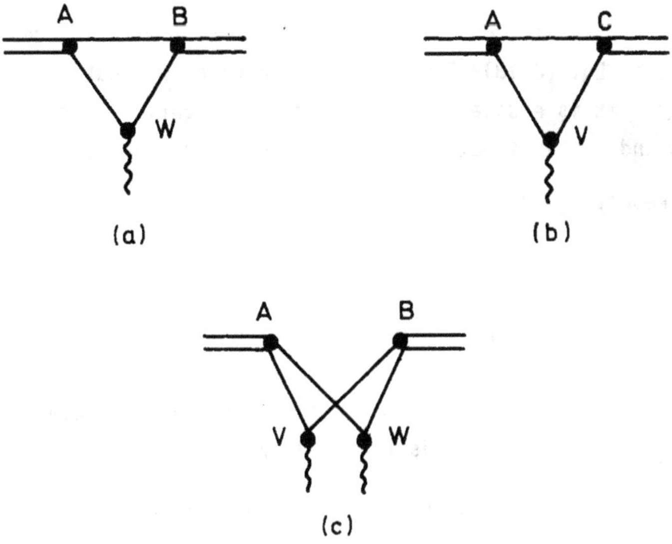

Fig. 6.3 Quark diagrams with universal vertex functions, V, W, G_A, G_B, G_C.

In view of such universality quark spectral integrals must be similar to the diagrams of the usual Lagrangian field theory. Practically the only difference is the integration over quark masses in the intermediate states, introduced in the spectral representation of the quark model.

Let us illustrate this idea with an example of the model Lagrangian, considered in the Sec. 5.4. Denoting this Lagrangian (Eq. (5.35)) by $L(m)$, we introduce then a sum of such Lagrangians with different m values:

$$L(m) \to \sum_m L(m) \quad . \tag{6.12}$$

In order to obtain the diagrams with spectral integrals, it is necessary to replace the sum in (6.12) by an integral:

$$\sum_m \to \int dm\, Z(m) \quad . \tag{6.13}$$

Since we are going to consider only the processes with sources of quarks (hadrons), in Eq. (6.13) $Z(m) = 1$ may be taken without any loss of generality: it is equivalent to some renormalization of the coupling constants and vertex functions of the transitions "hadron→quarks".

The resulting Lagrangian

$$L = \int dm\, (L_{QCD}(m) + L_H(m)) \tag{6.14}$$

does not include interactions of quarks with different masses with each other. It is easy to introduce such interactions in the Higgs part: the properties of the Lagrangian considered in the Sec. 5.4 will not be changed if $V(m)$ in (5.33) is replaced by

$$C(m)(\mathrm{Tr}\{\phi^+(m)\phi(m)\})^2 \to \int C(m, m')\mathrm{Tr}\{\phi^+(m)\phi(m)\}\mathrm{Tr}\{\phi^+(m')\phi(m')\}dm' \quad . \tag{6.15}$$

Hadrons are to be introduced here as sources of quarks, i.e. it is necessary to define the bare vertices "hadron→quarks", which are functions of quark and antiquark masses $G(m_1, m_2)$ for mesons. The total vertices $G(k_1^2, k_2^2; m_1, m_2)$ are then determined by a chain of quark-antiquark rescatterings.

Thus it is possible to construct spectral integrals over quark masses. Quark interactions are determined by a renormalizable Lagrangian which is essentially motivated by QCD and at the same time satisfies the requirements of the quark model. Some properties of this

Lagrangian may be varied in such a manner, that the resulting quark model is rather flexible in its details.

The only difference of such a quark model from a standard one is that the quark masses are considered as variables. It does not prevent one from applying the non-relativistic approximation. Indeed in this approximation spectral integrals for all processes may be written in terms of quark wave functions $\psi_{m_1 m_2}$ with different quark masses. For example, the meson form factor should be written as

$$F(\vec{q}^{\,2}) = \int dm_1 dm_2 \rho(m_1, m_2) \int \frac{d^3k}{(2\pi)^3} \psi_{m_1 m_2}(2\vec{k}) \psi^{*}_{m_1 m_2}(2\vec{k} + \vec{q}) \quad , \qquad (6.16)$$

where $\rho(m_1, m_2)$ is normalized quark density $\int dm_1 dm_2 \rho(m_1 m_2) = 1$, which is determined by the vertex functions $G(m_1, m_2)$. It is possible to estimate the width of the quark mass distribution $\rho(m_1, m_2)$, since in different processes the average quark mass values are different. For example, assuming $\rho(m_1, m_2) \sim \delta(m_1 - m_2)\exp\left[- \frac{(m_1 + m_2 - 2m_0)^2}{4\sigma^2} \right]$, we get

$$<m^N> \simeq m_0 + N \frac{\sigma^2}{2m_0} \quad . \qquad (6.17)$$

We may estimate m_0 and σ from the value of the quark magnetic momentum (it is proportional to m_q^{-1}) and from the mass formula of Glashow (Eq. (5.25)), in which quark mass enters linearly (the splitting item $\vec{\sigma}_1\vec{\sigma}_2/m_1 m_2$ is rather small and mass difference in it may be neglected). The non-strange quark mass determined from the nucleon magnetic momentum is about 335 MeV, while that from the mass formula (5.25) is 360 MeV. Thus, $m_0 \simeq 350$ MeV, $\sigma \simeq 90$ MeV. The value of σ seems to be quite reasonable for the strong interacting object and is not so large that it prevents us from introducing the notion of the "mass of the dressed quark". In the limit $\sigma \to 0$

$$\delta(m_1, m_2) = \delta(m_1 - m_0)\delta(m_2 - m_0) \qquad (6.18)$$

and spectral integrals turn into Feynman diagrams. A broad class of processes may be reasonably handled in the approximation (6.18).

We have considered the quark spectral integrals in order to clarify our position with respect to the quark model. This model is a semi-phenomenological approach. Starting from the main ideas of QCD as given a priori, one may carry out a self-consistent consideration of soft processes in the framework of it. The use of the approximations (e.g. (6.18)) which, if understood literally, ignore quark and gluon confinement, may be appropriate for some definite problems. In particular, such a problem is the description of high energy hadron collisions. These processes are the subject of the remaining part of the book.

The arguments considered allow one to assert that it is quite legal to use the well-known results of the theory of weakly bound system interactions in the framework of the quasinuclear quark model even when quark confinement is taken into account. Most of the quark model predictions for high energy hadron collisions are based on the impulse approximation. Hadron amplitudes in this case are proportional to the sums of the corresponding quark amplitudes (quark additivity); all non-additive contributions, which may come from the processes of multiple collisions of the constituents, are assumed to be parametrically small. When these contributions are estimated with the help of the technique presented in Chap. 4, they usually appear to reach 10-20% of the main contribution of the impulse approximation. However, in some special cases these corrections may dominate (see Sec. 5 of the present chapter).

6.2 Elastic Scattering of Quarks

Let us consider high energy hadron collisions, caused by the simplest underlying subprocess — elastic quark-quark scattering. In the case when polarization properties of the colliding particles are not measured, there are only 8 independent quark-quark elastic amplitudes for the light u, d, s quarks (uu, ud, us, $u\bar{u}$, $u\bar{d}$, $u\bar{s}$, ss, $s\bar{s}$; isotopic and charge invariance of the strong interaction assumed). The latter two of them do not enter any interaction process with a real (nucleon or antinucleon) target. The number of experimentally measurable hadron amplitudes is significantly larger, thus the quark model gives a set of relations for the amplitudes and cross sections of different processes.

6.2.1 *Total cross sections*

The elastic meson-nucleon and nucleon-nucleon amplitudes in the impulse approximation are determined by the diagrams of Fig. 6.4. This means that the amplitude of the process $12 \to 12$ may be written as

$$A_{12\to 12}(s, t) = \sum_{q_1, q_2} a_{q_1 q_2}(s_{qq}, t) F_1(-t) F_2(-t) \quad . \tag{6.19}$$

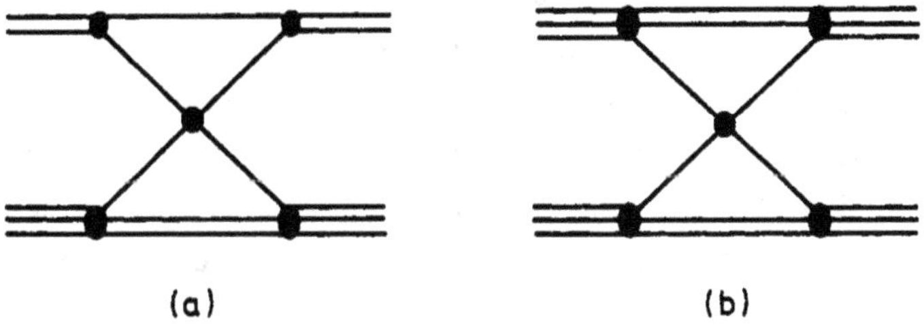

Fig. 6.4 Meson-nucleon (a) and nucleon-nucleon (b) elastic scattering in the impulse approximation of the quasinuclear quark model.

The sum in the r.h.s. of Eq. (6.19) runs over all the quark states q_1, q_2 involved in the particles 1 and 2; s_{qq} is the total energy squared of the colliding quarks in their c.m. frame. For nucleon-nucleon collisions $s_{qq} \simeq \frac{1}{9}(p_1 + p_2)^2$, for pion-nucleon ones $s_{qq} \simeq \frac{1}{6}(p_1 + p_2)^2$, t is the square of the momentum transferred, $t = (p_1 - p_1')^2$ (the required technique for high energy collisions of the composite systems is considered in Chap. 4). For the forward scattering $F_1(0) = F_2(0) = 1$, thus it is possible to write for the elastic pp forward amplitude the expression

$$A_{pp}(s, 0) = 4a_{uu}(\frac{s}{9}, 0) + 4a_{ud}(\frac{s}{9}, 0) + a_{dd}(\frac{s}{9}, 0) \quad . \tag{6.20}$$

The imaginary part of the elastic scattering defines the total cross section (see Appendix B). *Assuming that the forward amplitudes of the quark-quark scattering with definite flavours (e.g. ud) are the same for all hadron-hadron scattering processes (e.g.* pp *and* πp*), we obtain* the relations for the total cross sections of nonstrange hadrons[6,7]:

$$\sigma_{tot}(pp) + \sigma_{tot}(pn) + \sigma_{tot}(\bar{p}p) + \sigma_{tot}(\bar{p}n) = 3[\sigma_{tot}(\pi^+p) + \sigma_{tot}(\pi^-p)] \tag{6.21}$$

$$2[\sigma_{tot}(pp) + \sigma_{tot}(\bar{p}n)] - [\sigma_{tot}(pn) + \sigma_{tot}(\bar{p}p)] = 3\sigma_{tot}(\pi^+p) \quad . \tag{6.22}$$

The cross sections in both parts of Eqs. (6.21), (6.22) depend on the total energy of the colliding particles. Though at high energies this dependence is rather weak, it is preferable to compare (6.21), (6.22) with the experimental data at the same energies of quark-quark interactions, i.e. the nucleon-nucleon cross sections are to be taken at a laboratory momentum approximately 3/2 of that for the meson-nucleon cross section. The experimental ratios of the l.h.s. to r.h.s. of the Eqs. (6.21), (6.22) are shown in Fig. 6.5; there is agreement within the accuracy of 10-15% and it probably becomes better with energy increase. In the region $p_{lab} \sim 10^3$ GeV/c the direct data for πp collisions are now unavailable; these cross sections may be estimated with the help of dispersion relation analysis provided that the real part of the elastic amplitude is known. Looking at Fig. 6.5a, one can come to the conclusion that at $p_{lab} \sim 10^3$ GeV/c $\sigma_{tot}(\pi p)$ grows with the energy increase faster than $\sigma_{tot}(pp)$ thus making agreement with the quark model predictions better.

When the strange hadron collisions are considered, other relations of the type (6.21), (6.22) may be written. If the initial energies are not very high, the energy dependence of the total cross sections is not negligible. In this case it is necessary to take into account that the strange quark carries away a larger momentum fraction of the strange hadron, than non-strange quark does. For example, in the K^+ meson the $\bar{s}$-quark carries about 3/5 of the total momentum, while the u-quark about 2/5 of that. Thus, similarly to the Eq. (6.20), one may write:

$$A_{K^+p}(s,0) = 2a_{u\bar{s}}(\tfrac{s}{5}, 0) + a_{d\bar{s}}(\tfrac{s}{5}, 0) + 2a_{uu}(\tfrac{2s}{15}, 0) + a_{ud}(\tfrac{2s}{15}, 0) \quad . \tag{6.23}$$

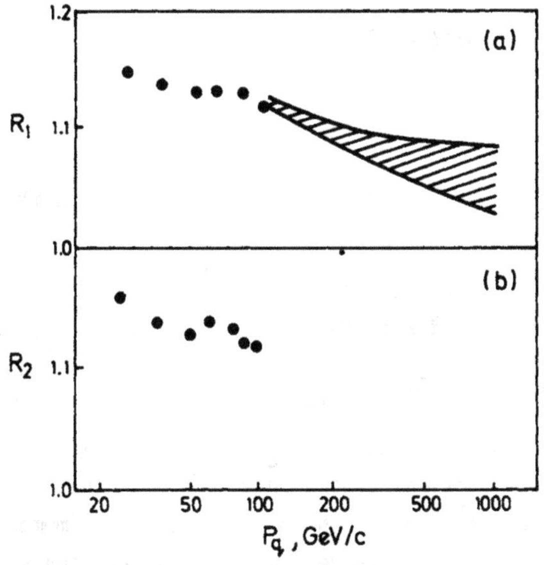

Fig. 6.5 The experimental states of the relations (6.21), (6.22):

$$R_1 = \frac{\sigma_{tot}(pp)+\sigma_{tot}(pn)+\sigma_{tot}(\bar{p}p)+\sigma_{tot}(\bar{p}n)}{3[\sigma_{tot}(\pi^+p)+\sigma_{tot}(\pi^-p)]} \qquad (a)$$

$$R_2 = \frac{2[\sigma_{tot}(pp)+\sigma_{tot}(\bar{p}n)] - [\sigma_{tot}(pn+\sigma_{tot}(\bar{p}p)]}{3\sigma_{tot}(\pi^+p)} \quad . \qquad (b)$$

The points are the data of the direct measurements[8), the shadowed region is the prediction of Refs. 9, 10, based on the dispersion relation analysis. $p_q \simeq p_\pi/2 \simeq p_N/3$ is the quark laboratory momentum.

In Λ- or Σ-hyperon the s-quark carries about 3/7 of the total momentum, while each of the non-strange quarks about 2/7 of that. As a result

$$A_{\Sigma^- p}(s,0) = 2a_{us}(\tfrac{s}{7}, 0) + a_{ds}(\tfrac{s}{7}, 0) + 4a_{ud}(\tfrac{2s}{21}, 0) + 2a_{dd}(\tfrac{2s}{21}, 0) \quad . \tag{6.24}$$

Combining the cross sections in such a manner that non-strange contributions cancel, we may obtain e.g. the relation:

$$\sigma_{tot}(pp;\ s) - \sigma_{tot}(\Sigma^- p;\ \tfrac{7}{6}s) = \sigma_{tot}(\pi^- p;\ \tfrac{2}{3}s) - \sigma_{tot}(K^- p;\ \tfrac{5}{6}s) + 2[\sigma_{tot}(K^+ p;\ \tfrac{5}{6}s) - \sigma_{tot}(K^+ n;\ \tfrac{5}{6}s)] \ . \tag{6.25}$$

This and some other similar relations are compared with the experimental data in Table 6.1.

At high initial energies $s \gtrsim 10^2\ GeV^2$ there are two simplifying factors. The cross section differences $\sigma_{tot}(pp) - \sigma_{tot}(pn)$ and $\sigma_{tot}(\pi^+ p) - \sigma_{tot}(\pi^- p)$, $\sigma_{tot}(K^+ p) - \sigma_{tot}(K^- p)$ decrease rapidly with the energy growth. Thus the relations (6.21), (6.22) may be rewritten as the Levin-Frankfurt[6)] relation $\sigma_{tot}(pp;s) = \frac{3}{2}\,\sigma_{tot}(\pi p;\ \frac{2}{3}s)$. Besides, all the total cross sections depend only weakly on the initial energy in this region, and we may neglect the difference in the momentum fractions carried by strange and non-strange quarks. However, the strange quark interaction amplitudes at these energies differ from those of the non-strange quarks quite noticeably. In the approximation when all the non-strange quark-quark amplitudes are taken to be equal to each other $a_{uu} = a_{ud} = \ldots = a_{\bar{u}u} \equiv a_{qq} \neq a_{sq}$, we obtain

$$\sigma_{tot}(KN) - \sigma_{tot}(\pi N) = \sigma_{tot}(\Lambda N) - \sigma_{tot}(pN) = \sigma_{tot}(\Sigma N) - \sigma_{tot}(pN) = \frac{1}{2}\,[\sigma_{tot}(\Xi N) - \sigma_{tot}(pN)] \ . \tag{6.26}$$

The hyperon cross sections were measured at $p_{lab} = 135$ GeV/c[11)]; for the proton target

$$\sigma_{tot}(pp) - \sigma_{tot}(\Sigma^- p) = 4.3 \pm 0.3 \text{ mb};$$

$$\frac{1}{2}\,[\sigma_{tot}(pp) - \sigma_{tot}(\Xi^- p)] = 4.55 \pm 0.15 \text{ mb} \tag{6.27}$$

while for the neutron target:

$$\sigma_{tot}(pn) - \sigma_{tot}(\Sigma^- n) = 5.6 \pm 0.4 \text{ mb};$$

$$\frac{1}{2}\,[\sigma_{tot}(pn) - \sigma_{tot}(\Xi^- n)] = 4.6 \pm 0.25 \text{ mb} \ . \tag{6.28}$$

The data for the meson beams at $p_{lab} \simeq 90$ GeV/c give:

Table 6.1 Experimental status of the quark model relations for the total cross sections of strange hadron collisions. The data are taken from Refs. 8, 11.

Relation	s, GeV^2	Left-hand side (mb)	Right-hand side (mb)
$\sigma_{tot}(pp;\ s) - \sigma_{tot}(\Sigma^- p;\ \frac{7}{6}s) = \sigma_{tot}(\pi^- p;\ \frac{2}{3}s) - \sigma_{tot}(K^- p;\ \frac{5}{6}s)$			
$+ 2[\sigma_{tot}(K^+ p;\ \frac{5}{6}s) - \sigma_{tot}(K^+ n;\ \frac{5}{6}s)]$	130	5.2 ± 0.3	3.0 ± 0.4
	240	4.3 ± 0.3	3.2 ± 0.2
$\sigma_{tot}(pp;\ s) - \sigma_{tot}(\Sigma^- n;\ \frac{7}{6}s) = \sigma_{tot}(\pi^- p;\ \frac{2}{3}s) - \sigma_{tot}(K^- p;\ \frac{5}{6}s)$	130	5.9 ± 0.4	3.9 ± 0.1
	240	5.5 ± 0.4	3.6 ± 0.1
$\sigma_{tot}(pp;\ s) - \sigma_{tot}(\Xi^- p;\ \frac{4}{3}s) = 2[\sigma_{tot}(\pi^- p;\ \frac{2}{3}s)$			
$- \sigma_{tot}(K^- p;\ \frac{5}{6}s) + \sigma_{tot}(K^+ p;\ \frac{5}{6}s) - \sigma_{tot}(K^+ n;\ \frac{5}{6}s)]$	200	9.1 ± 0.5	6.7 ± 0.4
$\sigma_{tot}(pp;\ s) - \sigma_{tot}(\Xi^- n;\ \frac{4}{3}s) = 2[\sigma_{tot}(\pi^- p;\ \frac{2}{3}s)$			
$- \sigma_{tot}(K^- p;\ \frac{5}{6}s)] + \sigma_{tot}(K^+ n;\ \frac{5}{6}s) - \sigma_{tot}(K^+ p;\ \frac{5}{6}s)$	200	9.1 ± 0.5	7.1 ± 0.3

$$\sigma_{tot}(K^{\pm}p) - \sigma_{tot}(\pi^{\pm}p) = 4.0 \pm 0.2 \text{ mb} \quad . \tag{6.29}$$

We consider such an agreement as quite satisfactory.

6.2.2 *Elastic cross sections at small momentum transferred*

The processes of the elastic lepton-nucleon or lepton-meson collisions are determined in the quark model by the diagrams of Fig. 6.6.

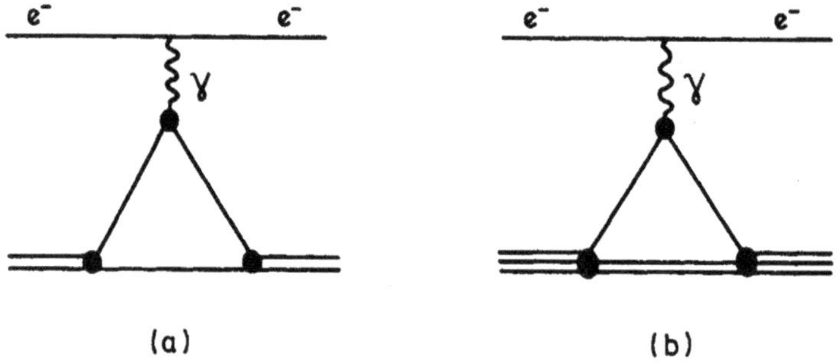

Fig. 6.6 Quark model diagrams for the elastic lepton-meson (a) and lepton-nucleon (b) scatterings.

The corresponding differential cross sections are proportional to hadron form factors squared, e.g.

$$\frac{d\sigma(ep)}{dt} \sim F_p^2(-t) \quad . \tag{6.30}$$

Similarly, the process of hadron-hadron collisions (Fig. 6.4) are determined by the same form factors[12]:

$$\frac{d\sigma_{el}(\pi p)}{dt} = |a_{qq}|^2 \; F_p^2(-t)F_\pi^2(-t) \quad . \tag{6.31}$$

$$\frac{d\sigma_{el}(pp)}{dt} = |a_{qq}|^2 \; F_p^4(-t) \quad . \tag{6.32}$$

Here $F_\pi(-t)$ and $F_p(-t)$ are pion and proton form factors, respectively, while $a_{qq} = a_{q\bar{q}}$ is the high-energy elastic quark-quark scattering amplitude.

In the region of moderately high energies we may neglect the dressed quark sizes as compared with those of hadrons, i.e. neglect the t-dependence of the amplitude $a_{qq}(s,t)$. In this case the behaviour of $d\sigma_{e\ell}/dt$ is determined by the form factors of the colliding hadrons. At $p_{lab} \simeq 10\text{-}20$ GeV/c such a behaviour was observed experimentally.

When p_{lab} increases, t-dependence of $a_{qq}(s,t)$ becomes more significant due to the growth of the dressed quark size. Regge parametrization of $a_{qq}(s,t)$ can be written[13,14] as (see also Sec. 2.3)

$$a_{qq}(s_{qq}, t) = A \exp(\alpha'_{\mathbb{P}} t \cdot \ell n \frac{s_{qq}}{s_0}) \tag{6.33}$$

where A determines the value of the $\sigma_{tot}(qq)$, s_0 is a model parameter, $\alpha'_{\mathbb{P}}$ is the slope of the Pomeron trajectory. It was shown in Refs. 15, 16 that at FNAL-ISR energies the pp elastic cross section is described reasonably by the Eqs. (6.32), (6.33).

The experimental data for the differential cross sections are usually parametrized in the form

$$\frac{d\sigma_{e\ell}(hp)}{dt} = A \exp(-b_{hp}|t|) \quad . \tag{6.34}$$

When the values of $d\sigma_{e\ell}/dt$ are measured with high accuracy, one may see that b_{hp} depends weakly on t[16]: the values of b_{hp} obtained from the data fit at $|t| \sim 0.05\ (\text{GeV/c})^2$ differ from those at $|t| \sim 0.2\ (\text{GeV/c})^2$ by $1\text{-}1.5\ (\text{GeV/c})^{-2}$. Such a behaviour of b_{hp} appears in the quark model. The proton and pion form factors are

$$F_p(-t) = \frac{1}{(1 - \frac{t}{m^2})^2} \qquad m^2 = 0.71\ \text{GeV}^2$$

$$F_\pi(-t) = \frac{1}{1 - \frac{t}{\mu^2}} \qquad \mu^2 = 0.40\text{-}0.55\ \text{GeV}^2 \quad . \tag{6.35}$$

Introducing (6.33), (6.35) and (6.36) into Eqs. (6.31), (6.32) and comparing the resulting expressions with Eq. (6.39), we get:

$$b_{pp} \simeq \frac{8}{m^2}(1 + \frac{t}{2m^2}) + 2\alpha'_{\mathbb{P}} \ln \frac{s_{qq}}{s_0} \quad . \tag{6.36}$$

$$b_{\pi p} \simeq \frac{4}{m^2}(1 + \frac{t}{2m^2}) + \frac{2}{\mu^2}(1 + \frac{t}{2\mu^2}) + 2\alpha'_{\mathbb{P}} \ln \frac{s_{qq}}{s_0} \quad . \tag{6.37}$$

If $b_{pp}(t)$ is normalized to the experimental points at the smallest $|t|$ values via the appropriate choice of s_0, we obtain

$$2\alpha'_{\mathbb{P}} \ln \frac{s_{qq}}{s_0} \simeq 1.2 \ (\text{GeV/c})^{-2} \quad . \tag{6.38}$$

Since the value of s_{qq} in πp collisions is about 3/2 of that in pp collisions, the last term in the Eq. (6.37) exceeds that in the Eq. (6.36) by approximately $2\alpha'_{\mathbb{P}} \ln 3/2 \simeq 0.25 \ (\text{GeV/c})^{-2}$. The calculated curves are shown in Fig. 6.7. They are in agreement with the experimental data[16,20].

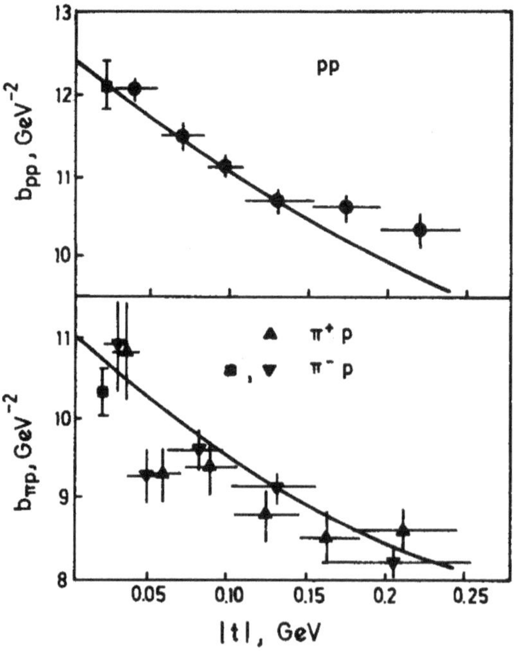

Fig. 6.7 t-dependence of the slope parameter for the elastic pp (a) p (b) scattering at p_{lab} = 200 GeV/c. The curves are calculated with the help of Eqs. (6.36), (6.37).

6.2.3 *The estimates of the dressed quark radius*

Considering hadron-hadron collisions in the impulse approximation of the quasinuclear quark model, it is possible to estimate the size of the dressed quark[21)] (see also Ref. 22). Such estimates were mentioned in Sec. 1.2.

The radius of the constituent quark may be estimated from the value of the total cross section of quark-quark interaction. At moderately high energies $\sigma_{tot}(qq) \simeq \sigma_{tot}(NN)/9 \simeq 4.5$ mb. Assuming that the total quark-quark cross section is determined by the geometrical size of the quark $\sigma_{tot}(qq) = 2\pi(2r_q)^2$ we get $r_q^2 \simeq 0.45\ \mathrm{GeV}^{-2}$.

Another way to estimate r_q^2 is connected with the consideration of the elastic quark-quark scattering amplitude which at high energies has the form of Eq. (6.33). If we assume that $|a_{qq}(t)|^2$ determines the size of the dressed quark in the same way as $F_p^4(-t)$ determines the size of the proton, we obtain the following estimate for r_q^2:

$$r_q^2(s) \simeq 3\alpha'_{\mathbb{P}} \ln \frac{s_{qq}}{s_0} \tag{6.39}$$

The estimated radius of the dressed quark grows with the energy increase; as we know such a growth is observed in high energy collisions (see Sec. 2.1). The slope of the Pomeron trajectory $\alpha'_{\mathbb{P}}$ is, according to different estimates, $\alpha'_{\mathbb{P}} \simeq 0.15\text{-}0.25\ (\mathrm{GeV}/c)^{-2}$. From the Eq. (6.38) we obtain then $s_0 \simeq 0.8\text{-}1\ \mathrm{GeV}^2$. Thus in the energy region $s \simeq 10^2\text{-}10^3\ \mathrm{GeV}^2$ the size of the dressed quark is about $r^2(10^2\ \mathrm{GeV}^2) \simeq 1\text{-}2\ \mathrm{GeV}^{-2}$, $r_q^2(10^3\ \mathrm{GeV}^2) \simeq 2\text{-}3.7\ \mathrm{GeV}^{-2}$.

Since the radius of the proton is $R_p^2 \simeq 17\ \mathrm{GeV}^{-2}$, for the ratio r_q^2/R_p^2 we obtain $1/3.5 \lesssim r_q^2/R_p^2 \lesssim 1/30$. Thus, estimations in the impulse approximation lead to a small value of r_q^2/R_p^2.

6.3 Flavour-Exchange Reactions

6.3.1 *Cross section*

We shall consider here binary reactions $a + b \to c + d$ in which hadron flavours in the final state differ from those in the initial state. A great variety of such reactions may be considered in the framework of the quark model, and a lot of relations may be obtained. However, only a few of these relations are tested experimentally with reasonable accuracy. Moreover, it is necessary to notice from the very beginning that practically all the data, which we consider here, are obtained at rather low initial laboratory momenta of a few GeV/c. The reason is that all flavour exchange cross sections fall down with the energy growth, so the higher the initial energy is, the more difficult it is to obtain the appropriate data. Because of that, all the comparisons of the quark model predictions with the data here should be treated with caution, remembering that the energies used are rather low.

An additional difficulty for inelastic collisions in the quark model is connected with the necessity to take into account the phase volume at low energies. When nonforward scatterings considered, the difference in particle form factors may be important too.

A detailed consideration of flavour exchange reaction in the framework of the quark model may be found in Ref. 23.

However there are relations for the inelastic reactions which do not depend on any properties of quark-quark interactions and are based only on the additivity assumption and on the impulse approximation for the quark model. They follow from the suppression of the t-channel exchanges, which are forbidden in the impulse approximation.

In this approximation flavour-exchange processes may be described by the quark diagrams of Fig. 6.8a (quark rearrangement) and Fig. 6.8b (quark annihilation-creation). Thus the processes with t-channel exchange of quantum numbers, which are impossible in the $q\bar{q}$ system, are forbidden (hereafter we shall denote forbidden reactions by the symbols $\not\twoheadrightarrow$). For instance in the forbidden reaction $\pi^- p \not\twoheadrightarrow \pi^+ \Delta^-$ charge

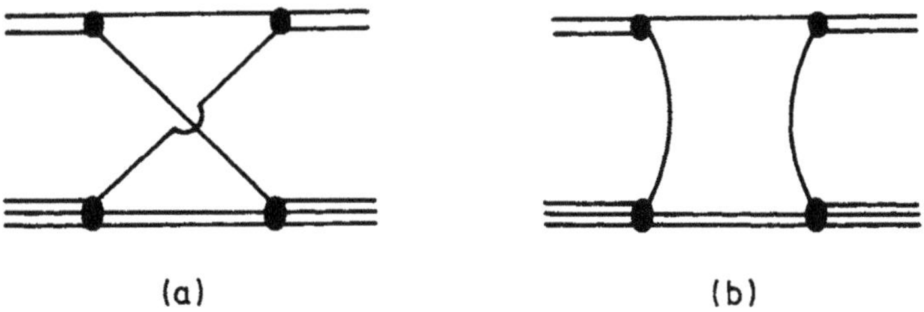

Fig. 6.8 Quark rearrangement (a) and quark annihilation-creation diagrams for flavour exchange processes in the quark model.

transfer is equal to $\Delta Q = 2$. There are two amplitudes in the reaction $\pi N \to \Delta N$, which correspond to the total isospin $I = 1/2$ and $3/2$. So if

$$A(\pi^- p \to \pi^+ \Delta^-) = \sqrt{\frac{2}{15}}\, A_{\frac{3}{2}} - \frac{1}{\sqrt{3}}\, A_{\frac{1}{2}} = 0 \tag{6.40}$$

we immediately get[24,25)]

$$\sigma(\pi^+ p \to \pi^+ \Delta^+) = \frac{2}{3}\,\sigma(\pi^+ p \to \pi^0 \Delta^{++}) = \sigma(\pi^- p \to \pi^- \Delta^+) = 2\sigma(\pi^- p \to \pi^0 \Delta^0) \quad . \tag{6.41}$$

Experimental data[26)] at $p_{lab} = 8$ GeV/c give

$$\sigma(\pi^+ p \to \pi^0 \Delta^{++}) = 0.12 \pm 0.02 \text{ mb}$$
$$\sigma(\pi^+ p \to \pi^+ \Delta^+) = 0.08 \pm 0.04 \text{ mb} \tag{6.42}$$

which is in agreement with (6.41) within the experimental accuracy. Other possible forbidden reactions of this type are considered in Ref. 25. It is interesting to note, that the cross sections of the forbidden reactions are suppressed in comparison to those of the allowed reactions by a factor 7-10, which confirms the accuracy of quark model predictions on the level 10-15%. All the measured differential cross sections of the forbidden reactions at $p_{lab} \gtrsim 5$ GeV/c have a pronounced minimum in the forward direction. The only exception is the annihilation $p\bar{p} \to \Xi\bar{\Xi}$ which has a nearly isotropic angular distribution at $p_{lab} = 7$ GeV/c[12)].

Another set of reactions, which are forbidden in the quark model is neutral meson production in πN or KN collisions, when the quark contents of the initial and final mesons are completely different[27]. These forbidden reactions are

$$\pi^{\pm}N \not\rightarrow (s\bar{s})N$$

$$\pi^{\pm}N \not\rightarrow (s\bar{s})\Delta$$

$$K^{\pm}N \not\rightarrow (d\bar{d})\Lambda(\Sigma)$$

$$K^{\pm}N \not\rightarrow (dd)\Sigma^* \tag{6.43}$$

and similar ones. As soon as $(s\bar{s})$ and $(d\bar{d})$ are not real observed mesons, the relations for the reactions with physical particles depend on their wave functions and mixing angles. For example, for pion-induced reactions we have

$$\frac{\sigma(\pi^- p \rightarrow \phi n)}{\sigma(\pi^- p \rightarrow \omega n)} = \frac{\sigma(\pi^+ p \rightarrow \phi\Delta^{++})}{\sigma(\pi^+ P \rightarrow \omega\Delta^{++})} = \left(\frac{1 - \sqrt{2}\ \mathrm{tg}\theta_{\phi\omega}}{1 + \sqrt{2}\ \mathrm{tg}\theta_{\phi\omega}}\right)^2 \tag{6.44}$$

where $\theta_{\phi\omega}$ is the $\phi\omega$ mixing angle. Thus the relation (6.44) enables us to estimate the mixing angle $\theta_{\phi\omega}$ independently of the mass formulae; the experimental data at a rather low energy $p_{lab} \simeq 3.5\text{-}4$ GeV/c[28] give the value of the ratio (6.44) about 1/40 : 1/60, in agreement with the "magic" mixing, when the ϕ-meson is built mainly of strange quarks. Similar relations may be written for other meson multiplets.

Further relations for the flavour exchange reactions are based on different assumptions about the properties of quark-quark amplitudes and quark-hadron vertices. Białas and Zalewski[29] proposed a classification of these relations according to the assumptions necessary to derive them. Due to this classification, some relations[29] follow from the broken U(3) symmetry for quarks alone and the same for antiquarks alone. Thus any meson-quark charge (or strangeness) exchange amplitude may be expressed in terms of four independent amplitudes:

$$T_1 = \frac{1}{2\sqrt{2}}\left[\langle u^{\uparrow}X|d^{\uparrow}Y\rangle + \langle u^{\downarrow}X|d^{\downarrow}Y\rangle\right] = T_{ud}\ ;$$

$$T_2 = T_{\bar{d}\bar{u}}\ ;\quad T_3 = T_{su}\ ;\quad T_4 = T_{\bar{u}\bar{s}}\ . \tag{6.45}$$

Here X and Y stand for arbitrary appropriate quark systems (in particular, particles). For example, charge exchange amplitudes are

$$<\pi^- X|\pi^0 Y> = T_1 - T_2$$

$$<K^- X|\bar{K}^0 Y> = \sqrt{2}\, T_2$$

$$K^0 X|K^+ Y> = \sqrt{2}\, T_1 \qquad (6.46)$$

etc., while for strangeness exchange amplitudes we get:

$$<K^- X|\pi^0 Y> = -T_3$$

$$<\pi^0 X|K^+ Y> = -T_4 \qquad (6.47)$$

and so on.

In order to obtain relations for measurable values from (6.46), (6.47) it is necessary to substitute nucleons for X and baryons for Y. Hereafter $\bar{\sigma}$ stands for any bilinear combination of the amplitudes; according to Ref. 30 it is related to the corresponding measured value σ by

$$\bar{\sigma} = s \frac{p_{in}}{p_{out}} \sigma \qquad (6.48)$$

where s, p_{in} and p_{out} are the total c.m. energy squared and the momenta of the incoming and outgoing particles, respectively. Thus for $\bar{\sigma}$ we get a set of relations like

$$\bar{\sigma}(\pi^- p \to \omega n) + \bar{\sigma}(\pi^- p \to \phi n) + \bar{\sigma}(\pi^- p \to \rho^0 n) = \bar{\sigma}(K^- p \to \bar{K}^{*0} n) + \bar{\sigma}(K^+ n \to K^{*0} p)$$

$$\bar{\sigma}(K^- p \to \omega\Lambda) + \bar{\sigma}(K^- p \to \phi\Lambda) = \bar{\sigma}(K^- p \to \rho^0\Lambda) + \bar{\sigma}(\pi^- p \to K^{*0}\Lambda) \quad . \qquad (6.49)$$

When ϕ is taken to be a pure $s\bar{s}$ system, the second of the relations (6.49) splits into two separate ones:

$$\bar{\sigma}(K^- p \to \omega\Lambda) = \bar{\sigma}(K^- p \to \rho^0\Lambda) \qquad (6.50a)$$

$$\bar{\sigma}(K^- p \to \phi\Lambda) = \bar{\sigma}(\pi^- p \to K^{*0}\Lambda) \quad . \qquad (6.50b)$$

The relation (6.49) remains valid when vector mesons in the final state are replaced by the mesons of any other nonet. For example, for 0^- mesons this will be

$$\bar{\sigma}(\pi^- p \to \eta n) + \bar{\sigma}(\pi^- p \to \eta' n) + \bar{\sigma}(\pi^- p \to \pi^0 n) = \bar{\sigma}(K^- p \to \bar{K}^0 n) + \bar{\sigma}(K^+ n \to K^0 p) \ . \tag{6.51}$$

Similarly, decuplet baryons may be taken instead of octet baryons (exact SU(6) wave functions are used):

$$\bar{\sigma}(\pi^+ p \to \eta \Delta^{++}) + \bar{\sigma}(\pi^+ p \to \eta' \Delta^{++}) + \bar{\sigma}(\pi^+ p \to \pi^0 \Delta^{++})$$
$$= \bar{\sigma}(K^+ p \to K^0 \Delta^{++}) + 3\bar{\sigma}(K^- p \to \bar{K}^0 \Delta^0) \ . \tag{6.52}$$

The relations (6.50) and (6.52) were tested in Refs. 31-33 and a good agreement with the data was reported.

Using standard methods of expanding baryon (and antibaryon) scattering processes into quark-quark amplitudes and assuming charge or isospin invariance of these interactions one comes to several identities for quark-quark amplitudes[29]. They give further relations for meson-baryon flavour exchange processes; these relations are in fact based on the assumption of equivalence of quarks constituting mesons and baryons and because of that they are very important for the quark model:

$$\bar{\sigma}(\pi^- p \to K^0 \Sigma^0) + \bar{\sigma}(\pi^- p \to K^{*0} \Sigma^0) = \frac{1}{3}\bar{\sigma}(\pi^- p \to K^0 \Lambda) + \frac{1}{27}\bar{\sigma}(\pi^- p \to K^{*0} \Lambda)$$

$$\bar{\sigma}(\pi^- p \to \omega n) = \frac{3}{8}\bar{\sigma}(\pi^+ p \to \pi^0 \Delta^{++}) + \frac{25}{24}\bar{\sigma}(\pi^+ p \to \omega \Delta^{++})$$

$$\bar{\sigma}(\pi^+ p \to \rho^0 \Delta^{++}) + \bar{\sigma}(\pi^+ p \to \omega \Delta^{++}) + \frac{27}{25}\bar{\sigma}(K^- p \to \bar{K}^0 N^{*0})$$
$$= \frac{24}{25}\bar{\sigma}(K^- p \to \bar{K}^{*0} n) + \bar{\sigma}(K^+ p \to K^{*0} \Delta^{++}) \ . \tag{6.53}$$

The agreement between prediction and experiment is again reasonably good except for the second relation. However, the measurements of the cross sections with double resonance production are often liable to large systematic errors.

A calculation for baryon-quark couplings, similar to that for mesons (Eqs. (6.45)-(6.47)) allows one to express baryon-quark cross sections as linear combinations of various kaon-quark cross sections. Eliminating the former from the obtained expressions, we come to addition relations of the type:

$$\bar{\sigma}(K^-p \to M^o\Lambda) = 3[\bar{\sigma}(K^-p \to M^o\Sigma^o) + \bar{\sigma}(K^-p \to M^o\Sigma^{*o})] \qquad (6.54)$$

where M^o is a neutral meson like π^o, η, ρ^o, etc. Similar relations for strangeness-exchange reactions are

$$\bar{\sigma}(\pi^-p \to M_s^o\Lambda) = 3[\bar{\sigma}(\pi^-p \to M_s^o\Sigma^o) + \bar{\sigma}(\pi^-p \to M_s^o\Sigma^{*o})] \quad . \qquad (6.55)$$

Here M_s^o is a neutral strange meson like K^o. The relations (6.54), (6.55) follow from the assumption of SU(6) symmetry for baryon wave functions. They are valid also for natural and unnatural parity exchange parts of the corresponding cross sections separately. The detailed test[34)] of (6.54) in a broad energy interval p_{lab} = 3-14 GeV/c shows that it is systematically broken: Λ production cross section is suppressed by a factor $2 \sim 3$ compared with the right hand side of (6.54). There are several possible explanations of this suppression, based on the account of the configuration mixings for baryon wave functions due to SU(6)-breaking interactions[35)] or quark-diquark clusterisation inside the baryon[36,37)]. In the latter case, the diquark (as it is claimed in Ref. 36) is not necessarily an elementary object like a quark; it may only express some asymmetry in the spatial part of the wave function, due to which one quark is removed sufficiently far from the other two. Such a picture of baryon structure is of course quite compatible with the quasinuclear quark model.

Using the mentioned expressions for meson-quark and baryon-quark amplitudes simultaneously, we obtain a number of relations connecting baryon-baryon (or baryon-antibaryon) and meson-baryon scattering processes. These relations are specific for the quark model and their detailed test is of great interest. However, some additional uncertainties appear when one tries to compare these relations with the data. First of all, strong (exponential) t-dependence of the hadron-hadron differential cross sections in the quark model is usually associated with hadronic form factors, which are essentially different for mesons and baryons. So, when non-forward scattering is considered, the relations for meson-baryon and baryon-baryon scattering cross sections should be corrected in order to take into account this difference; this procedure is not unique. Another source of the theoretical uncertainty is that meson-baryon and baryon-baryon collisions which we compare should be taken at different initial beam energies in order to obtain the same energies of quark-quark collision. Besides that, particle mass differences are also to be taken into account.

The experimental status of some of the relations between meson-baryon and baryon-baryon flavour exchange reactions is summarized in Table 6.2[29]. The general trend of these results is clear disagreement

Table 6.2 Beam momenta 3-8 GeV/c; see data references in Ref. 29.

Relation	Left-hand side (mb)	Right-hand side (mb)
$\bar{\sigma}(pp \to n\Delta^{++}) = \sigma(K^+p \to K^o\Delta^{++}) + \frac{25}{9}\sigma(K^+p \to K^{*o}\Delta^{++})$	2.2 ± 0.4	6.6 ± 1.8
$\bar{\sigma}(\bar{p}p \to \Delta^{++}\bar{\Delta}^{--}) = 8\bar{\sigma}(K^-p \to K^o\Delta^o)$	9.1 ± 0.6	1.0 ± 3.0
$\bar{\sigma}(\bar{p}p \to \Sigma^+\bar{\Sigma}^-) = 2\bar{\sigma}(\pi^-p \to K^o\Sigma^o) + \frac{2}{9}\bar{\sigma}(\pi^-p \to K^{*o}\Sigma^o)$	2.7 ± 1.1	4.0 ± 1.0
$\bar{\sigma}(\bar{p}p \to \Sigma^o\bar{\Lambda}) = \frac{3}{2}\bar{\sigma}(\pi^-p \to K^o\Sigma^o) + \frac{3}{2}\bar{\sigma}(\pi^-p \to K^{*o}\Sigma^o)$	2.4 ± 1.0	4.6 ± 2.0
$\bar{\sigma}(\bar{p}p \to \Sigma^o\bar{\Lambda}) = \frac{1}{2}\bar{\sigma}(\pi^-p \to K^o\Lambda) + \frac{1}{18}\bar{\sigma}(\pi^-p \to K^{*o}\Lambda)$	2.4 ± 1.0	1.8 ± 0.3
$\bar{\sigma}(\bar{p}p \to \Lambda\bar{\Lambda}) = \frac{3}{2}\bar{\sigma}(\pi^-p \to K^o\Lambda) + \frac{3}{2}\bar{\sigma}(\pi^-p \to K^{*o}\Lambda)$	2.6 ± 0.7	9.9 ± 1.3
$\bar{\sigma}(\bar{p}p \to \bar{n}n) = \bar{\sigma}(K^-p \to K^on) + \frac{25}{9}\bar{\sigma}(K^-p \to \bar{K}^{*o}n)$	7.56 ± 1.08	14.24 ± 3.1

of the quark model predictions with the data. However, in view of the mentioned theoretical uncertainties and specific prescriptions, made in Ref. 29 for kinematic corrections, it is not clear whether the observed disagreement should be considered as unavoidable. In Ref. 38 the relation

$$\frac{d\sigma}{dt}(pn \to np) + \frac{d\sigma}{dt}(p\bar{p} \to n\bar{n}) = \frac{d\sigma}{dt}(\pi^-p \to \pi^on) + 3\,\frac{d\sigma}{dt}(\pi^-p \to \eta n)$$
$$+ \frac{25}{24}\cdot\frac{25}{9}\left[\frac{d\sigma}{dt}(\pi^+p \to \rho^o\Delta^{++}) + \frac{d\sigma}{dt}(\pi^+p \to \omega\Delta^{++})\right] \qquad (6.56)$$

at p_{lab} = 8 GeV/c was tested; the value of the r.h.s. of (6.56) was shown to be about 3 times that of the l.h.s. The dominating contribution to the r.h.s. came from the reaction $\pi^+ p \to \rho^0 \Delta^{++}$, which was measured only at $|t| = 0.05\ (\mathrm{GeV/c})^2$. The cross section of the similar reaction $\pi^+ p \to \rho^+ p$ decreases by a factor 3 as $|t|$ changes from 0.05 to 0.02 $(\mathrm{GeV/c})^2$. Thus we do not see whether the results of Ref. 38 contradict the relation (6.56).

The consideration of the spin structure of charge exchange reactions $PN \to P'N'$ and $PN \to P'\Delta$, where P and P' are 0^- mesons, leads to another set of relations. Indeed, decuplet production amplitudes and spin-flip octet production amplitudes proceed through the same quark-quark interactions and so[24)]

$$\frac{d\sigma^f}{dt}(\pi^- p \to \pi^0 n) = \frac{25}{24}\frac{d\sigma}{dt}(\pi^+ p \to \pi^0 \Delta^{++}) = \frac{25}{8}\frac{d\sigma}{dt}(\pi^- p \to \pi^0 \Delta^0)$$

$$\frac{d\sigma^f}{dt}(\pi^- p \to \eta n) = \frac{25}{24}\frac{d\sigma}{dt}(\pi^- p \to \eta \Delta^0) = \frac{25}{8}\frac{d\sigma}{dt}(\pi^+ p \to \eta \Delta^{++})$$

$$\frac{d\sigma^f}{dt}(K^- p \to \bar{K}^0 n) = \frac{25}{8}\frac{d\sigma}{dt}(K^- p \to \bar{K}^0 \Delta^0)$$

$$\frac{d\sigma^f}{dt}(K^+ n \to K^0 p) = \frac{25}{24}\frac{d\sigma}{dt}(K^+ p \to K^0 \Delta^{++})$$

$$\frac{d\sigma^f}{dt}(K^- p \to \pi^- \Sigma^+) = \frac{1}{8}\frac{d\sigma}{dt}(K^- p \to \pi^- \Sigma^{*+}) \tag{6.57}$$

where $\frac{d\sigma^f}{dt}$ is the spin-flip part of the cross section. Since $\frac{d\sigma^f}{dt} \sim \sin^2\frac{\theta}{2}$ for small scattering angles, the differential cross sections of the reactions $PN \to P'\Delta$ must have a dip near the forward direction. Such a dip is observed in $\pi^+ p \to \pi^0 \Delta^{++}$ at $p_{lab} \simeq 6.8$ GeV/c[28)]. If spin-flip amplitudes dominate in charge exchange reactions $PN \to P'N'$ near the forward direction (it seems to be quite natural in the Regge pole scheme where ρ and ω exchange amplitudes are spin-flip dominant), the relations (6.57) may be checked experimentally at small $|t|$.

6.3.2 *Polarization phenomena*

All quark model relations, considered in the previous subsection are valid for any bilinear combination of the amplitudes, in particular for spin density matrix elements. The relations for density matrices may be divided into two groups[29]: those which relate the values, measured in the same reaction and those, which relate different reactions.

The relations belonging to the first group can be derived without assuming any equalities among quark-quark amplitudes in different processes. They may be tested experimentally without any arbitrary prescriptions for kinematic corrections; moreover, the experimental values to be compared with are usually obtained in one experiment and thus possible systematic uncertainties are significantly cancelled.

The polarization properties of the secondary particles might be expressed in terms of spin density matrices or in terms of statistical tensors (see Ref. 39). The obtained descriptions are equivalent, but the latter one is more concise.

When trying to compare the quark model prediction with experimental data, we encounter with the following uncertainty. Suppose, that in the reaction $a + b \to c + d$ the values of spin density matrix elements (or statistical tensors) of particles c and d are related in some given reference frame, e.g. Jackson frame for both particles c and d.

If the measured values refer to some other frame, e.g. helicity frame, there are two possible procedures. One may either transform the obtained relations from one frame to another or perform quark model calculations in the necessary frame from the very beginning. It turns out, that some predictions of the quark model depend on the frame chosen initially to perform the calculations. However, the relations, which follow only from the additivity assumption remain valid in any case.

Let us consider some simple examples. In the reaction $0^- + 1/2 \to 0^- + 3/2$ the polarization properties of the decuplet baryon may be exactly calculated starting from the impulse approximation only; for the density matrix in the helicity frame we get[40]

$$\rho_{\mu\nu} = \frac{1}{8}\begin{pmatrix} 3 & 0 & \sqrt{3} & 0 \\ 0 & 1 & 0 & \sqrt{3} \\ \sqrt{3} & 0 & 1 & 0 \\ 0 & \sqrt{3} & 0 & 3 \end{pmatrix} . \tag{6.58}$$

The data for the comparison with (6.58) which are shown in Table 6.3 are taken from Refs. 23, 40. They agree with quark model predictions rather well.

Table 6.3 Experimental data for spin density matrix elements and quark model predictions

Reaction	$\rho_{\mu\nu}$	Theory	Experiment	Beam lab. momentum
	$\rho_{3/2\ 3/2}$	0.375	0.22 ± 0.06	8 GeV/c
$\pi \quad \to \pi\Delta$	$\rho_{3/2 -1/2}$	0.215	0.132 ± 0.07	
	$\rho_{3/2\ 1/2}$	0	0.066 ± 0.01	
	$\rho_{3/2\ 3/2}$	0.375	0.40 ± 0.06	
$\pi N \to \pi\Delta$	$\rho_{3/2 -1/2}$	0.215	0.21 ± 0.08	4 GeV/c
	$\rho_{3/2\ 1/2}$	0	-0.03 ± 0.07	
	$\rho_{3/2\ 3/2}$	0.375	0.28 ± 0.06	
$K^+p \to K^0\Delta^{++}$	$\rho_{3/2 -1/2}$	0.215	0.21 ± 0.05	3 GeV/c
	$\rho_{3/2\ 1/2}$	0	0.04 ± 0.05	
	$\rho_{3/2\ 3/2}$	0.374	0.30 ± 0.15	
$K^-p \to \pi^-\Sigma^{*+}$	$\rho_{3/2 -1/2}$	0.215	0.25 ± 0.06	5.5 GeV/c
	$\rho_{3/2\ 1/2}$	0	0.00 ± 0.15	
	$\rho_{3/2\ 3/2}$	0.375	0.85 ± 0.09	
$K^-p \to \pi^-\Sigma^{*+}$	$\rho_{3/2 -1/2}$	0.215	0.16 ± 0.11	4.5 GeV/c
	$\rho_{3/2\ 1/2}$	0	0.16 ± 0.14	

In the reaction $0^- + 1/2 \to 1^- + 3/2$ the quark model relates the statistical tensors (or density matrices) of the secondary meson and baryon. Assuming only additivity, we get the following equalities for the components of the transversity statistical tensor[29]:

$$T^{0}_{-3\,1} = T^{0}_{3-1} = 0$$

$$T^{0}_{1\,1} = T^{0}_{1-1}$$

$$T^{m}_{3\nu\,\nu} = \sqrt{3}\,T^{m}_{\nu-\nu} \quad \left[m = \pm 1, \quad \nu = 2\cdot\left(\pm\frac{1}{2}\right)\right] \quad . \tag{6.59}$$

Since $T^{m}_{\Lambda\mu}$ are, generally speaking, complex values, the Eqs. (6.59) give 6 relations for the moduli $|T^{m}_{\Lambda\mu}|$ and 4 for the phases $\delta^{m}_{\Lambda\mu}$.

Supposing that two quark-quark spin-flip amplitudes with net transversity flip equal to 2 are equal, 3 more complex relations may be derived:

$$\sqrt{3}\,T^{1}_{-1\,1} = T^{-1}_{3\,1} = T^{1}_{-3-1} = \sqrt{3}\,T^{-1}_{1-1} \quad . \tag{6.60}$$

Lastly, if two quark-quark spin-flip amplitudes with net transversity flip equal zero are also equal to each other, 3 more relations appear:

$$T^{1}_{3\,1} = \sqrt{3}\,T^{-1}_{-1\,1} = \sqrt{3}\,T^{1}_{1-1} = T^{-1}_{-3-1} \quad . \tag{6.61}$$

The relations (6.59)-(6.61) lead to the following, less general relations among helicity density matrix elements of the vector meson and the baryon:

$$\rho_{11} + \rho_{1-1} = \frac{4}{3}\rho_{\frac{3}{2}\frac{3}{2}} + \frac{4}{\sqrt{3}}\rho_{\frac{3}{2}-\frac{1}{2}} \quad . \tag{6.62}$$

$$\rho_{11} = \frac{4}{3}\rho_{\frac{3}{2}\frac{3}{2}}$$

$$\rho_{1-1} = \frac{4}{\sqrt{3}}\mathrm{Re}\,\rho_{\frac{3}{2}-\frac{1}{2}}$$

$$\rho_{10} = \frac{4}{\sqrt{6}}\rho_{\frac{3}{2}\frac{1}{2}} \quad . \tag{6.63}$$

$$\mathrm{Re}\,\rho_{\frac{3}{2}\frac{1}{2}} = 0$$

$$\mathrm{Re}\,\rho_{10} = 0 \qquad (6.64)$$

High-statistics experiment[41)] enables one to test (6.59)-(6.61) for the reaction $K^-p \to \rho^-\Sigma^{*+}$ at p_{lab} = 4.2 GeV/c. The results are presented in Figs. 6.9-6.11; in every figure the values pictured by the different symbols are to be equal to each other. A good agreement is clearly seen for the relations (6.59) and (6.60) for the moduli; somewhat worse — for the phases. For the relation (6.61) there is some disagreement with the data (see Fig. 6.11); however, in Ref. 29 it is claimed that these predictions are most sensitive to the choice of the initial reference frame.

Helicity density matrix elements were measured in Ref. 42 for the reaction $\pi^+p \to \rho^0\Delta^{++}$ at p_{lab} = 16 GeV/c. These results are compared with the relations (6.62)-(6.64) in Fig. 6.12. Again the agreement for (6.62)-(6.63) is quite reasonable, while for (6.64) there is a clear disagreement.

Another group of relations may be derived if we assume the equivalence of similar quark-quark amplitudes in different processes. For example, the same arguments that lead to the relations (6.50a) and (6.50b) give for the vector meson spin density matrices

$$\rho_{ik}(K^-p \to \phi\Lambda) = \rho_{ik}(\pi^-p \to K^{*0}\Lambda) \qquad (6.65a)$$

$$\rho_{ik}(K^-p \to \omega\Lambda) = \rho_{ik}(K^-p \to \rho^0\Lambda) \qquad (6.65b)$$

in a good agreement with the data (see Ref. 29 and references given therein). Many other relations of this type may be derived, but only few of them were tested with reasonable experimental accuracy. It is interesting to note, that the relations (6.62)-(6.63) for the spin density matrices in the reactions $K^+n \to K^{*0}p$ and $pp \to n\Delta^{++}$ (i.e. the relations for meson-baryon and baryon-baryon reactions, which are free of kinematic corrections) are well satisfied by the data[43)] (see Fig. 6.13) at 6 GeV/c.

The properties considered of the flavour-exchange binary reactions

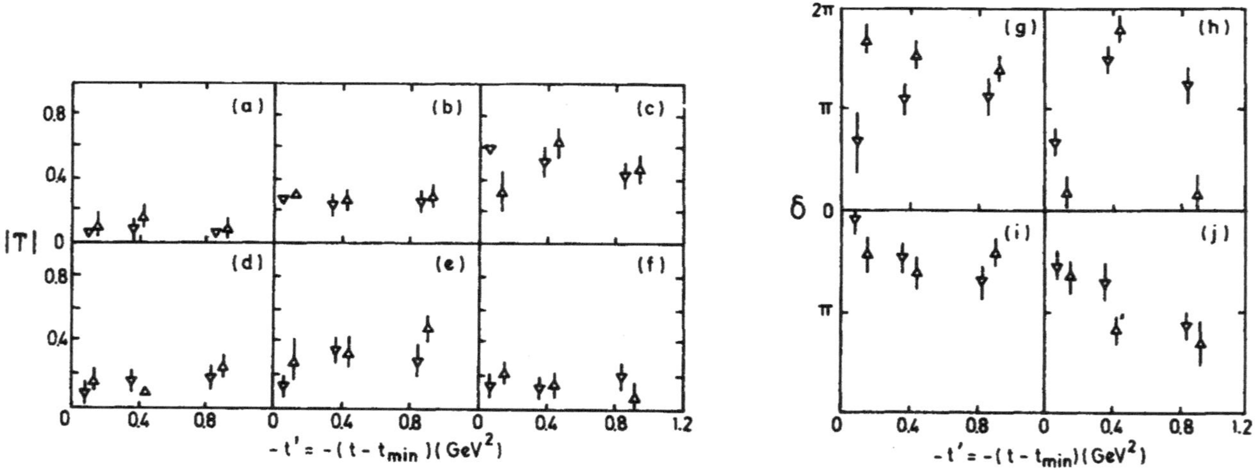

Fig. 6.9 Test of the relation (6.59) in the reaction $K^- p \to \rho^- \Sigma^{*+}$.

a) $|T^0_{-3\ 1}|(\nabla)$ and $|T^0_{3\ -1}|(\Delta)$; b) $|T^0_{1\ +1}|(\nabla)$ and $|T^0_{-1\ -1}|(\Delta)$;

c) $|T^1_{3\ 1}|(\nabla)$ and $\sqrt{3}|T^1_{1\ -1}|(\Delta)$; d) $|T^1_{-1\ 1}|(\nabla)$ and $(1/\sqrt{3})|T^1_{-3\ -1}|(\Delta)$;

e) $|T^{-1}_{3\ 1}|(\nabla)$ and $|T^{-1}_{1\ -1}|(\Delta)$; f) $|T^{-1}_{-1\ 1}|(\nabla)$ and $(1/\sqrt{3})|T^{-1}_{-3\ -1}|(\Delta)$;

g) $\delta^1_{-1\ 1}(\nabla)$ and $\delta^1_{-3\ -1}(\Delta)$; h) $\delta^0_{1\ 1}(\nabla)$ and $\delta^0_{-1\ -1}(\Delta)$;

i) $\delta^{-1}_{3\ 1}(\nabla)$ and $\delta^{-1}_{1\ -1}(\Delta)$; j) $\delta^{-1}_{-1\ 1}(\nabla)$ and $\delta^{-1}_{-3\ -1}(\Delta)$.

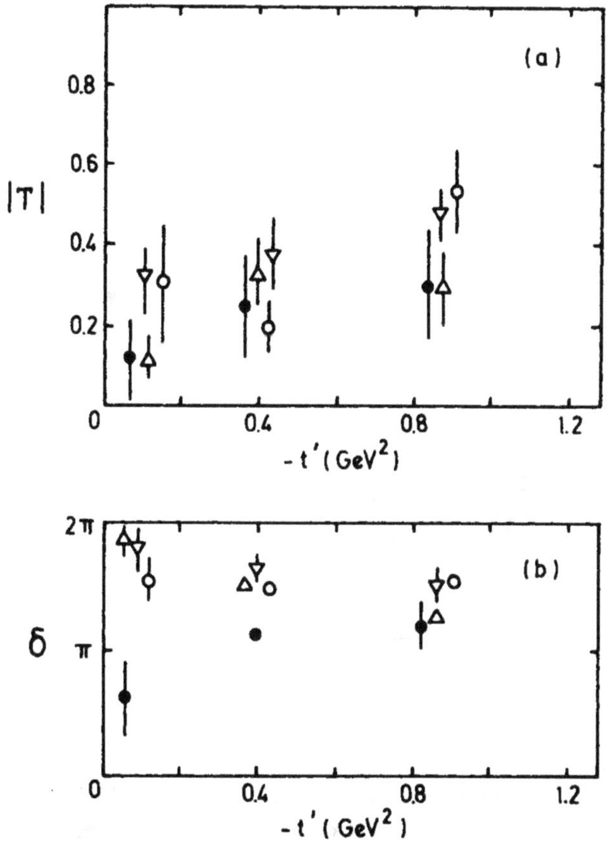

Fig. 6.10 Test of the relations (6.60) in the reaction $K^- p \to \rho^- \Sigma^{*+}$
a) $\sqrt{3}|T^{1}_{-1\ 1}|(\cdot)$, $|T^{-1}_{3\ 1}|(\Delta)$, $|T^{1}_{-3\ -1}|(\nabla)$ and $\sqrt{3}|T^{-1}_{1\ -1}|(0)$.
b) $\delta^{1}_{-1\ 1}(\cdot)$, $\delta^{-1}_{3\ 1}(\Delta)$, $\delta^{1}_{-3\ -1}(\nabla)$ and $\delta^{-1}_{1\ -1}(0)$.

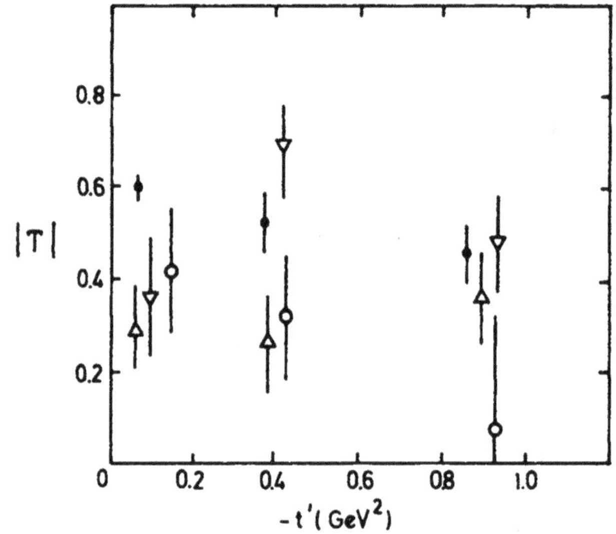

Fig. 6.11 Test of the relations (6.61) in the reaction $K^- p \to \rho^- \Sigma^{*+}$, $|T^1_{3\ 1}|(\cdot)$, $\sqrt{3}|T^{-1}_{-1\ 1}|(\Delta)$, $\sqrt{3}|T^1_{1\ -1}|(\nabla)$ and $|T^{-1}_{-3-1}|(0)$.

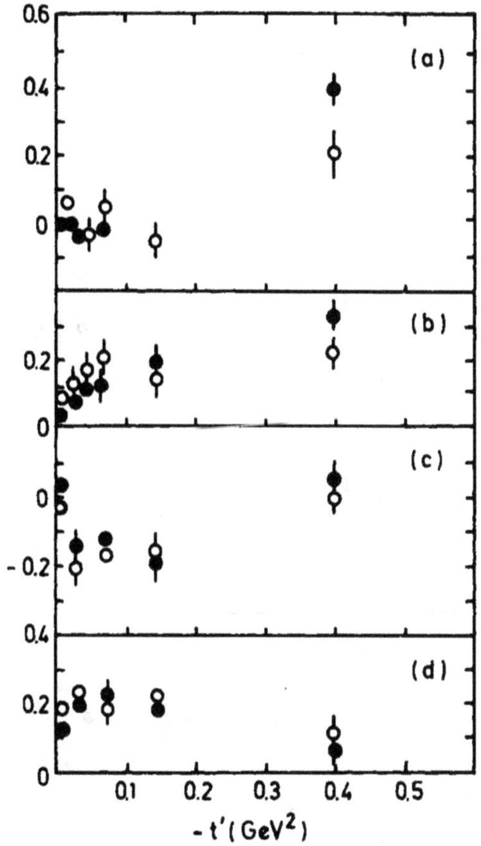

Fig. 6.12 Test of the relations (6.62)-(6.64) in the reaction $\pi^+ p \to \rho^0 \Delta^{++}$

a) $\rho_{11} + \rho_{1\,-1}$ (full circles) and $\frac{4}{3}\rho_{\frac{3}{2}\frac{3}{2}} + \frac{4}{\sqrt{3}}\rho_{\frac{3}{2}-\frac{1}{2}}$ (open circles).

b) ρ_{11} (full circles) and $\frac{4}{3}\rho_{\frac{3}{2}\frac{3}{2}}$ (open circles).

c) $\rho_{1\,-1}$ (full circles) and $\frac{4}{\sqrt{3}}\rho_{\frac{3}{2}-\frac{1}{2}}$ (open circles).

d) Re ρ_{10} (full circles) and $\frac{4}{\sqrt{6}}$ Re $\rho_{\frac{3}{2}\frac{1}{2}}$ (open circles).

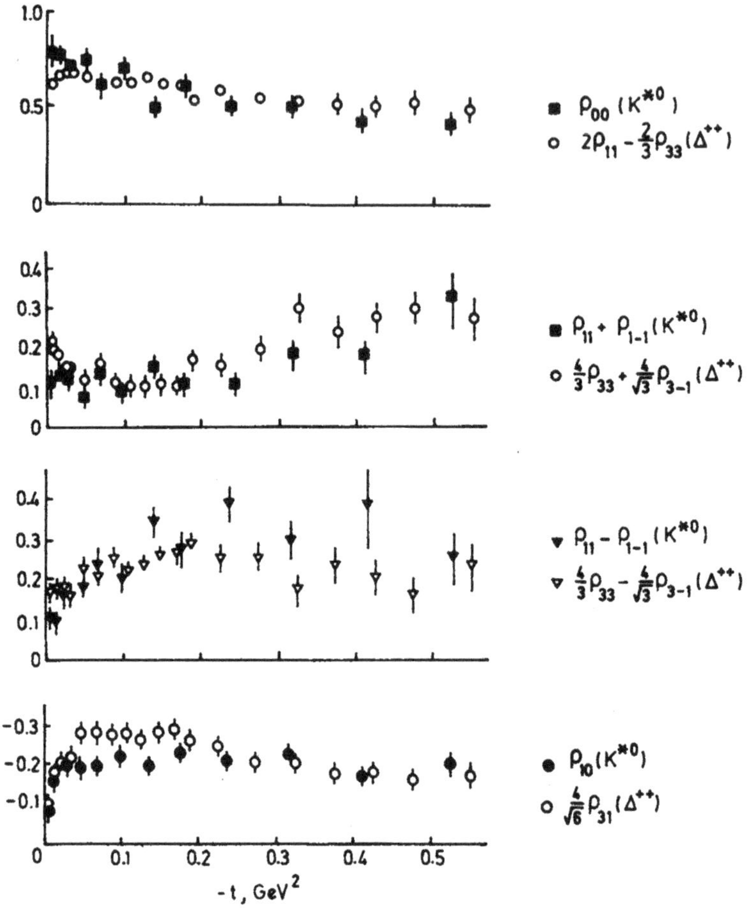

Fig. 6.13 Test of the relations (6.62), (6.63) in the reactions $K^+n \to K^{*0}p$ and $pp \to n\Delta^{++}$.

allow us to make the following conclusions:

1. The additive quark model gives a variety of relations for the flavour-exchange reactions. These relations may be classified into several groups according to the assumptions necessary to derive them.

2. Relations of the quark model, which are based only on the additivity assumption and impulse approximation, are in a good agreement with all experimental data, both for the cross sections and for the

polarization characteristics.

3. The predictive power of other relations, which are based on some specific assumptions about the properties of the quark-quark amplitudes, vary from a good agreement to a clear disagreement with the data.

4. Exact SU(6) symmetry of the quark-quark interaction is obviously violated by experiment; the same is true also for SU(6) broken on SU(3) level (i.e. by the strange quark interactions). However, if SU(6) breaking due to spin-spin and spin-orbital interactions is taken into consideration, the agreement with the data is significantly improved.

5. The procedure of the comparison of the quark model relations with the data suffers from some theoretical ambiguities. In particular, these are kinematical corrections for the measured cross section values and reference frame choice for the secondary resonance decay distributions. As a rule, the quark model predictions, which are less sensitive to these ambiguities, are in better agreement with the data; the worst disagreement is observed for the predictions, which contain many of the mentioned ambiguities.

6.4 Regge-Pole Exchanges in the Quark Model

The properties of the binary hadronic collisions, considered in the previous sections, give strong arguments in favour of the additivity of hadronic amplitudes in terms of the quark-quark amplitudes. On the other hand, it is well known that Regge pole phenomenology provides a unique quantitative tool for high energy hadron scattering processes (see Chap. 2). So one might conclude, that the Reggeon Born approximation (i.e. Regge pole exchange amplitudes) may be applied directly to the scattering of the individual quarks.

Standard assumptions of Regge Born approximation are the exchange degeneracy of the leading trajectories and spin non-flip dominance of the Pomeron exchange. These assumptions enables one to obtain some additional relations for the hadron-hadron cross sections and polarization parameters.

Let us consider the differences of the total cross sections first.

Using the notations

$$\Delta\sigma(p^{\pm}p) = \sigma_{tot}(\bar{p}p) - \sigma_{tot}(pp)$$

$$\Delta\sigma(\pi^{\pm}p) = \sigma_{tot}(\pi^{-}p) - \sigma_{tot}(\pi^{+}p) \tag{6.66}$$

etc. and assuming the universality of ρ, ω couplings to mesons and nucleons, we come to the following relations[44]:

$$\Delta\sigma(p^{\pm}p) + \Delta\sigma(p^{\pm}n) = 3[\Delta\sigma(K^{\pm}p) + \Delta\sigma(K^{\pm}n)] \tag{6.67a}$$

$$\Delta\sigma(p^{\pm}p) = 3\Delta\sigma(K^{\pm}p) - \Delta\sigma(\pi^{\pm}p) \quad . \tag{6.67b}$$

The experimental status of (6.67a,b) is shown in Fig. 6.14; the agreement is very good in a rather broad energy interval, when the right-hand side of (6.67a) and (6.67b) is taken at the beam energy which is 2/3 that of the left-hand side cross sections, as it is prescribed by the quasinuclear quark model.

Further relations for the polarization parameters $P_0(s,t)$ of the elastic scattering processes may be obtained when we assume that the Pomeron exchange contributes only to the spin non-flip amplitudes. If there are leading spin-flip contributions coming from exchange degenerate $\rho - A_2$ and $f - \omega$ exchanges, the resulting quark-antiquark polarizations vanish at small $|t|$, and we obtain the simple relations[45]

$$P_0(K^{+}p)\sigma_{tot}(K^{+}p) = P_0(\pi^{+}p)\sigma_{tot}(\pi^{+}p) \tag{6.68a}$$

$$P_0(pp)\sigma_{tot}(pp) = 2P_0(\pi^{+}p)\sigma_{tot}(\pi^{+}p) + P_0(\pi^{-}p)\sigma_{tot}(\pi^{-}p) \tag{6.68b}$$

together with other ones; however only these two are tested experimentally. The corresponding data are shown in Fig. 6.15; the agreement for (6.68a) is quite good and somewhat worse for (6.68b). However, it becomes better with the increase of the energy.

Similarly some consequences for the real parts of the forward scattering amplitudes of mesons and nucleons[45] can be derived. The equality of the real parts of $\pi^{+}p$ and $K^{+}p$ scattering amplitudes was noted in Ref. 46 and needs only exact SU(3) symmetry of Reggeon-hadron

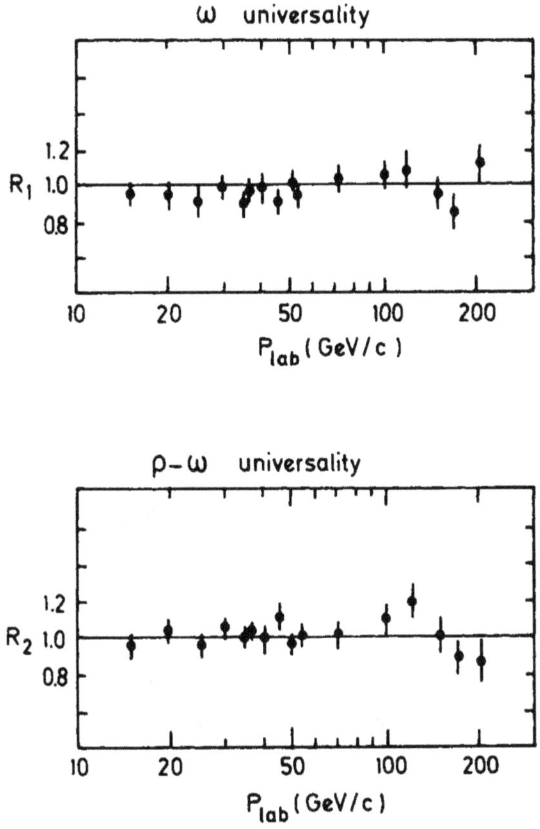

Fig. 6.14 Test of universality of ρ, ω couplings to mesons and nucleons:

$$R_1 = \frac{\Delta\sigma(p^{\pm}p) + \Delta\sigma(p^{\pm}n)}{3[\Delta\sigma(K^{\pm}p) + \Delta\sigma(K^{\pm}n)]}$$

$$R_2 = \frac{\Delta\sigma(p^{\pm}p)}{3\Delta\sigma(K^{\pm}p) - \Delta\sigma(\pi^{\pm}p)}$$

Quark model predicts $R_1 = R_2 = 1$.

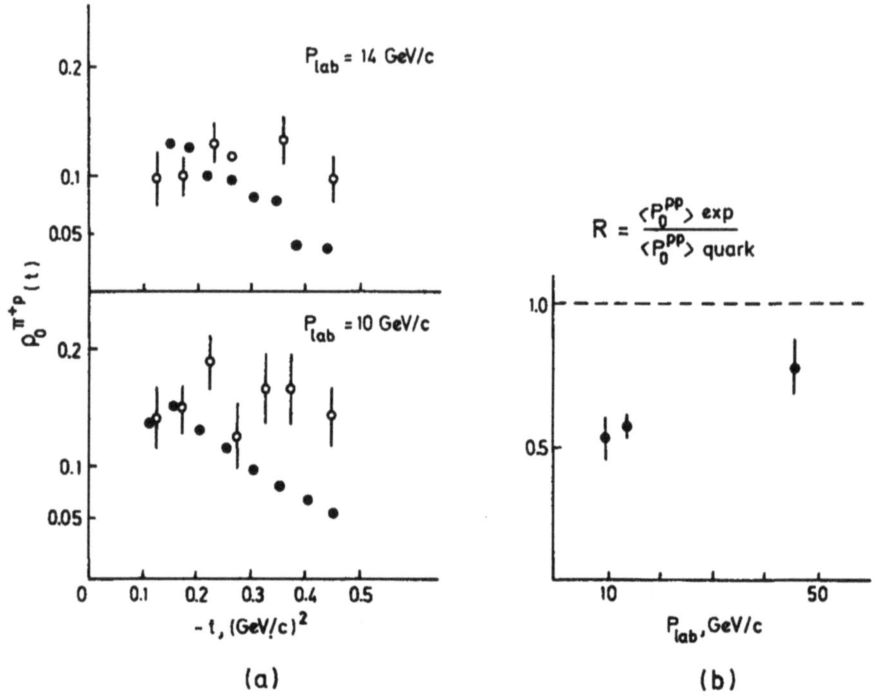

Fig. 6.15 Test of the relations (6.68) for the polarization parameter P_o.

a) $P_o(\pi^+p)$ — experimental data (full circles) and calculated values from the data on $P_o(K^+p)$ according to the Eq. (6.68a) (open circles);

b) $R = \frac{P_o(pp)_{exp}}{P_o(pp)_{quark}}$, where $P_o(pp)_{exp}$ is the experimental values, $P_o(pp)_{quark}$ is the calculated value with help of the Eq. (6.68b).

couplings, while the relations between meson-nucleon and nucleon-nucleon amplitudes

$$\rho(pp)\sigma_{tot}(pp) = 2\rho(\pi^+p)\sigma_{tot}(\pi^+p) + \rho(\pi^-p)\sigma_{tot}(\pi^-p) \tag{6.69a}$$

$$\rho(pp)\sigma_{tot}(pp) + \rho(np)\sigma_{tot}(np) = \frac{1}{3}\left[\rho(\pi^+p)\sigma_{tot}(\pi^+p) + \rho(\pi^-p)\sigma_{tot}(\pi^-p) \right] \tag{6.69b}$$

can also be obtained from the results of Ref. 12 by applying exchange degeneracy. Fig. 6.16 presents the available data for the real parts of pp, np and π^+p amplitudes compared with the quark model predictions; the agreement is quite reasonable.

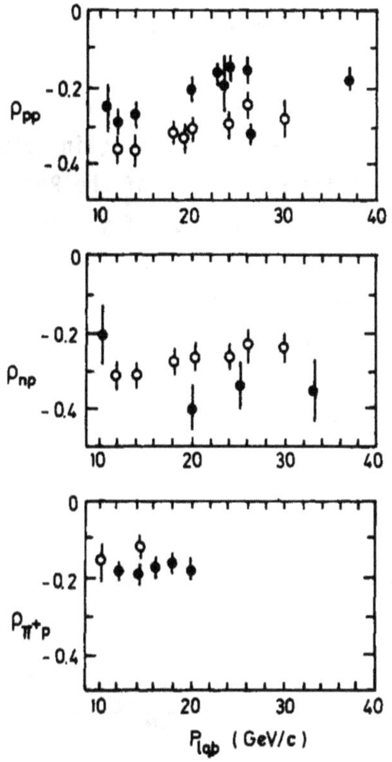

Fig. 6.16 Test of the relations (6.69) for the $\rho_{hp} = \dfrac{\text{Re } A_{hp}(s,o)}{\text{Im } A_{hp}(s,o)}$ ($h=p,n,\pi^+$). In each figure full circles are experimental values, open circles – calculated values with the help of the relations (6.69). ρ_{π^+p} is calculated with the help of K^+p data, assuming $\rho_{\pi^+p}\sigma_{tot}(\pi^+p) = \rho_{K^+p}\sigma_{tot}(K^+p)$.

Regge pole representation for quark-quark amplitudes assumes that Reggeon couplings for any hadronic processes can be calculated with the help of quark-Reggeon couplings. Such a formalism is presented in Ref. 47, where the quark-Reggeon couplings are determined by the electromagnetic and weak couplings of the quarks. Indeed, vector dominance of the quark electromagnetic currents relates ρ and ω couplings to those of the photon. Then, through exchange degeneracy and duality prescription for f-dominance of pomeron coupling, all the dominant natural parity exchanges (f, A_2, $\mathbf{P}$, etc) can be related to the photon coupling too. Similarly, unnatural parity exchanges (π and A_1) can be related to W-boson couplings; the latter are again related to γ coupling by CVC and PCAC. Thus the magnitudes of the Reggeon couplings are fixed by the vector-meson-photon constant $f\omega\gamma$. In Ref. 47 a lot of successful predictions both for particle decay rates and hadronic scattering amplitudes are made starting from this simple formalism. Two examples are given in Table 6.4 (meson decay widths) and Fig. 6.17 $\left(\rho(pp;\, s,o) = \dfrac{\mathrm{Re}\, A_{pp}(s,o)}{\mathrm{Im}\, A_{pp}(s,o)} \right)$.

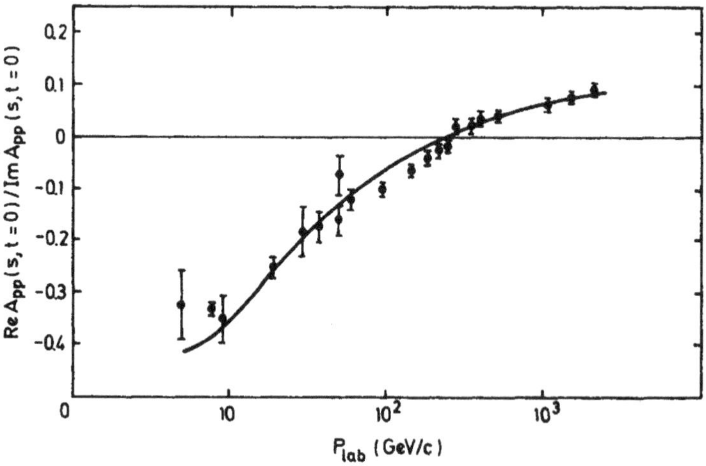

Fig. 6.17 ρ_{pp} value versus laboratory momentum and its description in the quark model with Reggeon exchanges[47].

Table 6.4 Meson decay widths, as predicted in Ref. 47.

Decay	Predicted width (MeV)	Experimental width[48]
$\rho \to \pi\pi$	264	155 ± 3
$f \to \pi\pi$	117	150 ± 17
$g \to \pi\pi$	56	48 ± 5
$h \to \pi\pi$	103	37 ± 5
$f \to K\bar{K}$	4.4	5.3 ± 0.6
$A_2 \to K\bar{K}$	5.9	5.5 ± 1.0
$\omega(1670) \to K\bar{K}$	6.6	-
$g \to K\bar{K}$	7.4	3 ± 1
$h \to K\bar{K}$	11.3	$1.5^{+0.1}_{-0.5}$
$\phi \to K\bar{K}$	3.47	3.5 ± 0.2
$f' \to K\bar{K}$	18.4	< 70 ± 10
$K^*(892) \to K\pi$	26	5.1 ± 1.0
$K^*(1430) \to K\pi$	16	45 ± 5

6.5 Multiple Quark Scatterings

The impulse approximation of the quasinuclear quark model is a typical first approximation in the composite system collision theory (see Chap. 4). The next approximations are related with the possibility of the repeated collisions of several constituent quarks in the process of hadron-hadron collision.

Here we shall consider the effect of the elastic quark rescatterings in hadron-hadron collisions. For definiteness, *pp* collisions are taken as an example.

We shall neglect the real part of high-energy quark-quark amplitude;

it is taken in the form of Eq. (6.33). The difference between a_{uu} and a_{ud} amplitudes is also neglected, and in the impulse approximation the pp scattering amplitude (Fig. 6.18a) may be written as:

$$A^{(1)}_{pp}(s,t) = 9a_{qq}(\tfrac{s}{9},\ t)F^2_p(-t) \quad . \tag{6.70}$$

$$a_{qq}(\tfrac{s}{9},\ t) = i \cdot 8\pi s\gamma e^{b_q(s)\cdot t} \tag{6.71}$$

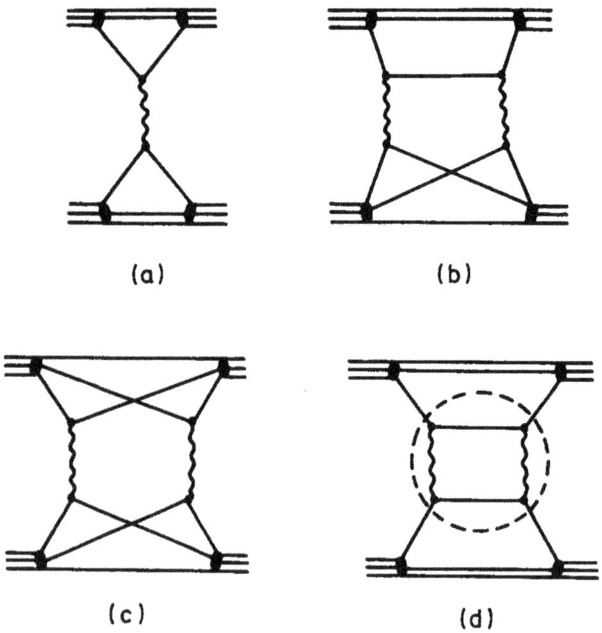

Fig. 6.18 Quark diagrams for the elastic pp scattering: a) impulse approximation; b, c) elastic shadow corrections; d) diagram, which renormalizes the impulse approximation quark-quark amplitude.

Here $\gamma = (\sigma_{tot}(pp))/72\pi$, $b_q(s) = \alpha'_{\mathbb{P}}\,\ell n(\frac{s}{9s_0})$. For the sake of simplicity we take a simple exponential parametrization for the proton form factor

$$F_p(-t) = \exp(\tfrac{1}{6}\,R^2\cdot t) \quad , \tag{6.72}$$

which proves to describe the data rather well at $|t| \lesssim 0.5\ (\text{GeV/c})^2$.

(It corresponds to the Gaussian parametrization of the quark wave function of the proton.)

The second approximation in the number of interactions is determined by two types of rescatterings, which are shown in Figs. 6.18b,c.

These amplitudes may be calculated easily introducing the trifactor of the proton, which in the non-relativistic approximation for the rest proton wave function can be defined as:

$$G_p(\vec{q}_1,\vec{q}_2,\vec{q}_3) = \int \frac{d^3k_1 d^3k_2 d^3k_3}{(2\pi)^9} \psi_p(k_1,k_2,k_3)\psi_p^*(\vec{k}_1+\vec{q}_1, \vec{k}_2+\vec{q}_2, \vec{k}_3+\vec{q}_3) \cdot \delta^3(\vec{k}_1+\vec{k}_2+\vec{k}_3) \quad . \tag{6.73}$$

In the Gaussian parametrization of ψ_p the trifactor G_p is:

$$G_p(\vec{q}_1,\vec{q}_2,\vec{q}_3) = \exp\left[-\frac{1}{6}R^2(\vec{q}_1^{\,2}+\vec{q}_2^{\,2}+\vec{q}_3^{\,2}-\vec{q}_1\cdot\vec{q}_2-\vec{q}_1\cdot\vec{q}_3-\vec{q}_2\cdot\vec{q}_3)\right] \cdot \tag{6.74}$$

In particular, $G_p(\vec{q},0,0) = F_p(\vec{q}^{\,2})$.

For small momentum transfer each of the colliding composite systems may be considered in its rest frame (see Chaps. 3, 4). Thus, using the trifactor (6.74), for the diagram in Fig. 6.18b we get:

$$A_{pp}^{(2b)} = -36i\,\frac{8\pi s\gamma^2}{R^2+4b_q(s)} \exp\left[(\frac{5}{24}R^2+\frac{1}{2}b_q(s)t\right] \,, \tag{6.75}$$

while the diagram in Fig. 6.18c gives:

$$A_{pp}^{(2c)} = -18i\,\frac{8\pi s\gamma^2}{R^2+2b_q(s)} \exp\left[(\frac{1}{12}R^2+\frac{1}{2}b_q(s)t\right] \,. \tag{6.76}$$

The total contribution of the double rescattering is

$$A_{pp}^{(2)} = A_{pp}^{(2b)} + A_{pp}^{(2c)} \quad . \tag{6.77}$$

Thus the double rescattering gives the contribution to the total amplitude of the opposite sign to the impulse approximation. This contribution is small at small $|t|$ (since the value of R^2 is large), but its role increases with the $|t|$ growth. The reason is that the contribution of the diagram 6.18c decreases at large $|t|$ significantly slower than that of the impulse approximation (Fig. 6.18a).

Before considering a possible role of multiple collisions at large $|t|$ let us discuss one more question. When quark rescatterings are considered in the Regge-pole exchange theory, one should probably consider the diagram in Fig. 6.18d along with the diagrams 6.18b,c. In the framework of the Glauber rescattering theory the diagram 6.18d contributes to the impulse approximation: the rounded blob in Fig. 6.18d renormalizes the quark-quark amplitude of Fig. 6.18a. Thus in fact the question is whether the amplitude of Eq. (6.71) describes quark-quark interaction adequately or it should be parametrized in some more complicated form.

Now let us return to the hadron scattering at large $|t|$ considering the role of multiple quark scatterings in this process. Since $A_{pp}^{(2)}$ has the opposite sign to $A_{pp}^{(1)}$ and decreases slower with $|t|$ growth, at some $t = t_0$ these contributions cancel, and a dip occurs in $d\sigma_{e\ell}(pp)/dt$ at $t = t_0$. The cross section in the minimum does not vanish; the depth of the dip is determined by the value of the real part of the amplitude or by some other contributions (e.g. spin-flip terms). The dip in $d\sigma_{e\ell}(pp)/dt$ was observed in ISR experiments at $|t| \simeq 1.4$ $(GeV/c)^2$ and $\sqrt{s}$ = 20-60 GeV[49]. There exist a number of papers in which this t region of the pp elastic scattering was described in the framework of the quark model.[50-55] As an example, in Fig. 6.19 the data for $d\sigma/dt$ at $\sqrt{s}$ = 53 GeV are shown in comparison with the calculations of Ref. 55. (These calculations in Ref. 55 were made in a more refined manner, by including inelastic screening effects for quarks). A confusing point for all calculations of this type is that at higher energy $\sqrt{s}$ = 540 GeV there is no experimental evidence for the dip in $d\sigma_{e\ell}(\bar{p}p)/dt$. This forces us to consider in more detail the validity of the calculations with the dressed quarks at large $|t|$ values.

We may treat the dressed quarks as quasiparticles only for soft collisions, i.e. in the region of small $|t|$ determined by the value of

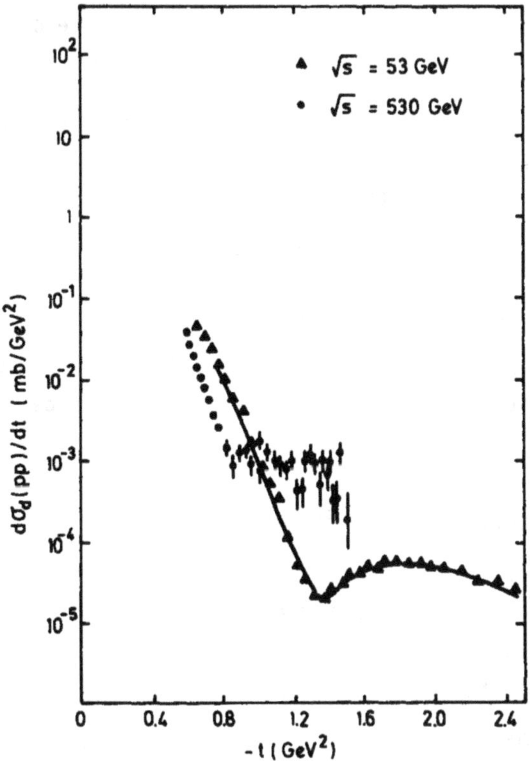

Fig. 6.19 Elastic differential cross section $d\sigma_{e\ell}/dt$ for pp collisions at $\sqrt{s}$ = 53 GeV and $\bar{p}p$ collisions at $\sqrt{s}$ = 540 GeV. The curve is the fit of Ref. 55 for $\sqrt{s}$ = 53 GeV.

r_q^2:

$$|t| \ll 6/r_q^2 \simeq (2\text{-}6)\ \text{GeV}^2 \qquad (6.78)$$

(assuming reasonable value of r_q^2 at ISR energies (1-3) GeV^{-2}). This estimate doesn't give an unambiguous answer in the considered case: if one should take $|t| \ll 6$ (GeV/c)^2, it is probably possible to consider the dressed quark scatterings at *$|t| \simeq 1.4$ (GeV/c)^2, but if $|t| \ll 2$* (GeV/c)^2 should be taken, such a consideration seems to be rather doubtful.

The estimate (6.78) is energy-dependent, the size of the dressed quark increases with the energy increase. The value of r_q^2 at SPS Collider energy exceeds that at ISR energy by the value $\Delta r_q^2 \sim 3$ GeV^2:

$$r_q^2(s = 3\cdot10^5 \text{ GeV}^2) - r_q^2(s = 10^3 \text{ GeV}^2) \simeq 3 \text{ GeV}^{-2} \quad . \tag{6.79}$$

Thus the estimate (6.78) for the allowed $|t|$ values at SPS energies is

$$|t| << 6/(r_q^2(\text{SPS})) \simeq 1\text{-}1.5 \ (\text{GeV/c})^2 \quad , \tag{6.80}$$

which is appreciably more restrictive than (6.78). Because of that the approach of the quasinuclear quark model may prove to be invalid at $|t| \sim 1.5$ (GeV/c)^2 at SPS energies.

6.6 Hadron Diffraction-Dissociation Processes

Hadron diffraction dissociation (DD) is an interesting class of processes, which lie on the boundary of binary processes and inelastic particle production. The diffraction-dissociation cross section (σ_{DD}) is a part of the total inelastic cross section ($\sigma_{in\ell}$).

Figures 6.20a,b present single diffraction dissociation in pp collisions: one of the colliding protons turns into some resonance or cluster of particles with a comparatively low effective mass M. The value of M^2 in DD processes must be essentially less than the total energy squared of the colliding particles $M^2/S << 1$. In other words, the $|X|$ value of the non-dissociated proton in the c.m. frame is to be close to 1: $1 - |X| << 1$. Another possible DD process is double diffraction dissociation of both colliding protons (Fig. 6.20c). The total cross section of the single DD in pp collisions (i.e. the sum of the processes Fig. 6.20a,b) is slightly less than the elastic cross section $\sigma_{DD}^{(1)}(pp)/\sigma_{e\ell}(pp) \simeq 0.75$. This estimate may vary by 10-15% depending on the M^2 values included into DD. The double DD cross section is not measured yet, it is known for some selected channels only.

The DD processes in the additive quark model are to a considerable extent determined by the elastic scattering of the dressed quarks[56,57]. Thus hadron DD processes are similar to those of the light nuclei. The

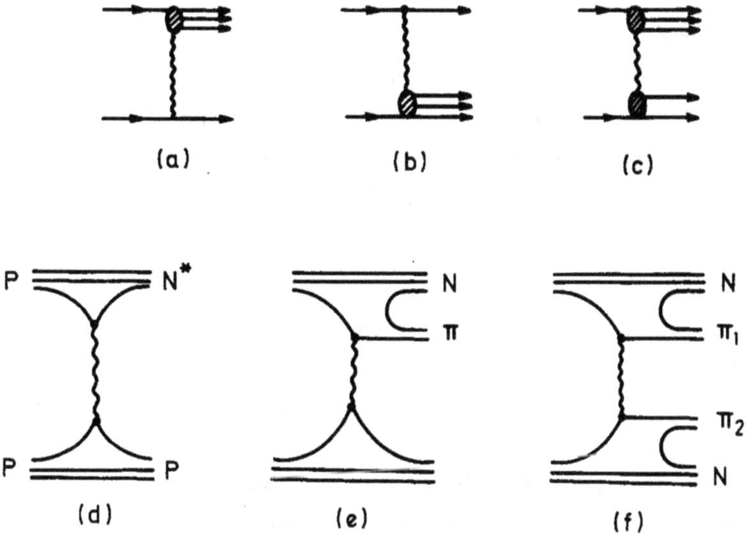

Fig. 6.20 (a-c) Processes of the single (a,b) and double (c) diffraction dissociation (DD). (d-f) DD processes in the quark model.

specific feature of the hadron diffraction dissociation is that it does not vanish in the forward direction, i.e. at $t = t_{min} \simeq -\frac{1}{S}(M_1^2 - m_1^2)(M_2^2 - m_2^2)$. Here m_1, m_2 are masses of the colliding hadrons, M_1, M_2 — those of the produced resonances or clusters (see Figs. 20a,b,c); since $t_{min} \to 0$ at $S \to \infty$, we may consider the forward DD collision as proceeding at approximately zero momentum transfer.

Let us note why the diffraction dissociation of the composite system must vanish at $t \to 0$. It is determined by the vertex "composite system → excited state" transition (see e.g. upper blocks in Figs. 6.20 d,e,f). In the impulse approximation this vertex is proportional to:

$$\int d^3r_1 d^3r_2 \psi_f^*(\vec{r}_1, \vec{r}_2 \ldots) e^{i\vec{q}\cdot\vec{r}_1} \psi_i(\vec{r}_1, \vec{r}_2, \ldots) \quad . \tag{6.81}$$

The wave functions of the ground and excited states of the composite system are orthogonal to each other. *Thus (6.81) vanishes for* $\vec{q} \to 0$ when $i \neq f$. The fact that the hadron DD amplitude does not vanish as $\vec{q} \to 0$ shows that in this case the ground states and the excited ones are not exactly orthogonal.

The problem of diffraction dissociation is studied in many papers (see e.g. Ref. 58 and references given therein). The possible mechanisms which may provide non-vanishing forward DD cross sections are usually considered in the framework of the eigenstate method[59,60] in the quark-parton picture of the hadron structure (see Refs. 61-67).

Non-orthogonality of the initial and final state wave functions in the additive quark model may reflect the existence of some inelastic transitions of the dressed quarks, which do not vanish at zero momentum transferred. These may be transitions into some excited states of quarks. We have already mentioned that the dressed quarks are rather complicated objects, which probably have no fixed masses. Instead, they are characterized by the distribution function $\rho(m_q^2)$. This distribution is characterized by the low-energy interaction of quark-gluon matter inside a hadron. High-energy interactions may disturb this distribution both in form and normalization. Simplifying the situation, one may argue that along with the elastic high energy quark-quark interaction

$$qq \to qq \tag{6.82}$$

there are inelastic collisions with quark excitation:

$$qq \to qq^* \tag{6.83}$$

The transitions (6.83) in this case provide the non-vanishing contribution to the DD cross section at $t \to 0$.

The nature of the interactions of the excited states q^* with the usual matter is an open question, and various speculations are possible here. Experimental data are usually described in the simple two-channel model. The quark scattering amplitude is in this case a 2×2 matrix

$$\begin{pmatrix} a_{11} & a_{12} \\ a_{21} & a_{22} \end{pmatrix} \tag{6.84}$$

where states 1 and 2 are qq and qq^*, respectively. Such a model gives a reasonable description of both elastic scattering and diffraction dissociation data in hadron-nucleon collisions. In Ref. 66 the value of the quark-nucleon cross section $qN \to qX$ was calculated in the framework of the two-channel model; we present here the explicit expression for this cross section (it includes both the elastic scattering and target diffraction dissociation), which will be useful in the description of the DD in hadron-nucleus collisions. This cross section is determined by the discontinuity of the diagrams in Fig. 6.21:

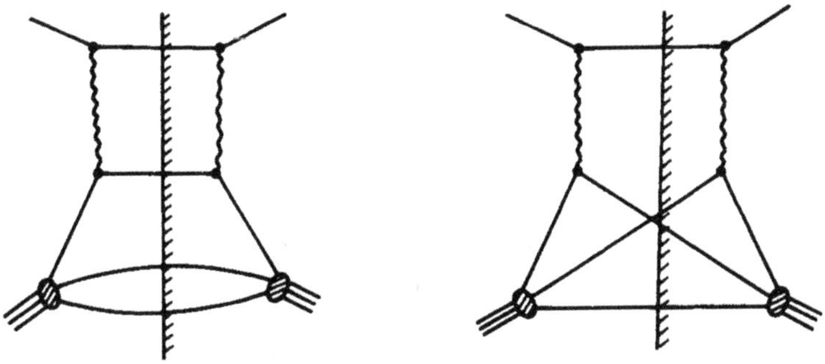

Fig. 6.21 Diagrams, the discontinuities of which determine the value of $d\sigma_{scat}/dt$ in the Eq. (6.85).

$$\frac{d\sigma_{scat}(qN)}{dk^2} = \frac{\sigma_{tot}^2(qN)}{16\pi}\left[\frac{1}{3}(1+g^2) + \frac{2}{3}\exp(-\frac{1}{2}R_N^2k^2)\right]\exp(-b_qk^2) \quad . \tag{6.85}$$

Here R_N is the nucleon radius,* b_q is the parameter of the quark-quark amplitude (see Eq. (6.71). At $p_{lab} \simeq 100\text{-}300\text{GeV/c}$ it is $b_q \simeq 1.5 \pm 0.5\ \text{GeV}^2$; $g = a_{21}/a_{11} \simeq 0.8\text{-}1.0$. In the following Sec. 8.3 when considering target diffraction dissociation in hadron-nucleus collisions we shall assume also that $a_{22}/a_{21} = g$. This is equivalent to the existence of some "passive" component in the nucleon, i.e. of some state of the quark-gluon matter which is devoid of high energy interaction. Such an assumption is common for all Refs. 61-67.

The diffraction dissociation is an interesting object for detailed studies, although it presents only a small part of the inelastic cross section. The attempts to describe DD processes quantitatively forced us to reject the simplest additive quark model and to introduce some excited states of quark-gluon matter like q^*. Explicit realizations of these excited states are by now rather naive, but they only form a first step in this direction. The future progress in this field may be related with the studies of the DD in hadron-nucleus collisions, where some information about the matrix elements a_{ik} in (6.84) may probably be obtained (see Sec. 8.3).

The elastic (or quasielastic) nature of the underlying subprocesses of the diffraction dissociation must reveal itself in the correlation phenomena[57)]. Let us consider, for example, the vertex of the diffraction dissociation of the proton (see Fig. 6.22): the fast proton with

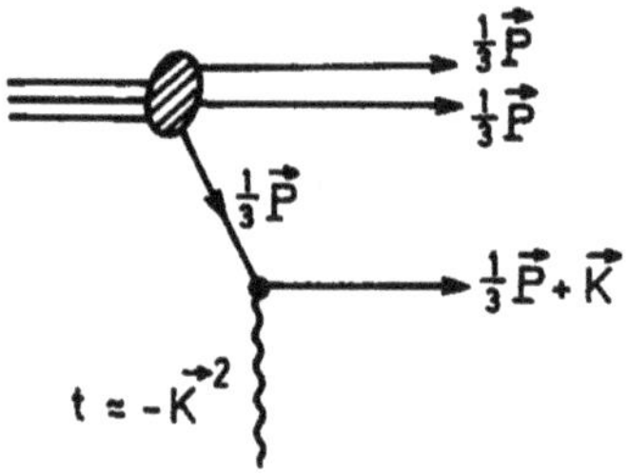

Fig. 6.22 Quark model diagram for proton disintegration in the DD process.

momentum P disintegrates as a result of the subprocess (6.82) or (6.83). If in the first approximation we neglect Fermi motion of the quarks inside the proton, the quark 4-momenta before and after the transition (6.82) are respectively $P_q = \left(\frac{1}{3P} + \frac{m_q^2}{\frac{2}{3}P}\ ,\ 0,\ 0,\ \frac{1}{3P}\right)$, $P'_q = \left(\frac{1}{3P} + \frac{m_q^2 + k_T^2}{\frac{2}{3}P}\ ,\ k_x,\ k_y,\ \frac{1}{3P}\right)$. Thus the following relation holds between the three-quark effective mass in the final state M^2 and the target momentum transferred:

$$M^2 = 9m_q^2 + 2|t| \quad . \tag{6.86}$$

A similar relation is valid also in the case of the transition (6.83) if

$9m_q^2$ is replaced by $6m_q^2 + 3m_{q*}^2$. Due to the Fermi motion of quarks the relation (6.86) is not exact; m_q^2 should be replaced by $m_q^2 + k_q^2$ and instead of an equality one obtains a correlation: the values of M^2 should be distributed in some way around the value $M^2 = \text{Const} + 2 \cdot |t|$. Besides that, hadronization processes may also diffuse the relation (6.86). Nevertheless, we may hope to observe maxima in the distributions $d^2\sigma/dM^2dt$ at fixed t values near the point $M^2 = \text{Const} + 2 \cdot |t|$. The spectra $d^2\sigma/dM^2dt$ at different $|t|$ and the maximum position versus $|t|$ for the process $pp \to pX$ are shown in Fig. 6.23; these data support

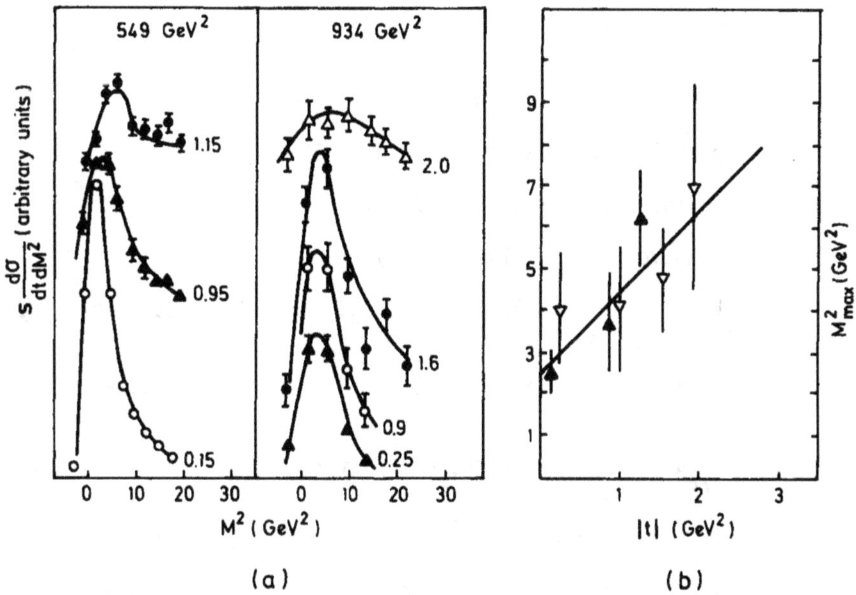

Fig. 6.23 (a) The differential cross section $\frac{d^2\sigma}{dtdM^2}$ of the DD process $pp \to pX$ at fixed $|t|$ values; the curves are eyefitted. (b) The position of the maximum in $\frac{d^2\sigma}{dtdM^2}$ from Fig. (a) versus $|t|$.

the existence of the correlation between M^2 and $|t|$ values.

It is possible to observe some other interesting correlations in diffraction-dissociation processes. For example, in the process of Fig. 6.20e the produced pion in the upper block must carry away almost the whole momentum transferred, while the final nucleon flies approximately in the same direction that the projectile, i.e. there is a correlation between the momenta transferred to the pion and to the target. In the double DD process of Fig. 6.20f the transverse momenta of the produced pions are correlated: $\vec{k}_T(\pi_1) \simeq -\vec{k}_T(\pi_2)$. The available experimental data for such processes are rather scarce and do not allow us to test these predictions.

References

1. G. Källen, Helv. Phys. Acta, 25, 417 (1952).
2. H. Lehman, Nuovo Cim. 11, 342 (1952).
3. Y. Nambu, Nuovo Cim. 16, 1064 (1957).
4. V.N. Gribov, ZhETF 34, 1310 (1958); 35, 416 (1958); Sov. Phys. JETP, 7, 903 (1958); 8, 287 (1959).
5. V.V. Anisovich, Yad. Fiz. 31, 1656 (1980); Sov. J. Nucl. Phys. 31, 859, (1980).
6. E.M. Levin and L.L. Frankfurt, Pis'ma v ZhETF 2, 105 (1965); JETP Lett. 2, 65 (1965).
7. H.J. Lipkin and F. Schneck, Phys. Rev. Lett. 16, 71 (1966).
8. A.S. Carroll et al., Phys. Lett. 61B, 303 (1976); 80B, 423 (1978).
9. J.P. Burq et al., Nucl. Phys. B217, 285 (1983).
10. J.P. Burq et al., CERN Courier 23, 137 (1983).
11. S.F. Biagi et al., Nucl. Phys. B186, 1 (1981).
12. J.J.J. Kokkedee and L. Van Hove, Nuovo Cim. 42, 711 (1966).
13. Ya. I. Azimov, E.M. Levin, M.G. Ryskin and V.A. Khoze, in Proc. of the IX LNPI Winter School on Physics, vol. 2, p. 5, Leningrad, 1974 (in Russian).

14. E.M. Levin and V.M. Shekhter, "Small Angle Elastic Scattering and Quark Model", Preprint LNPI-442, Leningrad, 1978.

15. A. Donnachie and P.V. Landshoff, Preprint M/C TH-83/13, DAMTP 83/9, 1983.

16. A. Schiz et al., Phys. Rev., D24, 26 (1981).

17. G.A. Jaroszkiewisz and P.V. Landshoff, Phys. Rev. D10, 170 (1974).

18. P.N. Kirk et al., Phys. Rev. D8, 63 (1973).

19. C.J. Bebek et al., Phys. Rev. D17, 1693 (1978).

20. J.P. Burq et al., Phys. Lett. 109B, 111, 124 (1982).

21. V.V. Anisovich, "Strong Interaction Processes at High Energies and Quark-Parton Model", in Proc. of the IX Winter School of Physics, vol. 3, p. 106, Leningrad, 1974 (in Russian).

22. V.V. Anisovich, M.N. Kobrinsky, J. Nyiri and Yu. M. Shabelski, "Quark Model and High Energy Collisions", Preprint KFKI 1982, 36, Budapest, 1982.

23. E.M. Levin and L.L. Frankfurt, Uspekhi Fiz. Nauk, 94, 243 (1968); Sov. Phys. Uspekhi, 11, 106 (1968).

24. C. Itzykson and M. Jacob, Saclay preprint, 1966.

25. P. Locher and H. Römer, Phys. Lett. 23, 496 (1966).

26. C. Becci and G. Morpurgo, Phys. Lett. 17, 352 (1965).

27. N. Cabibbo, L. Korwitz and Y. Ne'eman, Phys. Lett. 22, 336 (1966).

28. D.R.O. Morrison, Invited papers presented at the Conf. on High-Energy Two-Body Reactions, Stony Brook, April 1965.

29. A. Bialas and K. Zalewski, Nucl. Phys. B6, 449, 465, 478 (1968).

30. S. Meskov, G.A. Snow and G.B. Yath, Phys. Rev. Lett. 12, 87 (1964).

31. B. Gorczyca, Acta Phys. Pol. 33, 471 (1968).

32. A. Bialas, A. Guba and B. Muryn, Acta Phys. Pol. 32, 443 (1967).

33. G. Alexander, H.J. Lipkin and F. Scheck, Phys. Rev. Lett. 17, 412 (1966).

34. ACNO Collab., J.C. Kluyver et al., Nucl. Phys. B140, 141 (1978).

35. E. Hirsch, V. Karshon and H.J. Lipkin, Phys. Lett. 36B, 385 (1971).

36. ACNO Collab., M. Zralek et al., "Test of an Additive Diquark Spectator Model for Meson-Baryon Quasi-Two-Body Reactions", Nijmegen preprint HEN-164 (1979).

37. R.T. Van de Valle, Lectures presented at the 1979 Int. School on Subnuclear Physics, Erice, 1979.

38. V. Barger and L. Durand, Phys. Rev. 156, 1525 (1967).

39. A. Kotánski and K. Zalewski, Nucl. Phys. B4, 559 (1968).

40. J.D. Jackson, Phys. Rev. Lett. 15, 990 (1965).

41. ACNO Collab., M. Aguilar-Benitez et al., Nucl. Phys. B124, 89 (1977).

42. A. Honecker et al., Paper submitted to the EPS Int. Conf. on High Energy Physics, Palermo 1965; Quoted by P. Schremp, B. Schremp in Rapporteur Talk; CERN preprint TH-2054 (1975).

43. A.B. Wicklund et al., Paper presented to the XVII Int. Conf. on High Energy Physics, London, 1974. Quoted by R.C. Field in Proc. of the XVII Int. Conf. on High Energy Physics, London, 1974, part I, p. 185.

44. A.M. Wetherell, Inv. Talk at the EPS Int. Conf. on High Energy Physics, Palermo, 1975.

45. P. Kolár, B.Z. Kopeliovich and L.I. Lapidus, Czech. J. Phys. B26, 1294 (1976).

46. G. Altarelli, Phys. Rev. D1, 3200 (1970).

47. P.D. Collins and A.D.M. Wright, "Reggeon Exchange in the Quark Model", Univ. of Durham preprint, 1978.

48. Particle Data Group, Rev. Mod. Phys. 56, No. 2, Part 2, 1984.

49. E. Nagy et al., Nucl. Phys. B150, 221 (1979).

50. D.R. Harrington and A. Pagmanenta, Phys. Rev. 173, 1599 (1968).

51. E.M. Levin and V.M. Shekhter, "Quark Rescattering and Elastic Scattering Processes", in Proc. of the IX INPI Winter School on Physics, vol. 3, p. 28, Leningrad, 1974 (in Russian).

52. S. Wakaizumi and M. Tanimoto, Phys. Lett. 70B, 55 (1977).

53. A. Bialas et al., Acta Phys. Pol. B8, 855 (1977).

54. S.V. Goloskokov et al., Yad. Fiz. 33, 1349 (1981); Sov. J. Nucl. Phys., 33, 722 (1981).

55. E.M. Levin, Yu. M. Shabelski, V.M. Shekhter and A.N. Solomin, "Elastic Scattering with Large Transverse Momenta and Quark Model", Preprint LNPI-444 (1978).

56. V.A. Tsarev, Yad. Fiz. 28, 1054 (1978); Sov. J. Nucl. Phys. 28, 541 (1978).

57. V.V. Anisovich, E.M. Levin and M.G. Ryskin, Yad. Fiz., 29, 1311 (1979); Sov. J. Nucl. Phys. 29, 674 (1979).

58. K. Goulianos, Phys. Rep. 101C, 169 (1983).

59. E.L. Feinberg and I. Ya. Pomeranchuk, Doklady AN SSSR 93, 439 (1953) (in Russian).

60. M.L. Good and W.D. Walker, Phys. Rev. 120, 1857 (1960).

61. P. Grassberger, Nucl. Phys. B125, 83 (1977).

62. H.I. Miettinen and J. Pumplin, Phys. Rev. D18, 1696 (1978).

63. B.Z. Kopeliovich and L.I. Lapidus, Pis'ma v ZhETF 28, 664 (1978); JETP Lett. 28, 614 (1978).

64. L. Vegh, B.Z. Kopeliovich and L.I. Lapidus, Yad. Fiz. 35, 1514 (1982); Sov. J. Nucl. Phys. 35, 886 (1982).

65. Yu. M. Shabelski, Nucl. Phys. B132, 491 (1978).

66. V.M. Braun and Yu. M. Shabelski, Yad. Fiz. 34, 503 (1981); 35, 1247 (1982); 37, 1011 (1983); Sov. J. Nucl. Phys. 34, 280 (1981); 35, 731 (1982); 37, 599 (1983).

67. N.N. Nikolaev, ZhETF 81, 814 (1981); Sov. Phys. JETP, 54, 434 (1981).

68. M.G. Albrow et al., Nucl. Phys. B108, 1 (1976).

Chapter 7

MULTIPARTICLE PRODUCTION IN THE QUARK MODEL

7.1 Space-Time Picture and Spectator Mechanism

From the point of view of the quark model, any multihadron production process is a process of production of quarks and of their subsequent hadronization[1]. The quasinuclear structure of hadrons then leads to a specific space-time picture of the multiparticle production processes. It is essentially based on the notion of formation time of the secondary particles which was introduced into particle physics rather long ago[2,3].

Let us consider for definiteness the production of secondaries in deep inelastic lepton-induced collisions with sufficiently large Q^2 and W^2. In this process a virtual intermediate boson (or photon) strikes one of the quark-partons of the nucleon target. The removed parton flies in the direction of the boson (in the frame where $Q_0 = 0$), while the rest of the nucleon remains with unchanged momentum. In the quasinuclear quark model the remainder are the "wounded" dressed quark, from which a parton is removed, and two dressed quarks, which practically did not take part in the interaction (quark-spectators).

The process of hadronization of the flying quark-parton may be divided into three stages. At the first a jet of QCD quarks and gluons is produced. These interact with each other and form a jet of the dressed quarks. At the last stage the dressed quarks join together to form the experimentally observable hadrons.

In the framework of the quasinuclear model these stages of the multiple production process can be shown to be sufficiently well separated in time. The formation time of the dressed quark in its rest frame has to be determined by the quark radius r_q. If this quark is moving with a large momentum p, the corresponding time will be of the order of $r_q p/m$ (where $m = 0.3 - 0.4$ GeV in the mass of the dressed quark). This is obviously, the time of the existence of quark-partons and gluons. (Slow quark-partons and gluons cluster into dressed quarks earlier than fast ones.)

The time of existence of the cloud of the dressed quarks is determined by the time of hadron formation. As for any composite system in its rest frame this time is of the order of the constituent "collision" time, i.e. of the order $\tau_0 \sim R/v$. Here R is the radius of the system, v the average velocity of the constituents, which is of the order of $v \sim 1/mR$. Thus, $\tau_0 \sim mR_h^2$, and fast hadrons are formed from the dressed quarks during the time $\tau \sim pR_h^2$. Since $r_q/R_h \ll 1$ (see Secs. 1.2 and 6.2), the time of existence of the dressed quarks is essentially larger than their formation time: numerically the ratio of these two values is of the order of 3 - 6. Hereafter we shall consider this factor to be parametrically large.

A crucial point of our consideration is the assumption that the dressed quarks exist long enough to realize the possibility of interacting with each other repeatedly producing some additional pairs of dressed quarks. Thus it is quite natural to assume that the resulting cloud of the dressed quarks is just "prepared" for hadronization. This means that every quark has suitable colour and flavour partners for hadronization, which are sufficiently close to it in the rapidity scale. Indeed, if there is a given quark for which partners suitable to form a hadron are too far in the rapidity scale, some additional quark-antiquark pairs will be produced in such a manner, that the hadronization of all quarks would become possible without any suppression. The same is true for colour excitations: if a quark has no suitable partners to form a white hadron state, then new quarks are produced, and the wrong colour is transferred to another rapidity region. In this case again the hadronization of quarks, i.e. their transition into a white hadron state, takes place without any suppressing factors, and does not depend, for instance, on the probability of finding suitable partners for the quark. We call this property soft hadronization and soft colour neutralization of the dressed quarks.

Let us return to the multiparticle production in the deep inelastic lepton-nucleon scattering process. In the rest frame of the nucleon target the spectators are almost at rest, and therefore they are the first to form hadrons. The wounded quark and the slow component of quark-partons and gluons give rise to new pairs, of suitable colour; after this the quark-spectators q_j and q_k transform into a baryon B_{jk} (Fig. 7.1a), into a meson M_j and a baryon B_k (Fig. 7.1b), or into two mesons M_j and M_k (Fig. 7.1c). The newly produced dressed quarks on the average do not carry any definite quantum numbers, we will call them constituent sea quarks (and antiquarks) and denote them q and $\bar{q}$. If the SU(6) symmetry were not broken, q could be, with equal probabilities, any of the flavours with arbitrary spin projections ($s_z = \pm 1/2$).

After the hadronization of the quarks-spectators, their colour quantum numbers appear in the faster part of the jet. Furthermore, for a time of the order of pR_h^2 the faster constituent sea quarks form hadrons, while the quantum numbers not suitable for their hadronization move along the rapidity axis. The last quarks which form hadrons are those in the current fragmentation region. Since the process of hadronization goes on successively (upwards in Fig. 7.1), in the region of current fragmentation occur automatically those quantum numbers, which make possible the production of white hadron states. In this region the constituent sea quarks join the constituent quark q_i which carries the quantum numbers of the removed quark-parton. If at the same time the baryon charge is not transferred, a meson $M_i = q_i\bar{q}$ (Fig. 7.1a, b) will be produced; if such a transfer takes place, a baryon $B_i = q_iqq$ will be formed (Fig. 7.1c).

The hadronization process in e^+e^- annihilation is analogous to that in the deep inelastic lepton-nucleon scattering. The virtual

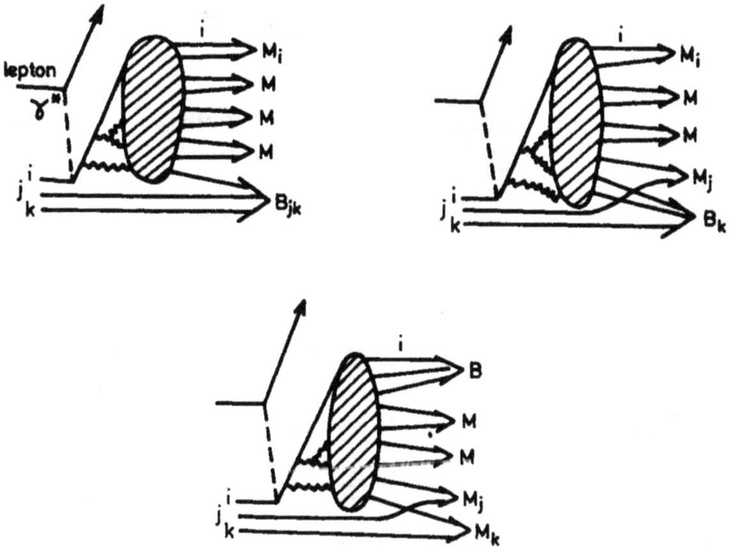

Fig. 7.1 Hadron production in deep inelastic scattering. QCD quarks and gluons are shown by thin lines, while constituent quarks by dense lines. In the traget nucleon rest frame the hadrons B_{ij}, B_k, M_k, M_j have small momenta; the hadrons B_i and M_i are the fastest ones.

γ-quantum produces a pair of point-like partons (quark-antiquark pair) which run in opposite directions. This pair induces new quark-partons and gluons, which, at sufficiently large distances, become dressed quarks. The constituent quarks then, joining each other, form hadrons. The particles which are produced this way form two jets of hadrons flying in opposite directions (if, of course, at the first stage hard gluons are not produced, which would then lead to additional jets).

The hadronization picture in hadron-hadron collisions is slightly more complicated. As an example let us consider the collision process of two nucleons. In the impulse approximation only one quark of the incident nucleon interacts; the two others remain spectators (see Fig. 7.2).

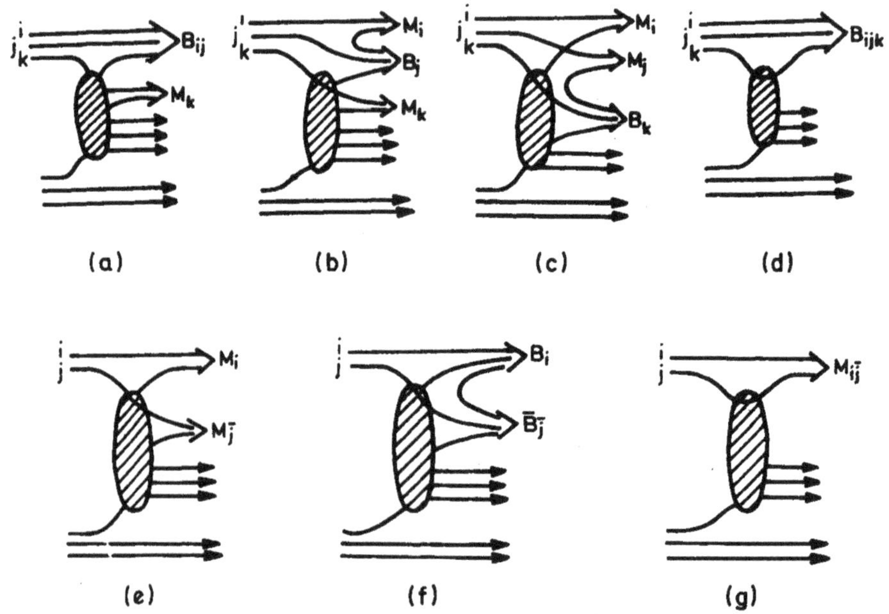

Fig. 7.2 Impulse approximation diagrams for inelastic nucleon-nucleon (a-d) and meson-nucleon (e-g) collisions.

The constituent quarks in the quasinuclear quark model are assumed to be some coherent clusters of QCD quarks and gluons. As a result of inelastic collision between the target and projectile quarks, the coherence of these clusters is broken and they disintegrate into quark-partons and gluons. The process of hadronization of the latters is the same as that in hard processes; a cloud of dressed quarks is produced first of all, and then they form the observable hadrons.

The notion of a quark-spectator in the quasinuclear quark model is, of course, somewhat different from the notion of a nucleon-spectator in the case of nuclear collisions. The hadronization process of the quark-spectators means inevitably that the spectators have to interact with each other and with the other quarks and gluons. In this sense, strictly speaking, these quarks are spectators rather conditionally.

The quark-spectator and the interacting quark differ by the value of the momentum transferred. In the quasinuclear model it is supposed that the interaction process changes the momentum of the quark-spectator only slightly.

As was already mentioned the quasinuclear quark model assumes that the final momentum distribution of the dressed quarks in the multiple production process (i.e. the momentum distribution just before the hadronization) is established through a large number of the repeated collisions of quarks, accompanied by the quark-antiquark pair production. Thus it seems to be quite probable that the final distribution of the dressed quarks (and that of the produced hadrons) over the carried momentum fraction of the jet does not depend on the origin of the process. In other words, the x-distributions of the secondaries in hard and soft multiple production collisions are some universal functions of x. This is, of course, not so for the quark-spectators: their x-distributions are determined by the wave functions of the colliding hadrons (see Sec. 3.2).

The assumption of soft hadronization leads to the prediction that x-distributions of the quark-spectators and that of the hadrons, formed by these quarks, are very close to each other. Indeed, quark-spectators produce their own partners for hadronization; thus the momentum loss, caused by quark-antiquark pair production, is probably compensated for when the newly produced quarks are joined by their "parents" to form a hadron. Such a spectator mechanism of particle production provides a possibility to calculate the inclusive distribution of fast secondaries. We shall consider it in more details in the following section.

7.2 Secondary Particles and Their Momentum Distributions

We consider once more the process of inelastic baryon-baryon collision, which is described by the diagrams of Fig. 7.2a-c in the impulse approximation. Let the initial baryon B_{ijk} consist of three dressed quarks q_i, q_j, q_k; quark q_k in Fig. 7.2 interacts with the target, while quarks q_i and q_j play the role of spectators. As a result of the inelastic collision of the quark q_k with the target, a large number of quark-antiquark pairs is produced. Among the new quarks there is one, q'_k, which carries the quantum number of the initial quark q_k; we will call q'_k a tagged quark. The spectators q_i, q_j and the tagged quark q'_k form hadrons, joining the newly produced (dressed) sea quarks. It follows from the assumption of the soft hadronization and the soft colour neutralization, that no suppression factors appear in the processes of Figs. 7.2a-c. If so, the momenta of secondary fragmentational hadrons would be determined mainly by the momenta of quark-spectators. Hence, the baryons which contain two spectators, q_i and q_j, carry about $x \sim 2/3$. The quark content of these baryons is $q_i q_j q$, where q runs over all possible flavour and spin states; we shall use the notation $B_{ij} = q_i q_j q$. In fact, B_{ij} is some ensemble of the real baryon states; its hadron content will be considered later on. The baryons $B_i = q_i qq$ and B_j and the mesons $M_i = q_i \bar{q}$, M_j, containing one of the spectators (see Fig. 7.2b, c) are produced in the region $x \sim 1/3$. Baryons and mesons B_k and M_k which are formed by sea quarks and antiquarks and the tagged quark q'_k, carry, as a rule, $x \sim 1/6 - 1/12$. Finally, the slowest part of the spectrum is formed by hadrons which consist only of sea quarks. These sea hadrons (which don't remember the quatum numbers of the incident particles) dominate the region $x < 0.1 - 0.05$. In true inelastic collisions the initial baryon B_{ijk} turns into different secondary particles:

$$B_{ijk} \to \tilde{\xi}_1 B_{ij} + \tilde{\xi}_2 (B_i + B_j) + \tilde{\xi}_3 (M_i + M_j) + \tilde{\xi}_4 B_k + \tilde{\xi}_5 M_k + (\text{sea hadrons}) \ . \tag{7.1}$$

The coefficients stand here for the probabilities of the production of the corresponding ensembles. They are bound by conditions (sum rules) which express the conservation of the total probabilities in the hadronization process for both the quark-spectators $(\tilde{\xi}_1 + \tilde{\xi}_2 + \tilde{\xi}_3 = 1)$ and the tagged quark $(\tilde{\xi}_4 + \tilde{\xi}_5 = 1)$, and the conservation of baryon number $(\tilde{\xi}_1 + 2\tilde{\xi}_2 + \tilde{\xi}_4 = 1)$.

Inelastic final states in hadron-hadron collisions may be produced via elastic quark scattering, which gives rise to the diffraction dissociation process. When the process of diffraction dissociation of the target $B_{ijk}N \to B_{ijk}X$ (see Fig. 7.2d) takes place, the secondary baryon of the same kind as the initial one is produced with $x \gtrsim 0.9$. The probability Δ of this process is about 10-12%. Particles, produced in the beam diffraction-dissociation processes, may be effectively included in the ensembles B_{ij}, B_i, M_i, etc.

As a result, the inelastic baryon-nucleon collision in a general case may be written the form:

$$\begin{aligned}
B_{ijk} \xrightarrow{N} \; & \Delta B_{ijk} && x \gtrsim 0.9 \\
& + (1-\Delta)\xi_1 B_{ij} && x \sim 2/3 \\
& + (1-\Delta)[\xi_2(B_i + B_j) + (1-\xi_1-\xi_2)(M_i + M_j)] && x \sim 1/3 \\
& + (1-\Delta)[(1-\xi_1-2\xi_2)B_k + (\xi_1+2\xi_2)M_k)] && x \sim 1/6 - 1/12 \\
& + (1-\Delta)N(s)[d \cdot M + B + \bar{B}] && x \lesssim 0.1 \qquad (7.2)
\end{aligned}$$

(here ξ_i's are redetermined coefficients which are expressed by $\tilde{\xi}_i$). The last term in (7.2) stands for the sea hadron contribution. The value $N(s)$ characterizes the total multiplicity of the sea secondaries; it depends on the total energy of the collision. The relative fraction of sea mesons is described by the value of d. Hereafter we assume that every ensemble B_{ij}, B_i, M_i, B, etc. is normalized to the unit average multiplicity, so that the coefficients in the right-hand side of (7.2) should be interpreted as average multiplicities of the corresponding secondaries.

Now we consider the case of a meson beam; each meson consists of a quark q_i and an antiquark $\bar{q}_j$. The quark q_i interacts with the nucleon target with the probability $\gamma = \sigma(q_iN)/[\sigma(q_iN) + \sigma(\bar{q}_jN)]$ where $\sigma(qN)$ (or $\sigma(\bar{q}N)$) is the total inelastic cross section of quark-nucleon (antiquark-nucleon) interaction; the probability of antiquark-nucleon interaction for a meson beam is then $1-\gamma$. The second constituent in both cases remains a spectator. Taking $\sigma(uN) = \sigma(dN) = \sigma(\bar{d}N) = 3\sigma(sN)/2$

$= 3\sigma(\bar{s}N)/2$, we have for the pion beam $\gamma = 1/2$, while for K^+, $\gamma = 3/5$ and for K^-, $\gamma = 2/5$. Four types of secondaries may be produced (see Figs. 7.2e-g): diffractive $(x \gtrsim 0.9)$, spectator $(x \sim 1/2)$, tagged $(x \sim 1/4 - 1/8)$ and sea $(x \lesssim 0.1)$ ones. Their average multiplicities depend on two phenomenological parameters: $\delta \simeq 0.08 - 0.1$ (the probability of the inelastic diffraction $M_{ij}N \to M_{ij}X$) and η. The value of η describes the ratio of secondary spectator (or tagged) mesons and baryons:

$$\begin{aligned}
M_{ij} \xrightarrow{N} \delta M_{ij} & \qquad (x \gtrsim 0.9) \\
+ (1-\delta)[\eta(\gamma\bar{B}_j + (1-\gamma)B_i) + (1-\eta)(\gamma M_j + (1-\gamma)M_i)] & \qquad (x \sim 1/2) \\
+ (1-\delta)[\eta((1-\gamma)\bar{B}_j + \gamma B_i) + (1-\eta)((1-\gamma)M_j + \gamma M_i)] & \qquad (x \sim 1/4\text{-}1/8) \\
+ (1-\delta)N(s) \cdot (d \cdot M + B + \bar{B}) & \qquad (x \lesssim 0.1)
\end{aligned} \tag{7.3}$$

The hypothesis of soft hadronization allows one to obtain some explicit expressions for x-distributions of the secondary particles. Let us consider this in more detail.

As it was mentioned in the previous section, x-distributions of the secondary hadrons, formed by quark-spectators, are determined mainly by the momentum fraction carried by spectators. The latter is determined by the wave function of the projectile hadron (see Sec. 3.2). Let us consider as an example the simple form of the nucleon (N) wave function with one parameter:

$$\Psi_N \sim \exp(-a_N s) \quad , \tag{7.4}$$

where $s = (k_1 + k_2 + k_3)^2$ is the sum squared of the 4-momenta of quarks. In the infinite momentum frame $s = \frac{m_{T1}^2}{x_1} + \frac{m_{T2}^2}{x_2} + \frac{m_{T3}^2}{x_3}$ where x_i is the fraction of the longitudinal hadron momentum carried by the quark; $m_T^2 = m_q^2 + k_T^2$, m_q is the mass of the dressed quark. The x-distribution of the dressed quark inside the nucleon is then

$$Q_N(x) = \int d\Omega_3 |\Psi_N|^2 \delta(x_1 - x) \quad , \quad \int_0^1 dx\, Q_N(x) = 1 \quad . \tag{7.5}$$

Integrating over the three-body "phase space"

$$d\Omega_3 = \frac{dx_1 dx_2 dx_3}{x_1 x_2 x_3} d^2k_{T1} d^2k_{T2} d^2k_{T3} \delta^2(\vec{k}_{T1} + \vec{k}_{T2} + \vec{k}_{T3}) \delta(x_1 + x_2 + x_3 - 1) \ , \tag{7.6}$$

we obtain the quark distribution in the nucleon in the following approximate form:

$$Q_N(x) \simeq C_N (1-x)^{3/2} \exp\left[-\frac{2a_N m_q^2 (1+3x)}{x(1-x)} \right] \quad , \tag{7.7}$$

where the normalization coefficient C_N has to be found from the condition (7.5). The value of the parameter a_N may be determined from the data on nucleon form factor; it turns to be about[4)] 2.5 GeV^{-2}. Thus, the obtained expression (7.7) gives the distribution of B_i, B_j and M_i, M_j from (7.2), i.e. the inclusive cross section of production of the hadrons, containing one quark-spectator of the beam nucleon N_{ijk}:

$$\frac{1}{\sigma_{inel}} \cdot \frac{d\sigma}{dx} (N_{ijk} N \to B_j X) = \frac{1}{\sigma_{inel}} \cdot \frac{d\sigma}{dx} (N_{ijk} N \to B_i X) = (1-\Delta)\xi_2 Q_N(x) \tag{7.8}$$

and

$$\frac{1}{\sigma_{inel}} \cdot \frac{d\sigma}{dx} (N_{ijk} N \to M_i X) = \frac{1}{\sigma_{inel}} \cdot \frac{d\sigma}{dx} (N_{ijk} N \to M_j X)$$

$$= (1-\Delta)(1-\xi_1-\xi_2) Q_N(x) \quad . \tag{7.9}$$

A similar expression may be obtained for the distribution of the baryons, containing a diquark-spectator[a]:

$$\frac{1}{\sigma_{inel}} \cdot \frac{d\sigma}{dx}\,(N_{ijk}N \to B_{ij}X) = (1-\Delta)\xi_1 D_N(x) \tag{7.10}$$

where

$$D_N(x) = \int d\Omega_3 |\Psi_N|^2 \delta(x_1 + x_2 - x) \simeq C_N x^{3/2} \exp\left[-\frac{2a_N m_q^2 (4-3x)}{x(1-x)}\right] \quad . \tag{7.11}$$

The distribution of the secondaries formed by quark-spectators in the case of meson beams may be found from the wave function

$$\psi_M \sim \exp(-a_M s) \quad , \qquad s = (k_1 + k_2)^2 \quad . \tag{7.12}$$

As a result we get:

$$\frac{1}{\sigma_{inel}} \cdot \frac{d\sigma}{dx}\,(M_{ij}N \to M_i X) = (1-\delta)(1-\eta)(1-\gamma)Q_M(x) \quad ,$$

$$\frac{1}{\sigma_{inel}} \cdot \frac{d\sigma}{dx}\,(M_{ij}N \to M_j X) = (1-\delta)(1-\eta)\gamma Q_M(x) \quad ,$$

$$\frac{1}{\sigma_{inel}} \cdot \frac{d\sigma}{dx}\,(M_{ij}N \to B_i X) = (1-\delta)\eta(1-\gamma)Q_M(x) \quad ,$$

$$\frac{1}{\sigma_{inel}} \cdot \frac{d\sigma}{dx}\,(M_{ij}N \to \bar{B}_j X) = (1-\delta)\eta\gamma Q_M(x) \quad , \tag{7.13}$$

---

[a] We call a diquark a pair of dressed quarks with small relative momentum inside the nucleon. It is not the same as the notion of diquark which is usually considered in hard collisions.

where

$$Q_M(x) \simeq C_M \exp\left[-\frac{2a_M m_q^2(1-x+\beta^2 x)}{x(1-x)}\right] \quad , \tag{7.14}$$

$$\int_0^1 dx Q_M(x) = 1 \quad . \tag{7.15}$$

Here C_M is determined by the normalization condition (7.15); β is the mass ratio of the quark-spectator and the interacting quark. β differs from unity for kaon beams; we take $m_s/m_u = m_s/m_d = 1.5$. The values of a_M, as determined from the data on meson radii, appear to be[4] $a_\pi \simeq 3.3\ \mathrm{GeV}^{-2}$, $a_k \simeq 2.8\ \mathrm{GeV}^{-2}$. The present level of our understanding of the hadron structure does not allow one to calculate the x-distributions of secondaries in hard processes like e^+e^- annihilation in the framework of the quasinuclear quark model. These distributions are to be determined from the fit of the experimental data or with the help of some additional model assumptions. However, once this distribution is given, one may calculate the distribution of the secondaries in the central region of hadron-hadron collisions (i.e. that of tagged and sea hadrons) — both distributions are described by universal functions of the momentum fraction of the jet carried by the secondary hadron.

So, if we introduce the distribution function $F(x)$ of the tagged hadrons in the quark jet (i.e. of those which carry the flavours of the quark-parton, initiating the jet), the distribution function $V_t^N(x)$ of the tagged hadrons in NN collisions may be written as

$$V_t^N(x) = \int_x^1 \frac{dx_1}{x_1} Q_N(x_1) F\left(\frac{x}{x_1}\right) \quad ; \quad \int_0^1 dx V_t^N(x) = 1 \quad . \tag{7.16}$$

Similarly, the distribution of sea hadrons in NN collisions may be written as

$$V_{sea}^N(x) = \int_x^1 \frac{dx_1}{x_1} Q_N(x_1) \Phi\left(\frac{x}{x_1}\right) \tag{7.17}$$

where $\Phi(x)$ is the sea hadron distribution in the quark jet. It is normalized to the sea particle multiplicity; using the notations of Eq. (7.2) for sea hadron contribution, we may write

$$\int_0^1 dx\Phi(x) = N(s) \quad . \tag{7.18}$$

A convenient variable for the particle spectra in the central region is not x, but y (rapidity). Integration in (7.18) assumes the integral over the whole rapidity space. The connection between x and y is given in Sec. 2.2. The distributions of the tagged mesons M_k, tagged baryons B_k and sea hadrons M, B and $\bar{B}$ may be expressed through $V_t^N(x)$ and $V_{sea}^N(x)$ with the help of Eq. (7.2) in a form similar to (7.13):

$$\frac{1}{\sigma_{inel}} \cdot \frac{d\sigma}{dx} (N_{ijk}N \rightarrow B_K X) = (1-\Delta)(1-\xi_1-2\xi_2)V_t^N(x) \quad ,$$

$$\frac{1}{\sigma_{inel}} \cdot \frac{d\sigma}{dx} (N_{ijk}N \rightarrow M_K X) = (1-\Delta)(\xi_1+2\xi_2)V_t^N(x) \quad ,$$

$$\frac{1}{\sigma_{inel}} \cdot \frac{d\sigma}{dx} (N_{ijk}N \rightarrow MX) = (1-\Delta)dV_{sea}^N(x) \quad ,$$

$$\frac{1}{\sigma_{inel}} \cdot \frac{d\sigma}{dx} (N_{ijk}N \rightarrow BX) = (1-\Delta) \frac{1}{\sigma_{inel}} \cdot \frac{d\sigma}{dx} (N_{ijk}N \rightarrow \bar{B}X) = (1-\Delta)V_{sea}^N(x) \ . \tag{7.19}$$

In the case of a meson beam $Q_N(x)$ in Eqs. (7.16), (7.17) should be substituted by $Q_M(x)$.

The distribution of the secondaries, produced in the target diffraction-dissociation processes, $V_{dif}(x)$, in our approach can be determined on purely phenomenological grounds. For example, when only the three-Pomeron contribution is taken into account (see Sec. 2.3), $V_{dif}(x)$ should be proportional to $(1-x)^{-1}$.

As to the beam diffraction dissociation, we shall assume that x-distributions for all types of secondaries in these processes are the same as those for true inelastic processes. Of course, it is a rather rough estimation, and these processes in fact should be considered separately; however, their contribution to the total spectra is of the order of 10-12% only, thus the resulting inaccuracy is not large.

7.3 Quark Statistics. Hadron Content of the Secondaries

7.3.1 *Statistical rules*

In the previous section it was shown, that the assumption of the quasinuclear hadron structure gives the possibility of expressing the inclusive distribution of secondaries in any multiple production process through several phenomenological parameters and two unknown functions $F(x)$ and $\Phi(x)$. However, the relations (7.8), (7.9), (7.10), (7.13), etc. refer to some "average" baryons or mesons, and up to now we have not established any correspondence between symbols B_{ij}, B_i, M_i, and so on and real, observable hadrons. In the following, we will see that in the framework of the quasinuclear quark model such a correspondence is established rather unambiguously with the help of the rules of quark statistics[5-7].

Let us consider the amplitude for secondary hadron production in quark jets. Assume that a tagged quark q_i picks up quarks and antiquarks from the sea and forms a meson $M_i = q_i\bar{q}$ (Fig. 7.3a) or a baryon $B_i = q_iqq$ (Fig. 7.3b). The inclusive cross sections of M_i and B_i production are determined by graphs of the Mueller-Kancheli type (see Figs. 7.3c, d and Appendix B) and are equal to

$$\frac{d\sigma_h(x, \vec{k}_T)}{dxd\vec{k}_T} = \int \prod_{i=1}^{n} \left(\frac{dx_i'}{x_i'} d^2k_{Ti}' \frac{dx_i''}{x_i''} d^2k_{Ti}'' \right) \delta(\Sigma x_i' - x)\delta(\Sigma x_i'' - x)$$

$$\cdot\ \delta^2(\Sigma\vec{k}_{Ti}' - \vec{k}_T)\delta^2(\Sigma\vec{k}_{Ti}'' - \vec{k}_T)\Psi_h^*(s_n')\Psi_h(s_n'')$$

$$\cdot\ F_n(x_1', \vec{k}_{T1}',\ldots,x_n', \vec{k}_{Tn}' \mid x_1'', \vec{k}_{T1}'',\ldots,x_n'', \vec{k}_{Tn}'') \quad . \tag{7.20}$$

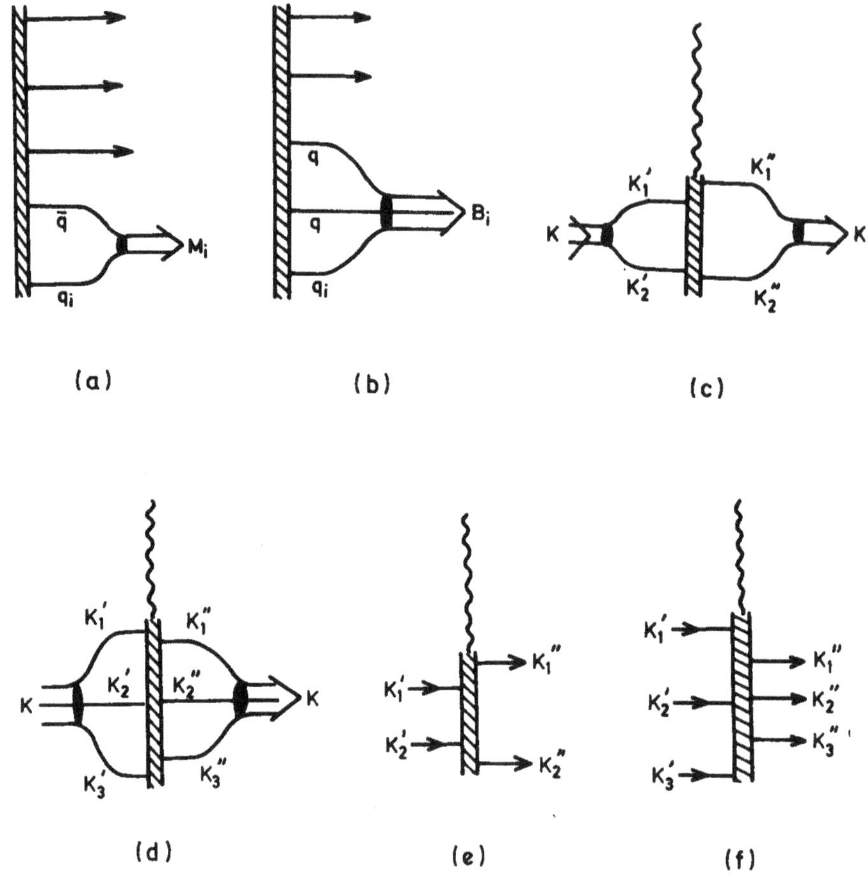

Fig. 7.3 Amplitudes (a, b) and inclusive cross sections (c, d) of meson M_i and baryon B_i production from the cloud of dressed quarks. Diagrams (e, f) represent the distribution functions F_n (see text, Eq. (7.20)).

Here $h = M_i$ or B_i, F_n are the distribution functions of a quark and an antiquark $(n = 2)$ or the three quarks $(n = 3)$, which correspond to Fig. 7.3e, f. These functions depend on the fractions of the total momentum (x_i', x_i'') of the jet carried away by the quarks and on the transverse momenta $(\vec{k}_{Ti}', \vec{k}_{Ti}'')$. We denote the wave functions of the meson or baryon by $\Psi_h(s)$. As above, we will suppose that the wave functions depend only on the invariant mass squared $s_n = (k_1 + \ldots + k_n)^2$ of the quarks (the consideration of more complicated energy dependences, possible in the systems of three quarks, would not change the result). The decrease of $\Psi_h(s)$ with the increase of s_n is characterized by the radius R_h^2 of the composite system, which is large compared to the characteristic quark sizes. Hence, all the values of x_i' and x_i'' have to be close to each other as well as those of $\vec{k}_{Ti}'$ and $\vec{k}_{Ti}''$. In other words, hadrons are formed by constituent quarks with small energies of their relative motion. However, the interactions of dressed quarks at small relative energies obey SU(6) symmetry. This means (see Chap. 5) that the s_n dependence of all wave functions ψ_h is the same within one SU(6) multiplet. Moreover, this means also, that in the limit of exact SU(6) symmetry the distribution functions F_n with similar x_i', x_i'', $\vec{k}_{Ti}'$, $\vec{k}_{Ti}''$ values do not depend on the quark quantum numbers like spin, isospin, strangeness. Because of this, inclusive cross sections of all mesons M_i (and baryons B_i) which belong to the same SU(6) multiplet have to be equal. Hence, we obtain relations of the following type (considering for the sake of definiteness $q_i = u$):

$$\frac{d\sigma_{\rho^+}(x, \vec{k}_T)}{dxd\vec{k}_T} \Bigg/ \frac{d\sigma_{\pi^+}(x, \vec{k}_T)}{dxd\vec{k}_T} = \frac{d\sigma_{K^{*+}}(x, \vec{k}_T)}{dxd\vec{k}_T} \Bigg/ \frac{d\sigma_{K^+}(x, \vec{k}_T)}{dxd\vec{k}_T} = 3 \quad ,$$

$$\frac{d\sigma_{\rho^0}(x, \vec{k}_T)}{dxd\vec{k}_T} \Bigg/ \frac{d\sigma_{\omega}(x, \vec{k}_T)}{dxd\vec{k}_T} = 1 \quad , \tag{7.21}$$

and (in the exact SU(6) limit):

$$\frac{d\sigma_{K^{*+}}(x, \vec{k}_T)}{dxd\vec{k}_T} \Bigg/ \frac{d\sigma_{\rho^+}(x, \vec{k}_T)}{dxd\vec{k}_T} = 1 \quad . \tag{7.22}$$

Hence, at each x and $\vec{k}_T$ value the ratios of the inclusive cross sections are equal to the ratios of the statistical weights of particles in the SU(6) multiplet.

In reality SU(6) symmetry is broken. We have already mentioned (Chap. 5) that the wave functions of hadrons belonging to the same SU(6) multiplet can be assumed to be equal with reasonable accuracy. Furthermore, we will make use of a very simple assumption, namely, we will suppose that the breaking of the symmetry between the u, d and s results only in the suppression of the probability of s-quark production as compared to that of u- or d-quarks by the factor $0 \leq \lambda \leq 1$. The limit $\lambda = 1$ corresponds to the exact symmetry. Thus, if the quark wave function changes are negligible, then the ratios (7.21) remain valid for the realistic case of broken SU(6), while instead of (7.22) we will have

$$\frac{d\sigma_{K^{*+}}(x, \vec{k}_T)}{dx d\vec{k}_T} \Bigg/ \frac{d\sigma_{\rho^+}(x, \vec{k}_T)}{dx d\vec{k}_T} = \lambda \quad . \tag{7.23}$$

Let us stress once more that the relative largeness of the radius R_h is relevant for the derivation of all statistical relations.

Relations like (7.21) may be tested against experimental data. The only difficulty is the necessity of separating decay products of different resonances (like ρ, K^*, etc.) from the prompt particles, since all quark statistical relations refer only to the latter. Due to this reason it is more convenient to test the statistical relation for kaons, which appear as decay products of other (i.e. non-strange) resonances rather rarely in comparison with pions.

Equation (7.21) and other similar relations assume, in particular, that the average multiplicity of any secondary meson is proportional to $2s_{q\bar{q}} + 1$, where $s_{q\bar{q}}$ is the total quark spin[8)]. In other words, we expect that the average number of secondary mesons with $s_{q\bar{q}} = 1$ is 3 times as large as that of mesons with $s_{q\bar{q}} = 0$, or 75% of all prompt secondary mesons have total quark spin $s_{q\bar{q}} = 1$ and only 25% of them have $s_{q\bar{q}} = 0$. All particles with $s_{q\bar{q}} = 1$ are resonances (see Sec. 5.1) and therefore (neglecting the decays of non-strange resonances like ϕ, f', etc. into K-mesons) about 75% of the secondary kaons have to be decay products of resonances with $s_{q\bar{q}} = 1$.

Experimental data on the production of K-resonances in pp[9] and Kp[10] collisions provide us with the possibility of testing the quark-statistical relation 3:1[8]. In the works mentioned the cross sections of K, $K^*(892)$ and $K^*(1430)$ production are determined. Due to the SU(6) classification, the first two of these particles belong to the multiplet with $L = 0$, while the tensor meson $K^*(1430)$ belongs to the $L = 1$ multiplet. The values of the cross sections are given in Table 7.1.

The meson multiplet with $L = 1$ contains four SU(3) nonets $J^P = 0^+$, 1^+, 1^+ and 2^+. The statistical weight of each of these nonets is proportional to $2J + 1$. This means that 5/12 of the particles belonging to the multiplet with $L = 1$ have to be tensor mesons, i.e. the total number of particles in this multiplet is 12/5 times the number of tensor mesons (T). 75% of all particles of the $L = 1$ multiplet are mesons with $s_{q\bar{q}} = 1$. All the vector mesons with $L = 0$ have also $s_{q\bar{q}} = 1$. Thus, the quark statistical rules predict $V + 9T/5$ for the contribution of resonances with $s_{q\bar{q}} = 1$ to the K-meson production. Neglecting higher SU(6) multiplets $L > 1$, this quantity has to give 75% of the total cross section of kaon production. As can be seen from the data in Table 7.1, the corresponding experimental value is, in each case, close to 75%.

Let us mention that since we did not take into account those possible resonances, which belong to the $L = 2$ SU(6) multiplet, all the kaons produced by decays of these resonances were added to the prompt particles. The contribution of resonances with $L = 2$ can be roughly estimated from the cross section of g-meson production; it turns out to be about 5-10%[11]. It is interesting to notice, that according to the most accurate measurement of K^- and K^0 production[10] the contribution of $s_{q\bar{q}} = 1$ resonances is somewhat less than the predicted 75% (see Table 7.1). Taking into account the resonances with $L = 2$, this value will increase. In this case the agreement between the prediction of quark statistics and the experimental data will be somewhat better.

Table 7.1 Production of kaons in K^-p collisions[10] (32 GeV/c) and in pp collisions[9] (405 GeV/c)

Reaction	$K^-p \to K^-X$		$K^-p \to \bar{K}^\circ X$		$K^-p \to K^+X$		$K^-p \to K^\circ X$		$pp \to K^\circ_s X$	
	mb	%	mb	%	mb	%	mb	%	mb	%
Inclusive production cross sections	8.3 ± 1.5	100	8.0 ± 0.5	100	1.6 ± 0.5	100	1.6 ± 0.5	100	7.4 ± 0.5	100
Total contribution of vector mesons (V)	4 ± 0.4	48	4.2 ± 0.3	53	1.1 ± 0.3	69	0.9 ± 0.2	56	3.4 ± 1	46
Total contribution of tensor mesons (T)	0.9 ± 0.2	11	0.8 ± 0.2	10	0.1 ± 0.01	5	0.1 ± 0.01	5	1.7 ± 0.8	23
Mesons with $s_{q\bar{q}} = 1$ (V + 9T/5)	5.6 ± 0.4	68 ± 5	5.6 ± 0.3	71 ± 4	1.3 ± 0.3	81 ± 19	1.1 ± 0.2	69 ± 13	6.5 ± 1.9	88 ± 26

Note: a) In comparison with the value given in Ref. 10 the inclusive cross section in $K^-p \to K^-X$ is decreased by the values of the diffraction dissociation cross section; b) The production of vector mesons from tensor meson decays is not taken into account. The prediction of quark statistics (neglecting the contribution of resonances with $L > 1$) is: $V + 9T/5 = 75\%$.

7.3.2 *Hadron ensembles*

Now we are able to identify the previously introduced notations B_{ij}, B_i, M_i, B, M with real hadrons. As we indicated before, in quark statistics the notion of sea quark means in fact a statistical average of the quarks u, d, s:

$$q = \frac{1}{2(2+\lambda)} u^{\uparrow} + \frac{1}{2(2+\lambda)} u^{\downarrow} + \frac{1}{2(2+\lambda)} d^{\uparrow} + \frac{1}{2(2+\lambda)} d^{\downarrow} + \frac{\lambda}{2(2+\lambda)} s^{\uparrow} + \frac{\lambda}{2(2+\lambda)} s^{\downarrow} \quad . \tag{7.24}$$

In the right-hand side of (7.24) the numerical coefficients before the quark symbols are probabilities; q might be $u^{\uparrow}$, $u^{\downarrow}$, $d^{\uparrow}$ or $d^{\downarrow}$ with the probability $1/(2(2+\lambda))$, and $s^{\uparrow}$ or $s^{\downarrow}$ with a somewhat lower probability $\lambda/(2(2+\lambda))$. Such a notation for q will be useful when we consider secondary hadrons.

According to (7.24), the baryons B_{ij}, B_i, B and the mesons M_i, M which figure in (7.2), (7.3), etc., are also average values of some ensembles of real baryons and mesons, which belong to different SU(6) multiplets.

Quark statistics enables us to calculate the relative production probabilities of different hadrons, belonging to the same SU(6) multiplet. However, the relations between probabilities of different multiplets cannot be obtained within the framework of quark statistics — they are determined by the dynamics of the hadronization process of the cloud of dressed quarks, i.e. by the features of the functions F_n in (7.20). Hence, these relations have to be obtained from experiment.

In Ref. 5 the dominance of the lowest SU(6) multiplets, i.e. of the meson 1 + 35-plet $(J^P = 0^-, 1^-)$ and of the baryon $\{56, 0^+\}$ multiplet $(J^P = 1/2^+, 3/2^+)$, was supposed. Now it is already clear that this is a rather rough approach. The analysis of experimental data shows that the contribution of the P-wave (L = 1) meson multiplet $(J^P = 0^+, 1^+, 1^+, 2^+)$ reaches 30% of the total meson production[10,11].

One can expect that the share of the D-wave $(L = 2)$ meson multiplet will be about 5-10% (see Refs. 8 and 11). Data on baryon production are much poorer still; as will be seen, the contribution of baryons belonging to the $\{70, 1^-\}$ SU(6) multiplet ($J^P = 5/2^-, 3/2^-, 3/2^-, 3/2^-, 3/2^-, 1/2^-, 1/2^-, 1/2^-, 1/2^-$) has to be significant also. Because of this, we will consider a general case, where the ensembles B_{ij}, B_i, B and M_i, M are superpositions of different SU(6) multiplets.

The features of the cloud of dressed quarks may be different in the fragmentation region and in the central one. If so, the SU(6) multiplet content of meson M_i, M and baryon B_{ij}, B_i, B ensembles will also be different. The decomposition of M_i and M into SU(6) multiplets can be written in the form

$$M = \sum_L \alpha_0^M(L)M(L) \quad ,$$

$$M_i = \sum_L \alpha_1^M(L)M_i(L) \quad , \tag{7.25}$$

where the index L enumerates different SU(6) multiplets. The coefficients $\alpha_0^M(L)$ and $\alpha_1^M(L)$ define the relative share of these multiplets; they obey the normalization condition

$$\sum_L \alpha_I^M(L) = 1 \quad , \quad I = 1, 0 \quad . \tag{7.26}$$

Similarly, for baryon ensembles the decompositions are

$$B = \sum_L \alpha_0^B(L)B(L) \quad ,$$

$$B_i = \sum_L \alpha_1^B(L)B_i(L) \quad ,$$

$$B_{ij} = \sum_L \alpha_2^B(L)B_{ij}(L) \quad . \tag{7.27}$$

with the normalization condition

$$\sum \alpha_I^B(L) = 1 \quad , \quad I = 0,1,2 \quad . \tag{7.28}$$

In the following we will investigate the hadron content of the ensembles for given L values.

We consider in detail the production of mesons and baryons with $L=0$. $M(0)$ is a statistical average of sea quark and antiquark pairs in S-wave states:

$$M(0) = (q\bar{q})_{L=0} = \sum_{k,\ell} \frac{\mathcal{L}_k \mathcal{L}_\ell}{4(2+\lambda)^2} (q_k\bar{q}_\ell)_{L=0} \quad . \tag{7.29}$$

The subscripts k, ℓ characterize the flavour of the quark as well as the projection of its spin: $u^\uparrow$, $u^\downarrow$, $d^\uparrow$, $d^\downarrow$, $s^\uparrow$, $s^\downarrow$. We take $\mathcal{L}_k = 1$ for non-strange quarks $(q_k = u, d)$ and $\mathcal{L}_k = \lambda$ for the strange one $(q_k = s)$. The state $(q_k\bar{q}_\ell)_{L=0}$ enters the ensembles $M(0)$ with probability $\mathcal{L}_k\mathcal{L}_\ell/|4(2+\lambda)^2|$. Supposing exact SU(6) symmetry $(\lambda = 1)$ we have $\mathcal{L}_k\mathcal{L}_\ell/|4(2+\lambda)^2| = 1/36$, i.e. $M(0)$ contains all states $(q_k\bar{q}_\ell)_{L=0}$ with equal probabilities. To go over from $(q_k\bar{q}_\ell)_{L=0}$ to real meson states $(\pi, \rho, K, \text{etc.})$, one must calculate the probabilities of these states to belong to $(q_k\bar{q}_\ell)_{L=0}$. In order to effect this the wave function $|q_k\bar{q}_\ell>_{L=0}$ has to be expanded in terms of the meson wave functions (see Appendix C); the squared coefficients of such an expansion give the probabilities we are looking for. For example,

$$(u^\uparrow\bar{d}^\uparrow)_{L=0} = \rho_1^+ \quad ,$$

$$(u^\uparrow\bar{d}^\downarrow)_{L=0} = \frac{1}{2}\pi^+ + \frac{1}{2}\rho_0^+ \quad ,$$

$$(u^\uparrow\bar{u}^\uparrow)_{L=0} = \frac{1}{2}\rho_1^0 + \frac{1}{2}\omega_1 \quad ,$$

$$(u^\uparrow\bar{u}^\downarrow)_{L=0} = \frac{1}{4}\pi^0 + \frac{1}{4}\rho_0^0 + \frac{1}{4}\omega_0 + \frac{1}{12}\eta^8 + \frac{1}{6}\eta^1 \quad ,$$

$$(s^\uparrow\bar{s}^\uparrow)_{L=0} = \phi_1 \quad , \quad \text{etc.} \tag{7.30}$$

The lower indices correspond to the spin projections of the vector mesons (1, 0, -1); η^8 and η^1 are pure octet and singlet states respectively. Denoting the mesons of the S-wave 1 + 35-plet by $h_{M(0)}$ and making use of Eq. (7.29) and relations of the type of (7.30), we can write the following decomposition for the sea meson ensemble $M(0)$ in terms of the real particles:

$$M(0) = \sum_h \mu_{h(0)} h_{M(0)} \tag{7.31}$$

The coefficients $\mu_{h(0)}$ are presented in Appendix D. Since we have no special interest in polarization effects in the particle production processes, the probabilities given in Appendix D are summed over all spin projections, e.g. $\mu_{\rho^+} = \mu_{\rho^+_1} + \mu_{\rho^+_0} + \mu_{\rho^+_{-1}}$.

The decomposition of $M_i(0)$ in terms of real hadrons is carried out similarly:

$$M_i(0) = (q_i\bar{q})_{L=0} = \sum_k \frac{\mathcal{L}_k}{2(2+\lambda)} (q_i\bar{q}_k)_{L=0} = \sum_h \mu_{h(0)}(i) h_{M(0)} \quad . \tag{7.32}$$

Analogous decompositions for the mesons of the next $(L = 1)$ multiplet are the following:

$$M(1) = (q\bar{q})_{L=1} = \sum \frac{\mathcal{L}_k \mathcal{L}_\ell}{4(2+\lambda)^2} (q_k\bar{q}_\ell)_{L=1} = \sum_h \mu_{h(1)} h_{M(1)}$$

$$M_i(1) = (q_i\bar{q})_{L=1} = \sum \frac{\mathcal{L}_k}{2(2+\lambda)} (q_i\bar{q}_k)_{L=1} = \sum_h \mu_{h(1)}(i) h_{M(1)} \tag{7.33}$$

The coefficients $\mu_{h(0)}(i)$, $\mu_h(1)$ and $\mu_{h(1)}(i)$ are also presented in Appendix D.

When deriving the relations (7.31-33) one can forget about the colour degrees of freedom of quarks since the symmetry properties of meson wave functions in the colour and the SU(6) indices are absolutely independent. In the case of baryons this is not so. The completely antisymmetric wave function of a three-quark system might be of mixed colour symmetry and mixed SU(6) symmetry as well. Such states, of course, do not correspond to real baryons, because they are not colour singlets. Hence, in the expansion of B_{ik}, B_i and B in terms of real hadrons one has to take into account the colour quantum numbers of quarks.

Let us return to the hadronization process of mesons. We will discuss the mechanism of colour neutralization, when a quark q_i picks up an antiquark $\bar{q}$ from the sea. This, however, does not mean that a meson is produced. Indeed, for the transition $q_i\bar{q} \to M_i$ the $q_i\bar{q}$ pair has to be in a white state. Without special colour corelations, $q_i\bar{q}$ will be white with the probability $W < 1$; $1 - W$ is the probability for $q_i\bar{q}$ to be in a coloured state. In the latter case this coloured excited state disintegrates and the quark q_i picks up again an antiquark $\bar{q}$ (see Fig. 7.4a). If the new $q_i\bar{q}$ pair happens to be again a coloured one, a repeated decay has to take place (Fig. 7.4b), etc. We assume (see Sec. 7.1) that the formation time of new quarks is much less than the lifetime of the cloud of dressed quarks, and thus the described "decay mechanism" has to provide the transition of the $q_i\bar{q}$ system into a white state with unit probability. This assumption corresponds to the hypothesis of soft colour neutralization.

The colour neutralization of baryon states has to take place in an analogous way (Fig. 7.4). Let us consider the case $L = 0$ and show how to calculate the hadron content of the "baryon" B_{ik}, B_i and B. The "baryon" $B_{ik}(0)$ is an average of S-wave baryons, containing the S-wave diquark $\{q_iq_k\}$ and the sea quark q. We will denote by braces wave functions symmetrized over SU(6) indices; e.g.

$$|\{q_iq_k\}\rangle = \frac{1}{\sqrt{2}}\,|(q_iq_k + q_kq_i)\rangle \qquad \text{for} \quad i \neq k \quad \text{and}$$

$$|\{q_iq_k\}\rangle = |q_iq_i\rangle \qquad \text{for} \quad i = k \quad .$$

Writing for $B_{ik}(0)$ a decomposition, analogous to (7.32) or (7.33),

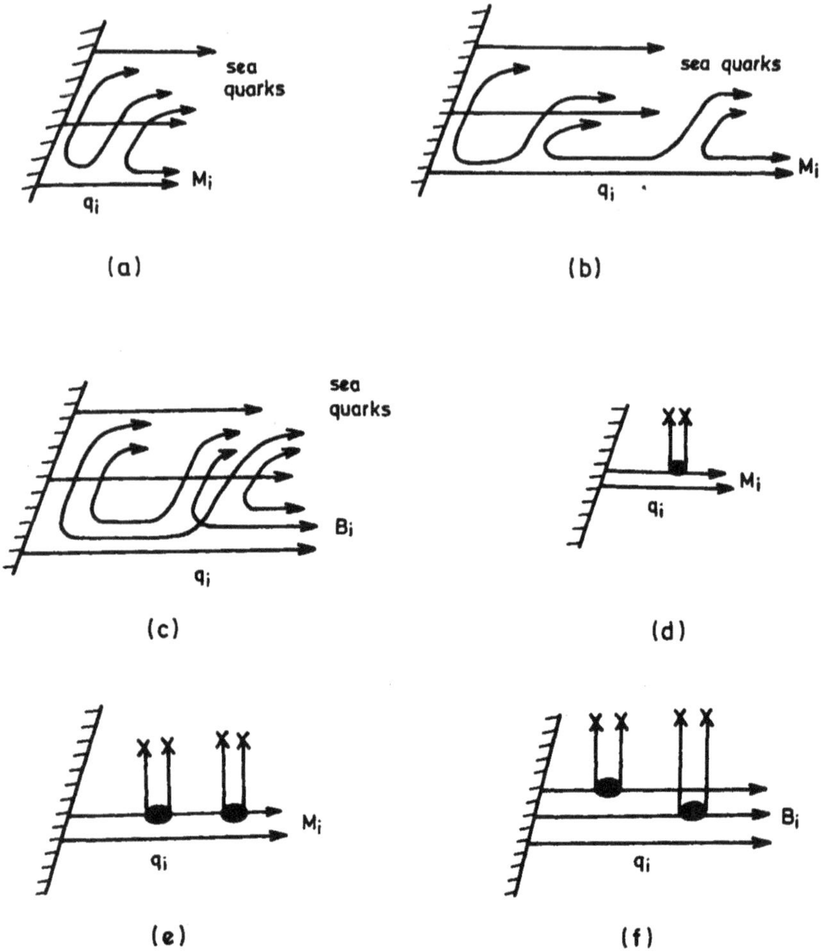

Fig. 7.4 The decay mechanism of soft colour neutralization in the processes of meson M_i (a, b) and baryon B_i (c) production; this mechanism may be interpreted as effective interaction represented by the diagrams (d-f) (see discussion in the text).

we obviously get a wrong result:

$$(\{q_i q_k\} q)_{L=0} = \sum_{\ell} \frac{\mathscr{L}_\ell}{2(2+\lambda)} (\{q_i q_k\} q_\ell)_{L=0} \quad . \tag{7.34}$$

Indeed, the state $(\{q_i q_k\} q_\ell)_{L=0}$ is not symmetric under the permutation of the index ℓ. It is easy to see that the problem is connected with colour neutralization: the states which are not symmetric in the SU(6) indices are coloured ones. This means that the ensemble (7.34) contains white states and coloured ones as well. Let us separate these states explicitly. (For definiteness we take $q_i = q_k = u^\uparrow$.)

$$(\{u^\uparrow u^\uparrow\}q)_{L=0} = \frac{1}{2(2+\lambda)}\Big[\{u^\uparrow u^\uparrow u^\uparrow\}_{L=0} + \frac{1}{3}\sum_{k\neq u^\uparrow} \mathscr{L}_k\{u^\uparrow u^\uparrow q_k\}_{L=0}$$
$$+ \frac{2}{3}\sum_{k\neq u^\uparrow} \mathscr{L}_k\{u^\uparrow u^\uparrow q_k\}_{70,L=0}\Big] \qquad (7.35)$$

The first two terms in the brackets correspond to the white states (completely symmetric in SU(6) indices), while the last term corresponds to the states of a coloured "70-plet". Because of soft colour neutralization the quark q_k does not stay in the coloured 70-plet but goes away into the quark sea, while the remaining diquark $u^\uparrow u^\uparrow$ picks up another sea quark. The iteration of this process leads to the final production of the white states

$$(u^\uparrow u^\uparrow u^\uparrow)_{L=0} \qquad \text{and} \qquad \sum_{k\neq u^\uparrow} \mathscr{L}_k\{u^\uparrow u^\uparrow q_k\}$$

only, whereas the proportion of the production probabilities remains the same as in (7.35):

$$(\{u^\uparrow u^\uparrow\}q_k)_{L=0} \longrightarrow \frac{3}{2(3+\lambda)}\Big[\{u^\uparrow u^\uparrow u^\uparrow\}_{L=0} + \frac{1}{3}\sum_{k\neq u^\uparrow} \mathscr{L}_k\{u^\uparrow u^\uparrow q_k\}\Big]$$
$$(7.36)$$

It is easy to see that the sum of probabilities at the right-hand side of (7.36) is equal to unity. Hence this expression is the required *ensemble* $B_{u^\uparrow u^\uparrow}(0)$.

The way described for the separation of white states leads to the correct answer in the sense that averaging over the quantum numbers of the initial diquark we obtain an ensemble which is not polarized and not aligned in any SU(6) index. This result is provided by the decay mechanism in which the initial diquark interacts with the cloud of dressed quarks via colour meson exchange. Note that if we tried to neutralize the coloured baryon by gluon (i.e. flavour singlet) exchange, this would correspond to the expansion of the last term in the right-hand side of (7.35) in terms of only $\{u^{\uparrow}u^{\uparrow}q_i\}_{q_i \neq u^{\uparrow}}$ states. Because of this the resulting baryons (after averaging over the initial diquark) would be aligned. Hence, the colour neutralization of a hadron state via gluon exchange is possible for the case of secondary mesons, but it leads to an unreasonable result for secondary baryons. One can say therefore that the colour neutralization in quark statistics is provided by repeated four-fermion interactions of a coloured hadron with the cloud of dressed coloured quarks. Using this language, the graphs of Fig. 7.4a-c can be rewritten as graphs of Fig. 7.4d-f. The interaction which leads to the colour neutralization has to be described by an effective Lagrangian of the type

$$G_{18}(k^2)(\bar{\psi}\frac{1}{\sqrt{6}}I_f\vec{\lambda}_c\psi)<\bar{\psi}\frac{1}{\sqrt{6}}I_f\vec{\lambda}_c\psi>$$

$$+\ G_{88}(k^2)(\bar{\psi}\frac{1}{2}\vec{\lambda}_f\vec{\lambda}_c\psi)<\bar{\psi}\frac{1}{2}\vec{\lambda}_f\vec{\lambda}_c\psi> \quad .$$

The matrices I and the Gell-Mann matrices $\vec{\lambda}$ act on the flavour (f) and colour (c) quark indices. The mean values $<\bar{\psi}\Gamma\psi>$ are "coloured mesons" (colour octet states) and singlet or octet states with respect to the flavour. The quark statistical rules require $G_{18}(k^2) = G_{88}(k^2)$: in this case there will be no alignment. The case of gluon exchange corresponds to $G_{18}(k^2) \neq 0$, $G_{88}(k^2) = 0$.

The above considerations allow us to formulate a prescription for building up baryon ensembles in the general case[7)]. For the "baryons" $B_{ik}(0)$, $B_i(0)$ and $B(0)$ this is the following. Ensembles

$$(\{q_iq_k\}q)_{L=0} \quad , \quad (q_iqq)_{L=0} \quad \text{and} \quad (qqq)_{L=0}$$

containing coloured states are constructed. Furthermore in these ensembles one leaves only white states with proper SU(6) symmetry (in the $L=0$ case this means states symmetric in SU(6) indices). The normalization has to be restored by multiplying the ensemble by a common factor. As a result, one has the following decompositions in terms of the 56-plet

$$B_{ik}(0) = \sum_{\ell} \frac{\mathscr{L}_\ell(1+\delta_{i\ell}+\delta_{k\ell})}{4+\mathscr{L}_i+\mathscr{L}_k+2\lambda} \{q_iq_kq_\ell\}_{L=0}$$

$$B_i(0) = \sum_{k\geq\ell} \frac{\mathscr{L}_k\mathscr{L}_\ell(1+\delta_{ik}+\delta_{i\ell})}{10+2\mathscr{L}_i+3\mathscr{L}_i^2+10\lambda+3\lambda^2} \{q_iq_kq_\ell\}_{L=0}$$

$$B(0) = \sum_{k\geq\ell\geq i} \frac{\mathscr{L}_k\mathscr{L}_\ell\mathscr{L}_i}{4(5+5\lambda+3\lambda^2+4\lambda^3)} \{q_iq_kq_\ell\}_{L=0} \tag{7.37}$$

Let us note that the numerical coefficients are probabilities for the states $\{q_iq_kq_\ell\}_{L=0}$ to belong to the corresponding baryon ensemble. Making use of the explicit form of the 56-plet wave functions (see Appendix C), the expansions of $\{q_iq_kq_\ell\}_{L=0}$ states into real hadrons can be carried out. We present here some of these relations:

$$\{u^\uparrow u^\uparrow d^\downarrow\}_{L=0} = \frac{2}{3}p + \frac{1}{3}\Delta^+ \qquad \{u^\downarrow s^\uparrow s^\uparrow\}_{L=0} = \frac{2}{3}\Xi^0 + \frac{1}{3}\Xi^{*0}$$

$$\{u^\uparrow u^\downarrow d^\uparrow\}_{L=0} = \frac{1}{3}p + \frac{2}{3}\Delta^+ \qquad \{u^\uparrow s^\uparrow s^\downarrow\}_{L=0} = \frac{1}{3}\Xi^0 + \frac{2}{3}\Xi^{*0}$$

$$\{u^\uparrow d^\downarrow s^\uparrow\}_{L=0} = \frac{1}{2}\Lambda + \frac{1}{6}\Sigma^0 + \frac{1}{3}\Sigma^{*0} \tag{7.38}$$

With the help of (7.37) and (7.38) the decompositions of $B_{ik}(0)$, $B_i(0)$ and $B(0)$ into baryons $h_B(0)$ of the S-wave 56-plet can be obtained:

$$B_{ik}(0) = \sum_h \beta_{h(0)}(ik)h_{B(0)}$$

$$B_i(0) = \sum_h \beta_{h(0)}(i) h_B(0) \quad , \quad B(0) = \sum_h \beta_{h(0)} h_B(0) \qquad (7.39)$$

The coefficients $\beta_{h(0)}(ik)$, $\beta_{h(0)}(i)$, and $\beta_{h(0)}$ (probabilities for $h_B(0)$ to enter the ensembles $B_{ik}(0)$, $B_i(0)$ and $B(0)$) are given in Appendix D. Also presented are the coefficients of the expansion of $B_{ik}(1)$, $B_i(1)$, and $B(1)$ into the baryons of the 70-plet[12].

The mechanism of colour neutralization just described leads to an important consequence: formulae (7.37) give the isotopic (and spin) alignment only if such an alignment is present in the initial quark states. As an example, consider the production of the $\Delta(1236)$ resonance in the target fragmentation region of the deep inelastic neutrino-nucleon collision process. The average cross section

$$\sigma(\Delta) = \sigma(\nu p \to \Delta) + \sigma(\nu n \to \Delta) + \sigma(\bar{\nu} p \to \Delta) + \sigma(\bar{\nu} n \to \Delta) \qquad (7.40)$$

is (except for small x_B values) defined by the transition of the di-quark (its isotopic states averaged) into Δ. Tables D.2, which are calculated on the basis of (7.37) give the production of the Δ-isobar without isotopic alignment; indeed

$$\sigma(\Delta^{++}):\sigma(\Delta^{+}):\sigma(\Delta^{0}):\sigma(\Delta^{-}) = 1:1:1:1 \quad .$$

If, on the contrary, the colour neutralization mechanism is not taken into account, we can come to the result that a symmetric quark state hadronizes into isotopically aligned particles. For example, formula (A.4) of Ref. 13 gives for the same $\sigma(\Delta)$ cross sections

$$\sigma(\Delta^{++}):\sigma(\Delta^{+}):\sigma(\Delta^{0}):\sigma(\Delta^{-}) = 1:\frac{16}{9}:\frac{16}{9}:1 \quad .$$

The production probabilities of the baryons belonging to the 70-plet are for the time being not well known. The parameters $\alpha_2^B(1)$, $\alpha_1^B(1)$, and $\alpha_0^B(1)$ in (7.27) can be estimated using the following simple considerations. We neglect the production of resonances higher than S- and P-wave states. The probability for production of a P-wave hadron is determined by the probability of finding a P-wave quark state in the cloud of dressed quarks. This probability, α_P, is characterized by the properties of the quark cloud itself, and it is reasonable to think that it will not depend on whether we consider a sea quark or an antiquark. Thus,

$$\alpha_2^B(1) \simeq \alpha_1^M(1) \simeq \alpha_0^M(1) = \alpha_P$$

$$\alpha_2^B(0) \simeq \alpha_1^M(0) \simeq \alpha_0^M(0) = \alpha_s \quad . \tag{7.41}$$

(Since we neglected the production of higher resonances $\alpha_P + \alpha_s = 1.$) For baryons containing two or three sea quarks, we have

$$\alpha_1^B(0) \simeq \alpha_0^B(0) \propto \alpha_s^2$$

$$\alpha_1^B(1) \simeq \alpha_0^B(1) \simeq 2\alpha_s\alpha_P \quad . \tag{7.42}$$

If P-wave mesons are produced with 30% probability, then, due to (7.41), (7.42), the probability of the production of baryons belonging to the 70-plet is 30% in the case of diquark fragmentation and 45% when the baryon is produced by the fragmentation of a quark or by sea quarks.

7.3.3 *Meson-baryon ratios*

Another type of the quark statistical relations is the relative probability of meson and baryon production in the quark jets. It is also based on the hypothesis of soft hadronization, considered in Sec. 1.

Let us return to the integral (7.20). At large R_h the wave functions $\psi_h(s_n)$ decrease very quickly with increase of s_n. Because of this, in the distribution function which is changing more slowly, one can take $x_i' = x_i'' = x/n$ and $\vec{k}_{Ti}' = \vec{k}_{Ti}'' = \frac{\vec{k}_T}{n}$. The remaining integrand is defined only by the quark wave functions, and does not depend, due to the Lorentz invariance, on x and $\vec{k}_T$ (this can be seen immediately in the rest frame of the secondary hadron, where $x = 0$, $\vec{k}_T = 0$). Hence, for the inclusive spectra of meson M_i or baryon B_i production one can write

$$\frac{d\sigma_h(x, \vec{k}_T)}{dxd\vec{k}_T} = \text{Const. } F\left(\frac{x}{n}, \frac{\vec{k}_T}{n}, \ldots\right) \qquad (7.43)$$

As we already considered, the hadronization process of the quark q_i proceeds through the repeated interactions of this quark with the others; these interactions are accompanied by the production of additional quark-antiquark pairs. Such a process may be interpreted as a chain of decays of the initial excited system "quark q_i + its neighbours" to the final state "meson M_i + the rest sea quarks" or "baryon B_i + the rest sea quarks". If so, it is natural to assume that both final distributions $F_2\left(\frac{x}{2}, \frac{\vec{k}_T}{2}; \frac{x}{2}, \frac{\vec{k}_T}{2} \middle| \frac{x}{2}, \frac{\vec{k}_T}{2}; \frac{x}{2}, \frac{\vec{k}_T}{2}\right)$ and $F_3\left(\frac{x}{3}, \frac{\vec{k}_T}{3}; \frac{x}{3}, \frac{\vec{k}_T}{3}; \frac{x}{3}, \frac{\vec{k}_T}{3} \middle| \frac{x}{3}, \frac{\vec{k}_T}{3}; \frac{x}{3}, \frac{\vec{k}_T}{3}; \frac{x}{3}, \frac{\vec{k}_T}{3}\right)$ are approximately proportional to the same function $F_{q_i}(x, k_T)$, which may be conditionally considered as some "initial distribution" of the dressed quark q_i in the jet. Thus,

$$\frac{d\sigma_{Mi}(x, \vec{k}_T)}{dxd\vec{k}_T} = W_1 F_{q_i}(x, \vec{k}_T) \quad , \quad \frac{d\sigma_{Bi}(x, \vec{k}_T)}{dxd\vec{k}_T} = W_2 F_{q_i}(x, \vec{k}_T) \qquad (7.44)$$

Of course, $W_1 + W_2 = 1$, for the total probability of hadronization for the quark q_i is equal to unity.

The next assumption, which allows us to calculate W_1 and W_2 explicitly, is connected with the underlying dynamics of multiple production. Namely, if sea quarks and antiquarks are produced without any correlations, the quark q_i interacts with a sea quark or a sea antiquark with equal probabilities. In this case W_1 is proportional to 1/2 — the relative probability of one sea antiquark production, while $W_2 \sim 1/2 \cdot 1/2 = 1/4$ — the relative production probability of two sea quarks. Thus, $W_1 = 2W_2$, i.e.

$$W_1 = 2/3 \quad , \quad W_2 = 1/3 \, . \qquad (7.45)$$

This means that within the framework of quark statistics the total probability for the transition of the quark q_i into the baryon B_i is equal to[7] 1/3, i.e. it coincides with the baryon charge of the quark. It is obvious, however, that in any realistic model of multiple production some quark-antiquark correlation is present. These are, for example, trivial kinematic correlations due to energy-momentum conservation. Besides, even in the limit $s \to \infty$ some correlations are possible: new quarks are always produced as $q\bar{q}$ pairs. Thus the absence of $q\bar{q}$ correlations means in fact that the average rapidity gap between the newly produced quark and its conjugate partner-antiquark is very large in comparison with the average rapidity gap in the sea. This is possible if the average quark density in the sea is large enough.

Thus, we see, that the relation (7.45) is in some sense an asymptotic one. There are, however, some model estimations[14,15] which show that the ratio $W_1 : W_2$ tends to its asymptotic limit 2:1 rather rapidly. A little later we shall see that the experimental data at the available energies confirm the relation (7.45). Here we wish to demonstrate how to obtain the relations of quark statistics by purely combinatorial considerations (see Refs. 5, 6). This derivation might not reveal the physics of the process; it is, however, much simpler.

Let us return to the hadronization processes of the incident quark q_i and of the quark-antiquark sea $(q\bar{q})$. We place all quarks in the sequence of increasing rapidities. For the sake of simplicity we suppose that hadrons are formed by quarks which are the nearest neighbours on the rapidity axis. From the hypothesis of quark statistics, the nearest neighbour of the quark q_i might be a quark or an antiquark with the same probability. Joining these quarks, q_i forms with $\bar{q}$ a meson ensemble $M_i = q_i\bar{q}$ or — with the same probability — a "diquark" ensemble q_iq. (In our terminology M_i, q_iq and other ensembles will be mesons, diquarks, etc.) *In the first case the hadronization* process of q_i is accomplished, in the second one further iterations have to be considered. Again, the next neighbour is either a quark q

or an antiquark $\bar{q}$. It follows that the diquark $q_i q$ forms with q a baryon B_i, or with $\bar{q}$ a system $q_i q\bar{q}$. By the accepted rule, the nearest neighbours on the rapidity axis in the latter system turn into a sea meson $M = q\bar{q}$, while the quark q_i remains in the cloud of quarks which are not hadronized yet. The stage of the hadronization process that we are considering can be symbolically written in the form

$$q_i(\frac{1}{2}q + \frac{1}{2}\bar{q})(\frac{1}{2}q + \frac{1}{2}\bar{q}) \longrightarrow (\frac{1}{2}q_i q + \frac{1}{2}q_i\bar{q})(\frac{1}{2}q + \frac{1}{2}\bar{q})$$

$$\longrightarrow (\frac{1}{2}q_i q + \frac{1}{2}M_i)(\frac{1}{2}q + \frac{1}{2}\bar{q}) \longrightarrow \frac{1}{4}q_i qq + \frac{1}{4}q_i q\bar{q} + \frac{1}{2}M_i(\frac{1}{2}q + \frac{1}{2}\bar{q})$$

$$\longrightarrow \frac{1}{4}B_i + \frac{1}{4}Mq_i + \frac{1}{4}M_i q + \frac{1}{4}M_i\bar{q} \qquad (7.46)$$

The numerical constants stand for probabilities. It is seen that already at this stage that M_i and B_i are produced in the proportion 2:1 (which is in accordance with (7.45)). To consider the hadronization of all quarks one has to write a chain of iterations:

$$q_i(\frac{1}{2}q + \frac{1}{2}\bar{q})(\frac{1}{2}q + \frac{1}{2}\bar{q})\ldots(\frac{1}{2}q + \frac{1}{2}\bar{q})\ldots \qquad (7.47)$$

and collect the quarks one by one into hadrons (from left to right). As a result of a sufficiently long chain the average multiplicities of produced hadrons turn out to be[7]:

$$< n_{M_i} > = \frac{2}{3} \quad , \quad < n_{B_i} > = \frac{1}{3} \qquad (7.48a)$$

$$< n_M > = 6N + \frac{1}{3} \quad , \quad < n_B > = < n_{\bar{B}} > = N \qquad (7.48b)$$

The results (7.48a) are identical with (7.45). The multiplicities of the sea mesons $M = q\bar{q}$, baryons $B = qqq$ and antibaryons $\bar{B} = \bar{q}\bar{q}\bar{q}$ are characterized by the quantity N which is determined by the length of the chain and depends on the total energy of the jet. In practical calculations N might be expressed, for example, in terms of the average multiplicity of all charged particles $< n_{ch} >$. If N is large enough, the ratios of the multiplicities of sea mesons and baryons are[5]:

$$< n_M > : < n_B > : < n_{\bar{B}} > = 6 : 1 : 1 \quad . \tag{7.49}$$

The relations (7.48) can be written in a form similar to Eqs. (7.2) and (7.3):

$$q_i + (q, \bar{q} \text{ sea}) \longrightarrow 1/3\ M_i + 2/3\ B_i + (6N + 1/3)M + N(B + \bar{B}) \quad . \tag{7.50}$$

The coefficients at the right-hand side of Eq. (7.50) are the average multiplicities of the corresponding secondaries.

The average multiplicities of secondaries produced in the hadronization process of the diquark $q_i q_j$ in the quark sea can be calculated in a similar way. In the form (7.50) the result is written as[7]:

$$q_i q_j + (q\bar{q} - \text{sea}) \longrightarrow \frac{1}{2} B_{ij} + \frac{1}{12}(B_i + B_j) + \frac{5}{12}(M_i + M_j)$$

$$+ (6N + \frac{1}{6})M + NB + N\bar{B} \tag{7.51}$$

The relation (7.49) is also based on the assumption of uncorrelated production of sea quarks and antiquarks. The same numerical estimations[14,15] show that the $< n_M > / < n_B >$ ratio tends to its asymptotic limit (7.49) rather slowly compared to the $< n_{M_i} > / < n_{B_i} >$ ratio. We shall consider this point once more a little later.

The relation (7.45) (or (7.48a), or (7.50)) refer to the hadronization process of the "isolated" quark q_i, i.e. of the quark q_i, which initially has no close neighbours in rapidity space. Thus it is most convenient to test this relation in quark jets, initiated in the hard processes such as e^+e^- annihilation into hadrons or deep inelastic lepton-hadron collisions in the current fragmentation region. In these processes the quark jet contains one tagged quark with the quantum numbers of the parton which has produced a cloud of sea quarks and antiquarks with the average quantum numbers of the vacuum. The tagged quark is usually one of the fastest in the jet. Still, it is hard to distinguish the fragmentational (tagged) hadrons M_i and B_i from sea hadrons with the same quantum numbers, if q_i is a u, d or s quark. The situation is simpler in cases when in the e^+e^- annihilation processes $c\bar{c}$ or $b\bar{b}$ pairs are produced by virtual γ^*. The production of c- and b-quarks in the sea is strongly suppressed, and practically the only source for charm and beauty hadrons are the tagged c and b quarks.

In e^+e^- annihilation there are two jets (initiated by a quark and an antiquark respectively), hence the total multiplicities of charm mesons and baryons are $\langle n_{B_c} \rangle = \langle n_{\bar{B}_{\bar{c}}} \rangle = 1/3$, $\langle n_{M_c} \rangle = \langle n_{M_{\bar{c}}} \rangle = 2/3$. The multiplicities of secondary particles in e^+e^- annihilation are usually measured in units of reduced inclusive cross sections: $R(h) = \sigma(h)/\sigma(\mu^+\mu^-)$. The cross section for the production of charm baryons might be estimated from the increase of the total baryon multiplicity passing the threshold of c-quark production. The experimental increase $R(B+\bar{B})$ is 0.92 ± 0.11[16] in the energy region $\sqrt{s} = 4.4 - 6.5$ GeV. This means that the predicted $R(B_c)$ value[17]

$$R(B_c) = R(\bar{B}_{\bar{c}}) = \frac{4}{9} \;, \tag{7.52}$$

is in a good agreement with experiment. Note that the growth of $R(B+\bar{B})$, due to the increase of only the sea baryon contribution, is small, of the order of 0.02-0.06. On the other hand, the total baryon production in the e^+e^- annihilation at $\sqrt{s} = 6.5$ GeV turns out to be somewhat less ($R(B+\bar{B}) = 1.8 \pm 0.2$[16]) than predicted by the rules of quark statistics, $R(B+\bar{B}) = 20/9$.

The quark statistical rules provide a possibility to calculate the hadron content of B_c and M_c. The decomposition of M_c and B_c into hadrons of the lowest S-wave multiplet is given in Appendix D. Taking into account the resonance decays at[b] $\lambda = 0.3$[18] we obtain[17] a quite good agreement between prediction and experiment:

$$R(\Lambda_c^+ + \bar{\Lambda}_c^-) = 0.62 \qquad 0.52 \pm 0.13^{16)}$$

$$R(D^+ + D^-) = 0.40 \qquad 0.75 \pm 0.32^{20)}$$

$$R(D^0 + \bar{D}^0) = 1.14 \qquad 1.2 \pm 0.4^{20)} \quad . \tag{7.53}$$

In the calculation[17] the contribution of $L = 1$ resonances was neglected. Besides, it was assumed that in hadrons containing c- and s-quarks the transitions with c-quark decay are significantly more probable than those with s-quark decay because of the much larger phase volume.

In conclusion let us make some remarks about the status of the relation (7.48b) (or (7.49)). As we shall see in the next section, the data at usual accelerator energies $s \sim 10^2 - 10^3$ GeV^2 show that the fraction of the secondary baryons is appreciably lower than predicted by (7.49): the $\langle n_M \rangle / \langle n_B \rangle$ ratio is about 10 instead of 6. Even at ISR energies the number of produced quarks is not enough to reveal the asymptotic behaviour of this ratio (see Refs. 14, 15). In other words, at these energies kinematic $q\bar{q}$ correlations are still large as far as the relation (7.49) is concerned. Recent data[21] from SPS Collider, where the $\bar{p}/\pi$ production ratio in the central region of $p\bar{p}$ collisions at $\sqrt{s} = 540$ GeV was measured, are in better agreement with (7.49) (see Fig. 7.5). Still, we don't know at present if the SPS value of $\bar{p}/\pi$ is an asymptotic one or if it will change with increase of energy. The answer will be provided by experimental data at higher energies.

---

b Due to our estimations the value of λ is about 0.3 for the quark jets of hard processes and about 0.2 for multiparticle production processes in hadron collisions. In a recent paper (Ref. 18) the estimation gives $\lambda = 0.3$, while in Ref. 19 $\lambda = 0.20 \pm 0.03$ is obtained.

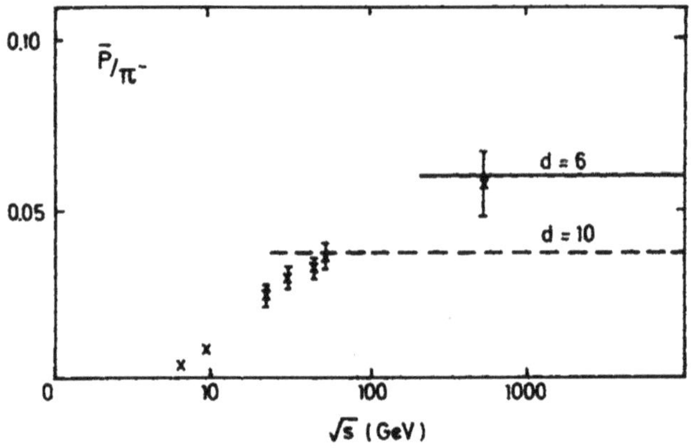

Fig. 7.5 Antiproton/pion yield ratio at $\eta^* = 0$ in $p\bar{p}$ and pp collisions versus total center-of-mass energy $\sqrt{s}$. The lines are the quark model predictions for different values of the parameter d in Eq. (7.2).

7.4 Total and Partial Multiplicities

The relations (7.2), (7.3) together with the decompositions of meson and baryon ensembles into real hadrons (Appendix D) allows one to calculate the average multiplicities of the secondary particles in multiple production processes.

The result of these calculations depend on several phenomenological parameters: Δ (or δ), which may be determined from the diffraction-dissociation cross section, λ (the suppression factor for the strange sea quark production), d, ξ_1, ξ_2 and η, $\alpha_I^M(L)$ and $\alpha_I^B(L)$. Moreover, the energy behaviour of the total sea multiplicity $N(s)$ cannot be calculated within the framework of the purely quark-statistical approach.

So let us begin with the value $N(s)$. In the impulse approximation of the quasinuclear quark model the available energy for the production of new particles is determined by the total energy of the colliding quarks in their center-of-mass system, $\sqrt{s_{qq}}$.

Let x_1 and x_2 be the fractions of the longitudinal momentum carried away by the interacting quarks of the projectile and target, respectively. Then quark center-of-mass system total energy squared s_{qq} is related with that of colliding hadrons, s, by the relation:

$$s_{qq} = x_1 x_2 s \quad . \tag{7.54}$$

We neglect in (7.54) the transverse motion of quarks inside the hadron. Now, if $Q_h(x)$ is the x-distribution of the interacting quark in the hadron h, the function $N(s)$ in pp collisions may be written as

$$N_{pp}(s) = \int_0^1 dx_1 \int_0^1 dx_2 Q_p(x_1) Q_p(x_2) N_{qq}(x_1 x_2 s) \tag{7.55}$$

where $N_{qq}(s_{qq})$ characterizes the multiplicity of hadron production in "quark-quark collisions". For πp collisions the corresponding result is

$$N_{\pi p}(s) = \int_0^1 dx_1 \int_0^1 dx_2 Q_\pi(x_1) Q_p(x_2) N_{qq}(x_1 x_2 s) \tag{7.56}$$

For Kp collisions it is necessary to account for the difference both in s and $u(d)$ quark x-distributions inside the kaon and in the cross sections of sq and qq inelastic collisions:

$$N_{Kp}(s) = \int_0^1 dx_1 \int_0^1 dx_2 Q_p(x_1) \left[\frac{\zeta}{1+\zeta} Q_K^s(x_2) + \frac{1}{1+\zeta} Q_K^q(x_2)\right] N_{qq}(x_1 x_2 s) \tag{7.57}$$

where $\zeta = \sigma_{inel}(sq)/\sigma_{inel}(qq) \simeq \frac{2}{3}$.

The unknown function $N_{qq}(s_{qq})$ can be parametrized in a convenient form; the values of parameters may then be determined from the fit of the experimental data on average charged multiplicity, e.g. in pp collisions. According to Eq. (7.2) we may write it in the form:

$$\langle n_{ch}\rangle_{pp} = 2\Delta + (1-\Delta)[2\xi_1 \langle n_{ch}(B_{ij})\rangle + 2\xi_2(\langle n_{ch}(B_i)\rangle + \langle n_{ch}(B_j)\rangle) + 2(1-\xi_1-\xi_2)(\langle n_{ch}(M_i)\rangle + \langle n_{ch}(M_j)\rangle) + 2(1-\xi_1-2\xi_2)\langle n_{ch}(B_k)\rangle + 2(\xi_1+2\xi_2)\langle n_{ch}(M_k)\rangle + N_{pp}(s)(d\langle n_{ch}(M)\rangle + \langle n_{ch}(B)\rangle + \langle n_{ch}(\bar{B})\rangle)] \quad . \tag{7.58}$$

Here $\langle n_{ch}(B_{ij})\rangle$, $\langle n_{ch}(B_i)\rangle$, etc. are the average charged multiplicities of the stable particles in the corresponding ensembles; they may be determined from the decompositions of Appendix D. It is necessary to emphasize once more, that every ensemble in the right-hand side of (7.2) or (7.3) is assumed to be normalized to one average prompt particle. Accounting for the decays of all relevant resonances and collecting the charged stable decay products we will obtain the average number $\langle n_{ch}\rangle$ we are looking for. The calculated $\langle n_{ch}\rangle$ depends, of course, on the values of the still undetermined parameters $\alpha_I^M(L)$, $\alpha_I^B(L)$; however, this dependence of $\langle n_{ch}\rangle$ is rather weak and to the first approximation $\lambda = 0.3$, $\alpha_0^M(0) = \alpha_1^M(0) = 0.7$ and $\alpha_I^B(0)$ from Eqs. (7.41), (7.42) seems to be quite suitable for our purpose.

The final expression for $\langle n_{ch}\rangle_{pp}$ depends also on the contribution of spectator and tagged hadrons. In the scaling limit this contribution is energy-independent; thus all variations of the relevant parameters would lead to some effective renormalization of $N_{qq}(s_{qq})$ by an additive factor.

The data for $\langle n_{ch}\rangle_{pp}$ in non-diffractive collisions at $s \sim 10 - 10^5\ GeV^2$ are shown in Fig. 7.6a; the curve is obtained from Eqs. (7.55) and (7.58) (at $\Delta = 0$) with

$$N_{qq}(s_{qq}) = -0.257 + 0.0649\,\ell ns + 0.00952\,(\ell ns)^2 \tag{7.59}$$

The same function $N_{qq}(s_{qq})$ has to determine the total multiplicity in the quark jets of the hard processes. In e^+e^- annihilation below the charm production threshold $\langle n_{ch}\rangle_{e^+e^-}$ may be easily

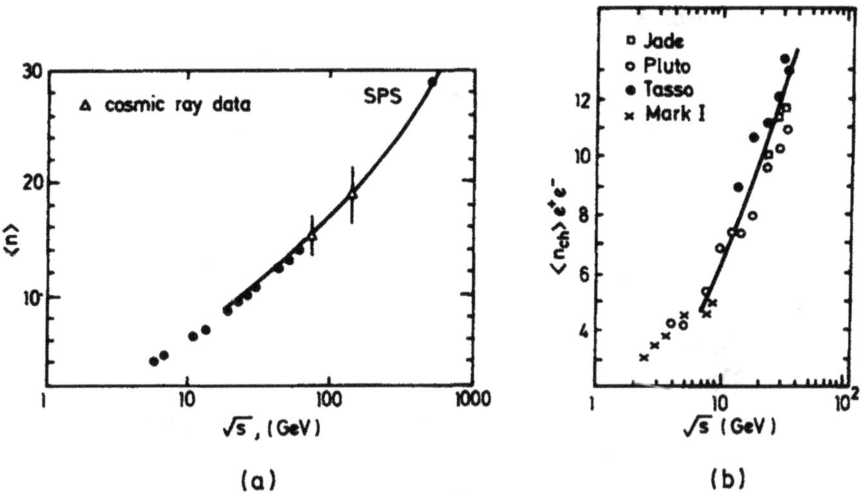

Fig. 7.6 (a) Average charge multiplicity in nondiffractive pp ($\bar{p}p$) collisions and its fit in the quark model (see Eqs. (7.55), (7.58), (7.59)).
(b) Average charge multiplicity in e^+e^- annihilation; the curve is calculated with the help of Eqs. (7.59), (7.60).

calculated using (7.48a):

$$\langle n_{ch} \rangle_{e^+e^-} = 2(1/3 \langle n_{ch}(B_i) \rangle + 2/3 \langle n_{ch}(M_i) \rangle) + N_{qq}(s)(d \langle n_{ch}(M) \rangle + \langle n_{ch}(B) \rangle + \langle n_{ch}(\bar{B}) \rangle) \quad (7.60)$$

It is interesting to compare (7.60) with the data at energies higher than the charm-production threshold. So far we don't know the decay channels of charm (or beauty) hadrons exactly; on the other hand, due to the large masses of these hadrons it is evident, that sea hadron multiplicities in c- or b-quark initiated jets are less than those in u, d, s-quark initiated jets. Experimental values of $\langle n_{ch} \rangle_{e^+e^-}$ near $c\bar{c}$ and $b\bar{b}$ thresholds are rather smooth functions of energy; this may mean that the decay products of the fragmentational particles with heavy flavours effectively reproduce the missing sea hadron multiplicity. In other words, one may assume that the total secondary multiplicity in the jet depends only on the available energy for particle production, not on the flavour content of the quark-partons, initiated by the jet. The comparison of (7.60) with the data at 4 GeV $\lesssim s \lesssim$ 30 GeV in Fig. 7.6b supports this assumption.

There are many attempts to compare the main properties of the multiparticle production in hard and soft processes directly considering the experimental data (see e.g. Refs. 22-25). The authors of these works usually try to separate leading secondary particles in hadron-hadron inelastic collision; the properties under consideration (multiplicity, jet characteristics, particle distributions, etc.) are then referred to the "available" energy $E_a = \sqrt{s} - E_{lead}$, where E_{lead} is the energy of the separated leading particles. Such an approach is very similar to the one used in the additive quark model: the "available energy" is usually close to the total energy of the colliding quarks $\sqrt{s_{qq}}$. However, experimentalists are always confronted with some uncertainties in the procedure of separation of leading particles while the expression (7.54) of the quark model is quite definite. In any case, the results obtained here and in the papers cited give arguments in favour of the assumption that the properties of the sea secondaries produced in the multihadron production process do not depend on the "source" of the process — on hard or soft collisions.

Let us now calculate the average multiplicities of the secondary hadrons of a given sort: partial average multiplicities. Using the same considerations as for $\langle n_{ch} \rangle$, we may write the expression for the average multiplicity of the prompt hadron h in h_1h_2 collision in the form:

$$\langle n_h \rangle_{h_1h_2} = \Delta_{h_1} \cdot \delta_{hh_1} + \Delta_{h_2} \cdot \delta_{hh_2} + \sum_{i=1}^{5} (a_i^{hh_1} + a_i^{hh_2})$$

$$+ (1 - \Delta_{h_1} - \Delta_{h_2}) N_{h_1h_2}(s)(da^{hM} + a^{hB} + a^{h\bar{B}}) \quad . \qquad (7.61)$$

The first two terms in the right-hand side of Eq. (7.61) stand for the diffraction-dissociation contribution: Δ_{h_i} is the corresponding probability, while δ_{hh_i} is Kronecker's symbol. The coefficients $a_i^{hh_1}$ and $a_i^{hh_2}$ are the products of three factors: 1) the coefficient at the corresponding ensemble symbol in the right-hand side of Eq. (7.2) or (7.3) ($i = 1$ for B_{ij} and thus $a_1^{hh_k} \equiv 0$ if h_k is a meson; $i = 2$ for spectator baryons; 3 for spectator mesons; $i = 4$ and 5 for the tagged baryons and mesons correspondingly; 2) the coefficient $\alpha_I^B(L)$ or $\alpha_I^M(L)$ from the decomposition of the corresponding ensemble into SU(6) multiplets (Eqs. (7.25), (7.27)); 3) the coefficient $\mu_{h(L)}$ or $\beta_{h(L)}$ from the expansion of the corresponding SU(6) ensemble into real hadrons (see Appendix D). The coefficients a^{hM}, a^{hB} and a^{hB} are also the products of the factors 2) and 3).

Let us again consider for example pp collisions. For the secondary proton multiplicity $\langle n_p \rangle_{pp}$ the first two terms give the contribution $2\Delta_p$; $a_3^{pp} = a_5^{pp} = a^{pM} = a^{p\bar{B}} = 0$, while $a^{pB} = \alpha_0^B(0) \cdot \beta p$. In order to calculate a_1^{pp}, a_2^{pp} and a_4^{pp} it is necessary to obtain the explicit form of the baryon ensembles B_{ij}, $B_i + B_j$ and B_k in the case of the proton beam. To make the calculations simpler, we assume the beam to be totally polarized; it is unimportant from the point of view of the final answer. The spin-flavour part of the proton wave function in this case is (see Appendix C):

$$|p\rangle_{+\frac{1}{2}} = \Big|\frac{\sqrt{2}}{3}(u^\uparrow u^\uparrow d^\downarrow + u^\uparrow d^\downarrow u^\uparrow + d^\downarrow u^\uparrow u^\uparrow) - \frac{\sqrt{2}}{3}(u^\downarrow d^\uparrow u^\uparrow + d^\uparrow u^\downarrow u^\uparrow + u^\uparrow u^\downarrow d^\uparrow + u^\downarrow u^\uparrow d^\uparrow + u^\uparrow d^\uparrow u^\downarrow + d^\uparrow u^\uparrow u^\downarrow)\Big\rangle \quad . \tag{7.62}$$

We assume that the third quark (i.e. the last one in every term of (7.62)) is an interacting one. Thus the quark $d^\downarrow$ interacts with the probability 2/9 (the first term in the right-hand side of (7.62)), the quark $u^\uparrow$ interacts with the probability 5/9 (the next four terms), the quark $d^\uparrow$ with the probability 1/9 (the next two terms) and the quark $u^\downarrow$ with 1/9 (the last two terms). Thus the ensembles of the tagged hadrons for the proton beam in (7.2) are:

$$B_k = 2/9 \cdot B_{d\downarrow} + 5/9 \cdot B_{u\uparrow} + 1/9 \cdot B_{d\uparrow} + 1/9 \cdot B_{u\downarrow}$$

$$M_k = 2/9 \cdot M_{d\downarrow} + 5/9 \cdot M_{u\uparrow} + 1/9 \cdot M_{d\uparrow} + 1/9 \cdot M_{u\downarrow} \quad . \qquad (7.63)$$

Since the hadron content of the ensembles does not depend on the polarization of the initial quark, we obtain:

$$B_k = 1/3 \cdot B_d + 2/3 \cdot B_u$$

$$M_k = 1/3 \cdot M_d + 2/3 \cdot M_u \quad . \qquad (7.64)$$

Deleting the interacting quark from the wave function (7.62), we shall in each case obtain the corresponding state of quark-spectators. Thus with the probability 2/9 the spectators are in the state $|u^\uparrow u^\uparrow>$; with 5/9, in the state $\frac{1}{\sqrt{5}}\,|2(u^\uparrow d^\downarrow + d^\downarrow u^\uparrow) - \frac{1}{\sqrt{2}}(u^\downarrow d^\uparrow + d^\uparrow u^\downarrow) > \equiv |ud_0>_p$ (the subscript 0 means the quark spin projection, the subscript p refers to the initial proton beam); with 1/9, in the state $1/\sqrt{2}\,|u^\uparrow u^\downarrow + u^\downarrow u^\uparrow) > \equiv |uu_0>$ and with 1/9, in the state $1/\sqrt{2}|u^\uparrow d^\uparrow + d^\uparrow u^\uparrow) > \equiv |ud_1>$. The hadron content of $B_{u^\uparrow u^\uparrow}$ and B_{uu_0} ensembles is identical, and so we get for the proton beam:

$$B_{ij} = 5/9 \cdot B_{ud_{0p}} + 1/9 \cdot B_{ud_1} + 1/3 \cdot B_{uu} \quad . \qquad (7.65)$$

Similarly,

$$B_i = B_j = 2/3 \cdot B_u + 1/3 \cdot B_d \quad ; \qquad M_i = M_j = \frac{2}{3} M_u + \frac{1}{3} M_d \quad . \qquad (7.66)$$

We can now calculate the coefficients a_1^{pp}, a_2^{pp} and a_4^{pp}:

$$a_1^{pp} = (1-\Delta_p)\xi_1\alpha_2^B(0)\left[\frac{5}{9}\,\beta_p(ud_{0p}) + \frac{1}{9}\,\beta_p(ud_1) + \frac{1}{3}\,\beta_p(uu)\right]$$

$$a_2^{pp} = 2(1-\Delta_p)\xi_2\alpha_1^B(0)\left[\frac{2}{3}\,\beta_p(u) + \frac{1}{3}\,\beta_p(d)\right]$$

$$a_4^{pp} = 2(1-\Delta_p)(1-\xi_1-2\xi_2)\alpha_1^B(0)\left[\frac{2}{3}\,\beta_p(u) + \frac{1}{3}\,\beta_p(d)\right] \quad . \qquad (7.67)$$

Previously we characterized sea particles as those which do not remember the quantum numbers of the initial particles. In fact this is true only approximately. The reason is that the average quantum numbers (charge and strangeness) of the ensembles B_{ij}, B_i, M_i, etc. in (7.2) or (7.3) are not equal to those of the quarks (q_i, q_j, q_k or q_i, $\bar{q}_j$) of the beam (target). As the total charge and strangeness of the secondaries are equal to those of the colliding particles, this means that sea particles have some nonzero average quantum numbers, which depend on the initial quantum numbers. In other words, there is some "leakage" of the charge and strangeness of the beam (target) into the sea.

One source of this leakage is the relative suppression of the strange sea quark production. As a result an "average" sea quark carries nonzero average charge: $Q(q) = \frac{1}{2+\lambda} \cdot (Q(u) + Q(d) + \lambda Q(s)) = (1-\lambda)/(3(2+\lambda))$; $Q(\bar{q}) = -Q(q)$. Hence, the average charge of any ensemble $M_i = q_i\bar{q}$ or $B_i = q_iqq$ is not equal to that of the quark q_i: for example $Q(M_u) = (1+\lambda)/(2+\lambda)$. However, one can see that at $\lambda = 1$ (exact SU(6) symmetry) $Q(M_i) = Q(q_i)$.

Another source of leakage is the mechanism of soft colour neutralization for the baryon states, which was discussed in Sec. 7.3. This mechanism may be considered, for instance, as repeated decays of coloured "baryon" states with emission of white sea mesons. As a result of these decays some fraction of the initial quark (spectator or tagged) quantum numbers is transferred to the sea mesons. This leakage occurs even in the case of exact SU(6) symmetry. One can see from Table D.5 of Appendix D that, for example, when $\lambda = 1$ the average charge of the baryon ensemble B_u when only 56-plet baryons are produced is $Q(B_u)_{56} = 6/7$, while for 70-plet $Q(B_u)_{70} = 22/35$; the same for B_{uu} ensemble are $Q(B_{uu})_{56} = 3/2$, $Q(B_{uu})_{70} = 6/5$ and so on. The missing quantum numbers are transferred through the considered "decays" to sea mesons. So for the sake of simplicity we shall assume that only sea meson quantum numbers depend on the value of leakage, while sea baryons and antibaryons remain symmetric. Thus the content of sea meson ensemble M is not determined by the Eqs. (7.29) or (7.33), but slightly differ from them. The exact expressions for the leakage are given in Appendix D.

The expression (7.61) gives the average multiplicity of the prompt particles. In order to obtain the observable values one has to account for all the relevant decays of the produced resonances.

Up to now we considered ξ_1, ξ_2 and η in Eqs. (7.2) and (7.3) as free parameters. The rules of quark statistics (7.50) and (7.51) give some extremal estimates for the values of these coefficients. Equations (7.50) and (7.51) are valid in the case of the hadronization of the "isolated" initial quark. The question is whether quark-spectators and tagged quark may be regarded as isolated. In other words, if the hadronization processes of quark-spectators and tagged quark proceed independently, the coefficients ξ_1, ξ_2 and η are determined by Eqs. (7.50) and (7.51):

$$\xi_1 = 1/2 \quad , \quad \xi_2 = 1/2 \quad , \quad \eta = 1/3 \quad . \tag{7.68}$$

However, if these processes are correlated, the values of ξ_1, ξ_2 and η may differ from (7.68) significantly. This question may hardly be solved on the basis of average multiplicity data only. Indeed, for the baryon beam, $\langle n_h \rangle$ does not depend on the value of ξ_2 at all — the hadron content of the spectator and tagged ensembles are identical. More preferable is the situation for the meson beam: the parameter η is responsible for the number of secondary baryons in the meson fragmentation region. The data give arguments in favour of the presence of some correlations in the process of hadronization, η seems to be less than 1/3. The detailed analysis of the inclusive spectra (see the next section) allows one to estimate η more accurately. The situation for ξ_1 and ξ_2 is less definite. The value $\xi_1 = 1/2$, predicted by the rules of quark statistics (7.51) seems to fit data rather well.

Some data for average multiplicities in pp, πp, and Kp collisions are shown in Figs. 7.7-7.9. The curves correspond to the quark model calculations. In all cases the agreement is quite reasonable;

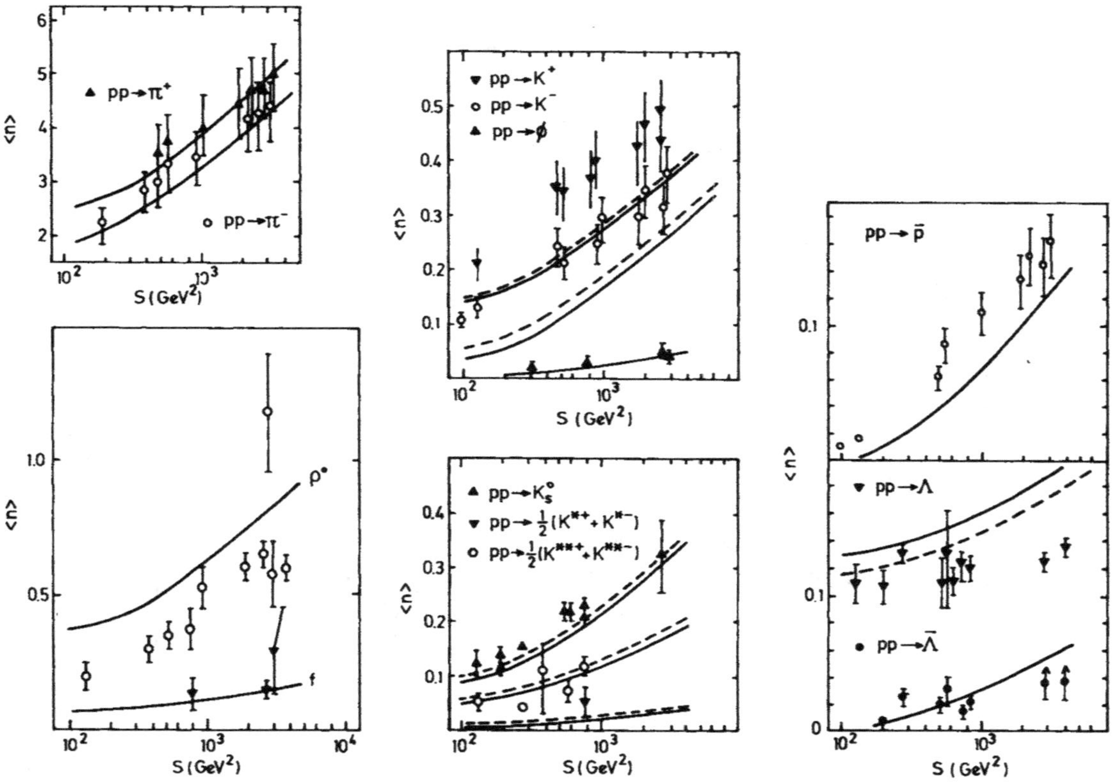

Fig. 7.7 Average particle multiplicities in pp collisions. Solid lines correspond to $\xi_1 = 1/2$, $\xi_2 = 1/12$ in Eq. (7.2), the dashed ones to $\xi_1 = 1/2$, $\xi_2 = 0.215$. Full list of references for experimental data is given in Ref. 35.

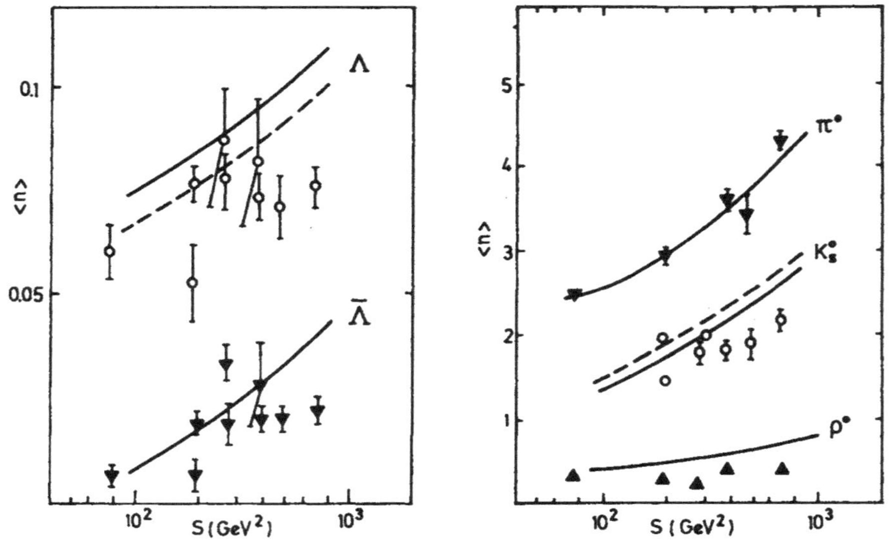

Fig. 7.8 Average particle multiplicities in πp collisions. Solid lines correspond to $\xi_1 = 1/2$, $\xi_2 = 1/12$ in Eq. (7.2), the dashed ones to $\xi_1 = 1/2$, $\xi_2 = 0.215$. Full list of references for experimental data is given in Ref. 35.

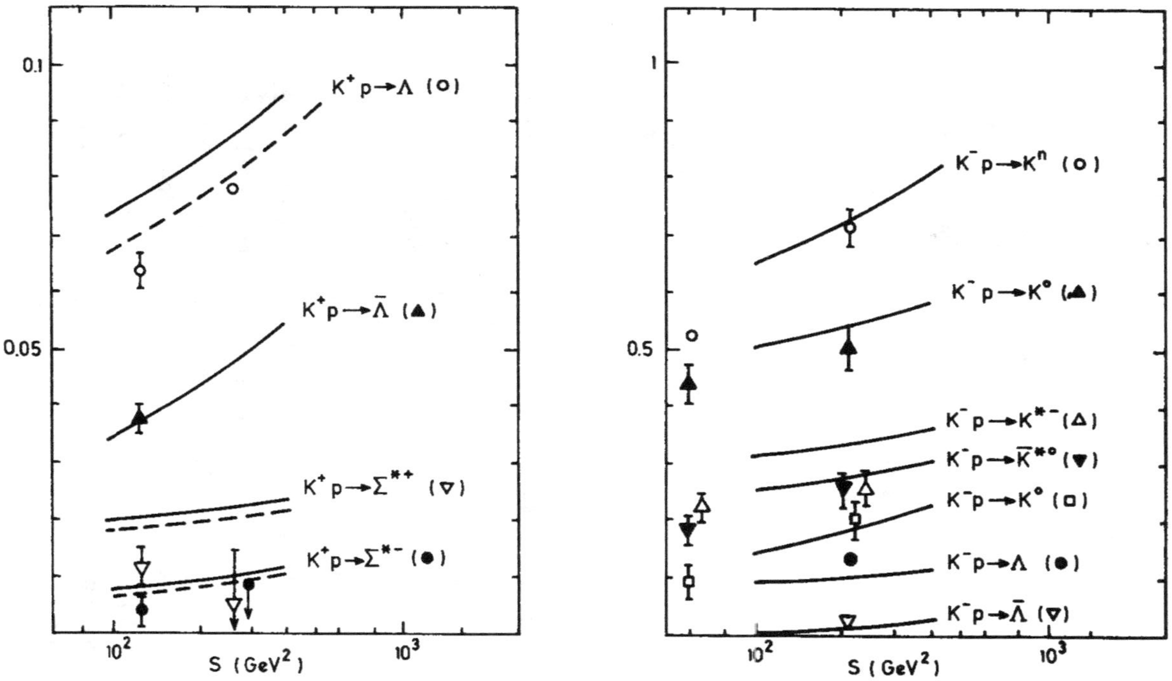

Fig. 7.9 Average particle multiplicities in Kp collisions. Solid lines correspond to $\xi_1 = 1/2$, $\xi_2 = 1/12$ in Eq. (7.2), the dashed ones to $\xi_1 = 1/2$, $\xi_2 = 0.215$. Full list of references for experimental data is given in Ref. 35.

the difference between the calculated and the measured values is, as a rule not more than 20%. Still, for $\langle n_{\rho^0} \rangle$ the predictions in many cases lie systematically higher than the experimental data. A reason for this can be that we neglected the contribution of the meson multiplet with $L = 2$ which is in fact produced with a probability of 5-10%. Estimating this contribution we see that it lowers the calculated ratio $\langle n_{\rho^0} \rangle / \langle n_{\pi^0} \rangle$ by the same 5-10%. Besides, one has to remember that there are several difficulties in the determination of the $\langle n_{\rho^0} \rangle$ production cross section because of background in the $\pi^+\pi^-$ mass spectrum. Taking into account the reflection of the ω-meson in the ρ^0 background one usually comes to an increase of the extracted value for the ρ^0 multiplicity[26-28].

Note here that the value of $\langle n_{ch}^{frag} \rangle$ of "purely fragmentational" charged particle multiplicity (i.e. of spectator and tagged hadrons) as predicted by our model is rather large: $\langle n_{ch}^{frag} \rangle \simeq 5.9$ for pp collisions, 5.5 for πp collisions and 5.4 for Kp collisions. This means that the predictions of the model may be compared with the experimental data only for sufficiently high initial energies, when $\langle n_{ch} \rangle > \langle n_{ch}^{frag} \rangle$ and sea hadron multiplicity is therefore non-negative. The corresponding energies are $s \geq 200\ \text{GeV}^2$ for pp collisions and $s \geq 150\ \text{GeV}^2$ for πp and Kp collisions. The extrapolation of the theoretical curves to the region of smaller s values though sometimes possible may lead to disagreement with the data.

The comparison of the predictions of the quark statistics for the multiplicity of secondary particles in hadron-hadron collisions with the experimental data in the pp, πp and Kp interactions permits us to conclude that this approach gives a good description for the main relations between the particles produced. The experimental data support the assumption due to which secondary particles are produced in SU(6) multiplets.

7.5 Inclusive Spectra

The inclusive distributions of the secondaries were already considered in Sec. 7.2, where all necessary expressions have been obtained. Since the hadron content of the meson and baryon ensembles is also established, these expressions allow one to calculate the prompt particle spectra explicitly.

The expressions of Sec. 7.2 contain two unknown functions $F(x)$ and $\Phi(x)$ which describe the secondary hadron x-distributions in the quark jets: $F(x)$ is the distribution of the tagged hadrons, while $\Phi(x)$ is that of the sea hadrons. They may be determined from the experimental data.

From the point of view of quark statistics, quark jets have the simplest structure in the hard processes of e^+e^- annihilation into hadrons and deep inelastic scattering in the region of current fragmentation. It is therefore important to understand how, within the framework of quark statistics, the spectra of hadrons produced in these processes is handled. The inclusive spectra of hadrons in quark jets are usually investigated in terms of the variable $z = $ (hadron energy)/(jet energy) (the notation z is used as a rule in deep inelastic processes while in e^+e^- annihilation this variable is denoted by x_E). The approach using quark statistics cannot be applied to the description of spectra at $z \to 1$. Indeed, the aim of this approach is only to deal with multiparticle-production processes. However the particles which carry away almost all the total momentum of the jet are not produced under the conditions of multiparticle production. The rapidity gap between these particles and their neighbour is much larger than the average value (Fig. 7.10a); see also Eqs. (2.15) and (2.23). Their spectrum can be characterized by the behaviour of the hadron form factors at large momentum transfers (see Refs. 29, 30, 31). The typical distribution of hadrons in multiparticle production processes (in the case when only u, d, s quarks are produced) is demonstrated in Fig. 7.10b. Obviously quark statistics tries to describe the inclusive spectra only at $z \lesssim 0.7$; for the region $z \gtrsim 0.7$ this approach is not justified. By the same reason it is not possible to handle inclusive spectra of hadrons with large p_T within the framework of quark statistics. Indeed, in this case spectra are formed by jets with large

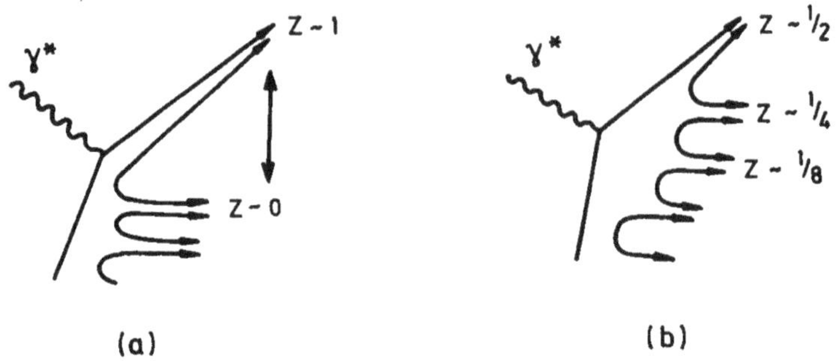

Fig. 7.10 Hadron production in quark jets:
(a) production of the particle with $z \to 1$;
(b) typical hadron distribution in multiple production processes.

intervals on the rapidity axis (of the type shown in Fig. 7.10a). Such momentum configurations are the most probable in these processes because of the rapid decrease of quark-quark or quark-gluon cross sections with increase of the momentum transferred.

The most accurate data for inclusive spectra in hard scatterings are obtained in e^+e^- annihilation. Below the charm production threshold these spectra are determined by the hadronization processes of u, d, s quarks and antiquarks. Due to the early scaling observed in the spectra $\frac{s}{\beta}\frac{d\sigma}{dx_E}$ $(e^+e^- \to \text{hadrons})$ (here $\beta = v/c$, the hadron velocity, $x_E = z = 2E/\sqrt{s}$) there is reason to hope that the rules of quark statistics may be applied here at comparatively low initial energies.

In Sec. 7.2 it was shown that the secondary baryons and mesons in the quark jets have the same z-distributions. This may be true, of course, only in the region where the model is applicable, i.e. at $z \lesssim 0.7$, while at $z \to 1$ these distributions must have another z-behaviour due to the difference in Q^2-dependence of the meson and baryon form factors. It is, however, necessary to take into account somehow the particle production in the region $z > 0.7$ also: the decay products of resonances with $z > 0.7$ might appear in the region $z < 0.7$. For this purpose distribution functions $F^M(z)$ and $F^B(z)$ of Ref. 32 were introduced, which were different only as $z \to 1$:

$$F^B(z) \sim \frac{F^M(z)}{1 + \frac{\kappa^2}{(1-z)^2}} \qquad \text{where} \qquad \kappa \simeq 0.1 \tag{7.69}$$

Such a parametrization provides us with the correct behaviour $F^M(z) \sim (1-z)$, $F^B(z) \sim (1-z)^3$ as $z \to 1$, which is determined by the asymptotes of the meson and baryon form factors. (The latter behave like Q^{-2} and Q^{-4} respectively.) Because of the smallness of κ in Eq. (7.69) the functions $F^M(z)$ and $F^B(z)$ practically coincide in the region $z < 0.7$. Thus in the jet initiated by the u-quark-parton the distribution of the prompt secondary hadron h from the L-th SU(6) multiplet may be expressed in the form:

$$\begin{aligned}\frac{1}{\sigma_{inel}} \cdot \frac{d\sigma}{dz} (u \to h(L)) = \frac{1}{3}\, \alpha_1^B(L) \beta_{h(L)}(u) F^B(z) + \frac{2}{3}\, \alpha_1^M(L) \mu_{h(L)}(u) F^M(z) \\ + d\alpha_0^M(L) \mu_{h(L)} \phi(z) \\ + (\alpha_0^B(L) \beta_{h(L)} + \alpha_0^{\bar{B}}(L) \beta_{\bar{h}(L)}) \Phi(z)\end{aligned} \tag{7.70}$$

Of course for the secondary baryon (or antibaryon) only the first and the last terms are different from zero in the right-hand side of Eq. (7.70) while for the secondary meson, only the second and the third terms are different from zero.

In e^+e^- annihilation below the charm production threshold there are two jets with two tagged quarks; their flavours are distributed proportionally to the squares of the quark charges, so that in each jet

$$\begin{aligned}B_i &= \frac{1}{3} B_u + \frac{1}{3} \bar{B}_{\bar{u}} + \frac{1}{12} B_d + \frac{1}{12} \bar{B}_{\bar{d}} + \frac{1}{12} B_s + \frac{1}{12} \bar{B}_{\bar{s}} \\ M_i &= \frac{1}{3} M_u + \frac{1}{3} M_{\bar{u}} + \frac{1}{12} M_d + \frac{1}{12} M_{\bar{d}} + \frac{1}{12} M_s + \frac{1}{12} M_{\bar{s}} \quad .\end{aligned} \tag{7.71}$$

and

$$\frac{1}{2\sigma_{e^+e^- \to \text{hadrons}}} \cdot \frac{s}{\beta} \frac{d\sigma}{dx_E} (e^+e^- \to h(L))$$

$$= \frac{1}{3} \alpha_1^B(L) F^B(x_E) \Big[\frac{1}{3} \beta_{h(L)}(u) + \frac{1}{3} \beta_{h(L)}(\bar{u}) + \frac{1}{12} \beta_{h(L)}(d)$$

$$+ \frac{1}{12} \beta_{h(L)}(\bar{d}) + \frac{1}{12} \beta_{h(L)}(s) + \frac{1}{12} \beta_{h(L)}(\bar{s})\Big]$$

$$+ \frac{2}{3} \alpha_1^M(L) F^B(x_E) \Big[\frac{1}{3} \mu_{h(L)}(u) + \frac{1}{3} \mu_{h(L)}(\bar{u}) + \frac{1}{12} \mu_{h(L)}(d)$$

$$+ \frac{1}{12} \mu_{h(L)}(\bar{d}) + \frac{1}{12} \mu_{h(L)}(s) + \frac{1}{12} \mu_{h(L)}(\bar{s})\Big]$$

$$+ d\alpha_0^M(L) \mu_{h(L)} \Phi(x_E) + \alpha_0^B(L) \beta_{h(L)} \Phi(x_E) + \alpha_0^{\bar{B}}(L) \beta_{\bar{h}(L)} \Phi(x_E) \quad . \tag{7.72}$$

The factor 2 in the denominator of the left-hand side of (7.72) appears because the data in e^+e^- annihilation are usually presented for the sum of two jets.

The fit of the data for $\pi^\pm$, $K^\pm$, K^n, p, and ρ^0 production was carried out with the help of (7.72) in Ref. 32; it is shown in Fig. 7.11a.

The same functions $F(z)$ and $\Phi(z)$ (Fig. 7.11b) determine the production of secondaries in the current fragmentation region for deep inelastic leptoproduction. Here only one jet is produced, and the expression for inclusive spectra is similar to Eq. (7.70) with the only difference that the flavour content of quark-partons which initiate the jets is determined by the structure function of the target.

EMC data for p/h^+ and $\bar{p}/h^-$ ratios at $0.4 < z < 0.6$ in μp deep inelastic collisions[33] (here $h^\pm = \pi^\pm + K^\pm + p^\pm$) are shown in Fig. 7.12 together with the quark model curves, obtained from the already known functions $F(z)$ and $\Phi(z)$. The agreement is quite reasonable, and we may conclude that the rules of quark statistics are confirmed by the data in the different z regions.

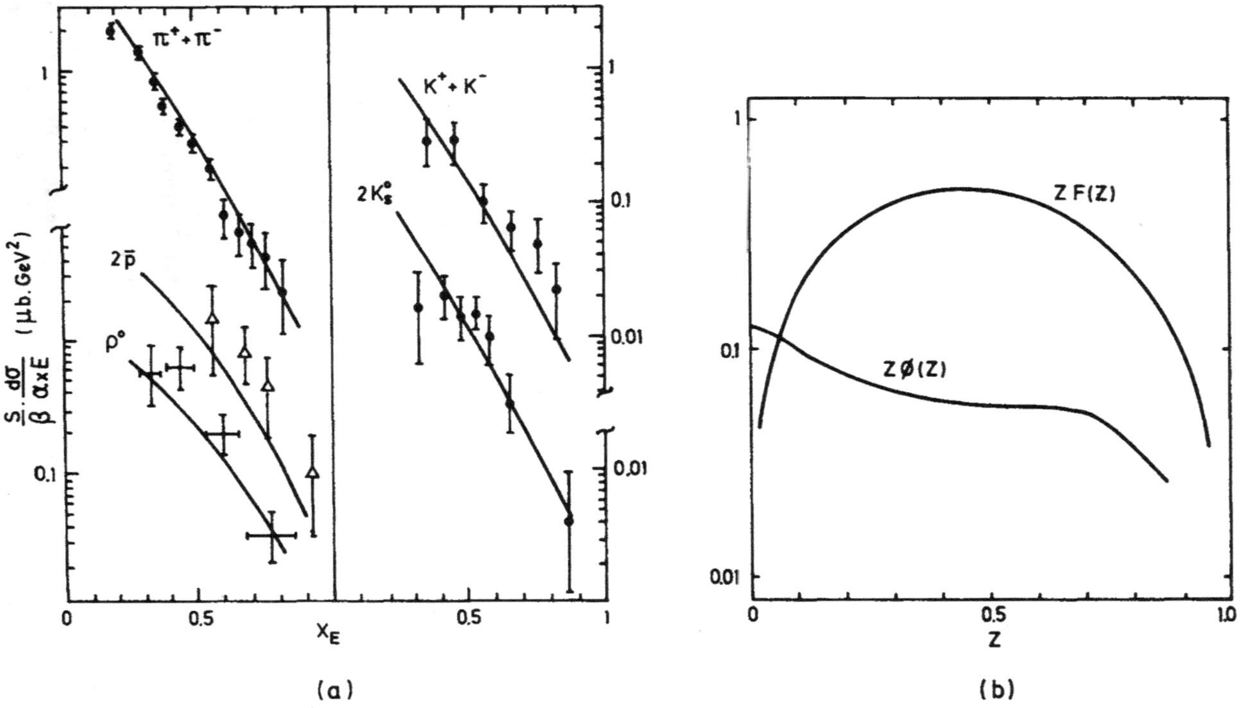

Fig. 7.11 (a) Secondary particle spectra in e^+e^- annihilation below the charm-production threshold and their description in the quark model.
(b) Fitted functions $zF(z)$ and $z\Phi(z)$.

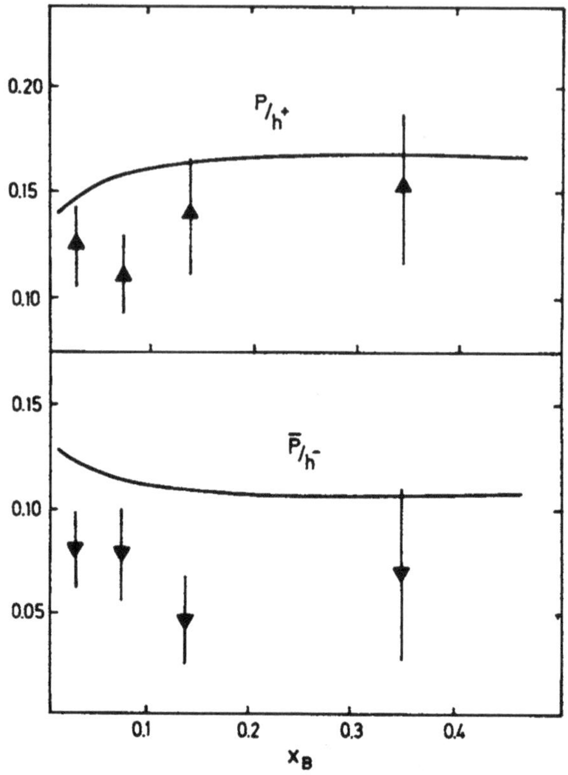

Fig. 7.12 The ratios p/h^+ and $\bar{p}/h^-$ (where $h^\pm = \pi^\pm + K^\pm + p^\pm$) in deep inelastic μp collisions at[33] $0.4 \leq z \leq 0.6$. The curves are the quark model predictions, calculated with the help of functions $F(z)$ and $\Phi(z)$ in Fig. 11.

Let us emphasize once more, that all the expressions for quark statistics refer to the prompt particles, and it is necessary to account for all the relevant decays when comparing the theoretical predictions with the data. The expressions for the inclusive spectra of the resonance decay products are presented in Appendix E. In order to demonstrate the significance of the higher resonance contributions, let us compare the data for the $\rho^0/\pi^\pm$ production ratio in e^+e^- annihilation[34] (Fig. 7.13) with our calculations with (dashed curve) and

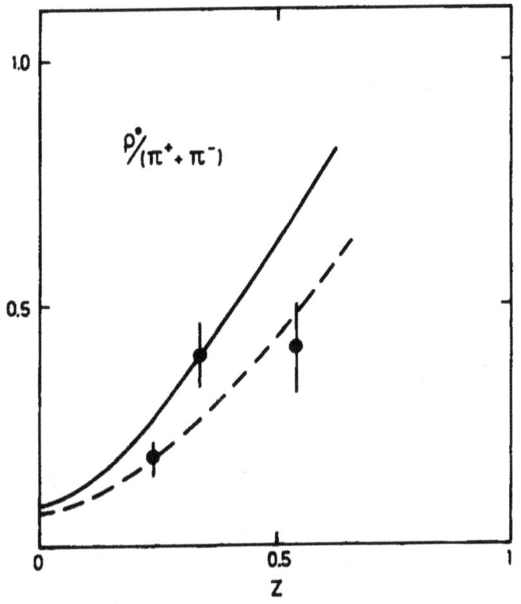

Fig. 7.13 $\rho^0/(\pi^+ + \pi^-)$ spectrum ratio[34] in e^+e^- annihilation as a function of x_E. The solid curve represent the quark model prediction without accounting for P-wave meson production, the dashed one takes these mesons into account.

without (solid curve) P-wave meson production accounted for. In the first case the agreement is quite reasonable, while in the second one at $z \gtrsim 0.5$ the calculations differ from the data appreciably.

Now we may turn our attention to the inclusive spectra in hadron-hadron inelastic collisions. The prompt inclusive cross section of hadron h_2 production in the forward hemisphere of the center-of-mass system of the reaction initiated by the hadron h_1 is determined by the sum

$$\frac{1}{\sigma_{inel}(h_1N)}\cdot\frac{d\sigma}{dx}(h_1N\to h_2X)$$

$$= \Delta_{h_1}\delta_{h_1h_2}V_{dif}(x) + a_1^{h_1h_2}D_{h_1}(x) + a_2^{h_1h_2}Q_{h_1}(x) + a_3^{h_1h_2}Q_{h_1}(x)$$

$$+ a_4^{h_1h_2}V_t^{h_1}(x) + a_5^{h_1h_2}V_t^{h_1}(x) + (1-\Delta_{h_1})$$

$$\cdot\,(da^{h_2M} + a^{h_2B} + a^{h_2\bar{B}})V_{sea}^{h_1}(x) \quad . \qquad (7.73)$$

The coefficients $a_i^{h_1h_2}$ and a^{h_2M}, a^{h_2B}, a^{h_2B} in the right-hand side of Eq. (7.73) are just the same as those in the right-hand side of Eq. (7.61) for the average multiplicities. It is obvious, that the expression in the right-hand side of (7.73), being integrated over x from 0 to 1, just gives the average multiplicity of h_2 in the forward hemisphere of the h_1N collision.

The detailed comparison of the Eq. (7.73) with the data in pp, $\pi^{\pm}p$, and $K^{\pm}p$ collisions is given in Refs. 4, 35. Here we shall restrict ourselves by a brief discussion of the results obtained.

First of all we note, that the rough parametrization (7.4) or (7.12) for the hadron wave function is by no means suitable in the whole x region. The parameters a_N, a_π, a_k determine only the half-width of the quark distribution. Hence, the calculations of the Refs. 4 and 35 are able to describe the data only in the region of x, where the corresponding secondary is produced with considerable probability. In the region of x, where spectra fall by 1-2 orders of magnitude, significant deviations of the calculated values from the data may be expected. For example, the exponential parametrization (7.4), (7.12) obviously gives a wrong asymptotic behaviour for the spectra as $x\to 1$ (let us remind ourselves that the quark statistical approach cannot be applied in this region).

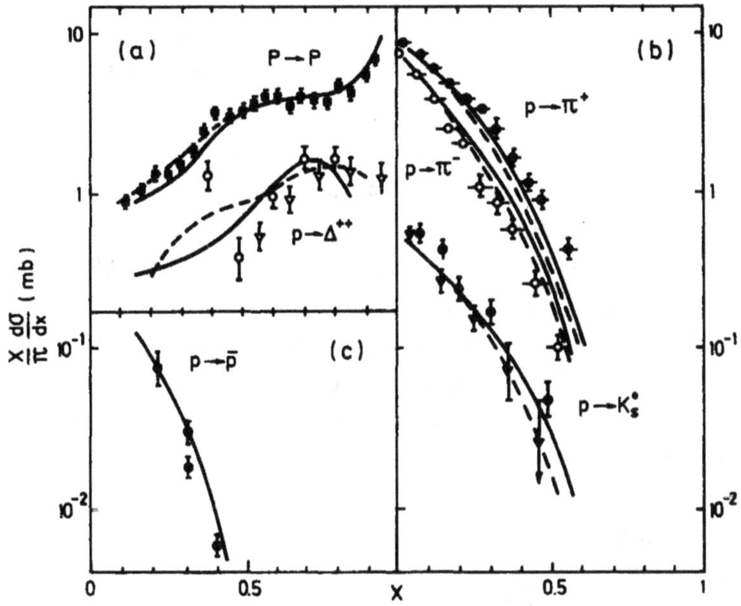

Fig. 7.14 Inclusive spectra of secondary particles in pp collisions. Solid lines are quark model calculation with $\xi_2 = 1/12$ in Eq. (7.2); dashed ones, the same with $\xi_2 = 0.215$. Full list of references for experimental data is given in Ref. 35.

In Fig. 7.14 some secondary spectra in pp collisions are shown. The value of $\xi_1 = 1/2$ in (7.2) is determined rather unambiguously from the experimental spectrum[36] $pp \to pX$ of $x \simeq 2/3$ (Fig. 7.14a); this value suits the data for Δ^{++} production[37,38] as well. As to the value of ξ_2, it remains rather ambiguous. We consider two extreme cases: $\xi_2 = 1/12$ as it is predicted by the quark statistical rules for the independent hadronization of the diquark by Eq. (7.51) (full curves in Fig. 7.14) and $\xi_2 = 0.215$ (dashed curves) which corresponds to the equal fractions of secondary tagged baryons for baryon and meson beams (see the discussion for the pion beam below). The difference between the predicted p and Δ^{++} spectra for $\xi_2 = 1/12$ and $\xi_2 = 0.215$ is rather small. The secondary π^- spectrum is described rather well

(Fig. 7.14b), while the π^+ spectrum at $x > 0.3$ is significantly underestimated. We considered the proton to be a pure $\{56, 0^+\}$ state with a symmetric coordinate part of the wave function (Sec. 5.1). Thus the x-distributions of u and d quarks in the proton are, in this case, identical and the production ratio π^+/π^- for the prompt particle in the proton fragmentation region has to be exactly $\pi^+/\pi^- = 2$. The contribution of the heavy baryon resonance decays (like N(1400), N(1690), and so on) would slightly increase this ratio, while that due to the production of meson resonances (like ρ^0, ω, A_2, etc.) would decrease. As a result, the calculated value of the observable π^+/π^- ratio for the symmetric proton wave function would hardly exceed 2, while the experimental value at $x > 0.3$ is appreciably larger. However, there are some arguments favouring an u-d asymmetry in the proton wave function. Such an asymmetry appears in potential models as a result of configurational mixing due to hyperfine spin-spin interactions (see Sec. 5.3). Another evidence for this asymmetry was mentioned in Sec. 6.3, when flavour exchange reactions were considered. At last, there is a well-known asymmetry in the distributions of u and d quark-partons in the proton structure function; it may also reflect the asymmetry in the dressed quark distributions.

Turning to the particle production in meson beams (Figs. 7.15, 7.16) we may note that for the double charge exchange production (like $\pi^+p \to \pi^-X$ or $\pi^-p \to \pi^+X$) there is a significant discrepancy with the data at $x \sim 0.8$. The large x region for these reactions is dominated by the diffraction dissociation processes, which are often described by the triple Reggeon graphs (see Sec. 2.3). As was previously mentioned, we effectively accounted for triple pomeron contribution (PPP) in our calculation, but did not take into account contributions like PRR, where R is some nonvacuum Reggeon. Thus, reactions $\pi^{\pm}p \to \pi^{\pm}X$, which may proceed through PPP diagrams, are described at $x \sim 0.8$ rather well, while the reactions $\pi^{\pm}p \to \pi^{+}X$, which may proceed through PRR diagrams (e.g. $\pi^-p \to \rho^0X \to \pi^+\pi^-X$ through π or A_1 exchange in the meson vertex) disagree with these data.

The data for Λ, $\bar{\Lambda}$ hyperon production in pp, πp, and Kp

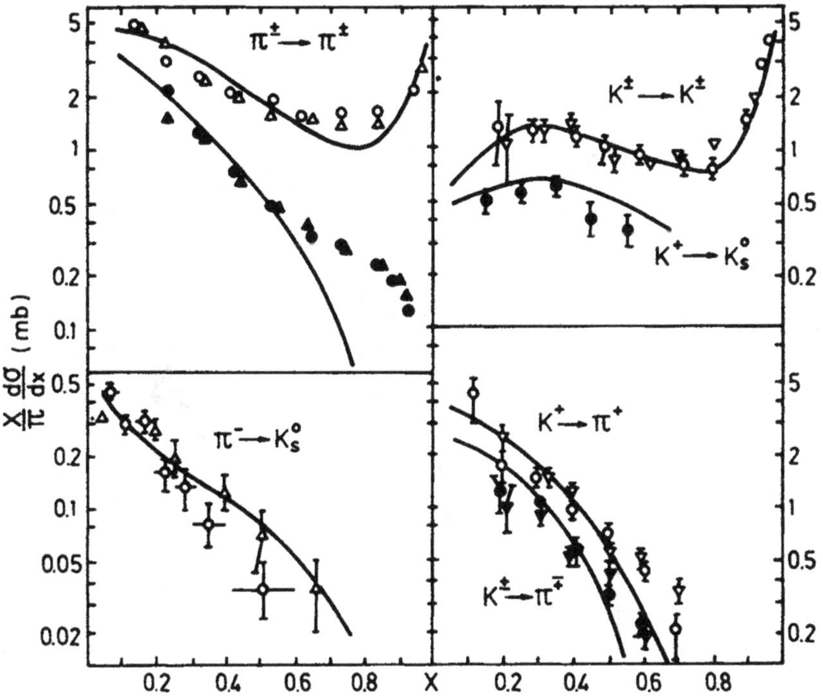

Fig. 7.15 Secondary meson spectra in p and Kp collisions. The curves are the quark model predictions.

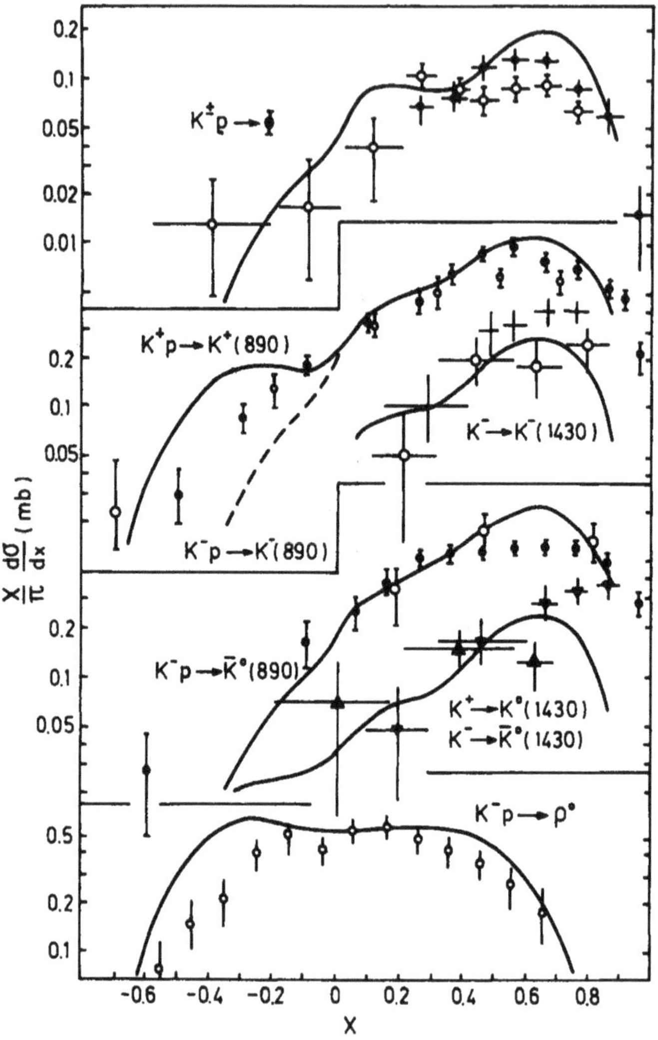

Fig. 7.16 Secondary meson resonance spectra in Kp collisions. The curves are the quark model predictions.

collisions are presented in Fig. 7.17. As in Fig. 7.14, the solid curve corresponds to the value $\xi_2 = 1/12$; in the region $x \sim 0.2 - 0.3$ of the proton hemisphere it slightly underestimates the data. The dashed curve for $\xi_2 = 0.215$ is in better agreement with the data. Baryon production in the meson fragmentation region allows one to estimate the parameter η in (7.3) rather accurately; it appears to be $\eta = 0.07$. This means that there is a significant correlation in the hadronization processes of the spectator and the tagged quark for the meson beam.

Recently the data on $\Lambda(1520)$ production (P-wave 70-plet resonance) in pp collisions were reported[39]; they were compared with our calculations in Fig. 7.18. The experimental points refer to the fixed value $p_T = 0.65$ GeV/c, while the theoretical curve is calculated for the spectrum integrated over p_T. The experimental values of the ratio $d\sigma/dx(\Lambda(1520))/(d\sigma/dx(\Lambda))$ are systematically higher than the theoretical curve by 40-50%, but the observed x-dependence at $x \gtrsim 0.6$ is very similar to the predicted one. The discrepancy may be due to slightly different p_T dependences of Λ and $\Lambda(1520)$ cross sections; another reason is the experimental uncertainty in the branching fraction $\Lambda(1520) \to \Lambda\pi\pi$, which is used in Ref. 39 to account for the unobserved modes of $\Lambda(1520)$ decay. Besides that, the fractions of P-wave baryons $\alpha_I^B(1)$ used in our calculations were not taken from the experimental data, but determined by simple qualitative considerations (see Eqs. (7.41), (7.42)). They are probably rather rough, and this may give an additional discrepancy in Fig. 7.18.

One can conclude, that the described structure of multiparticle production processes at high energies — namely, the formation of clouds of dressed quarks and their hadronization — enables us to give a quantitative description of the production for all secondary particles (mesons, baryons, meson and baryon resonances).

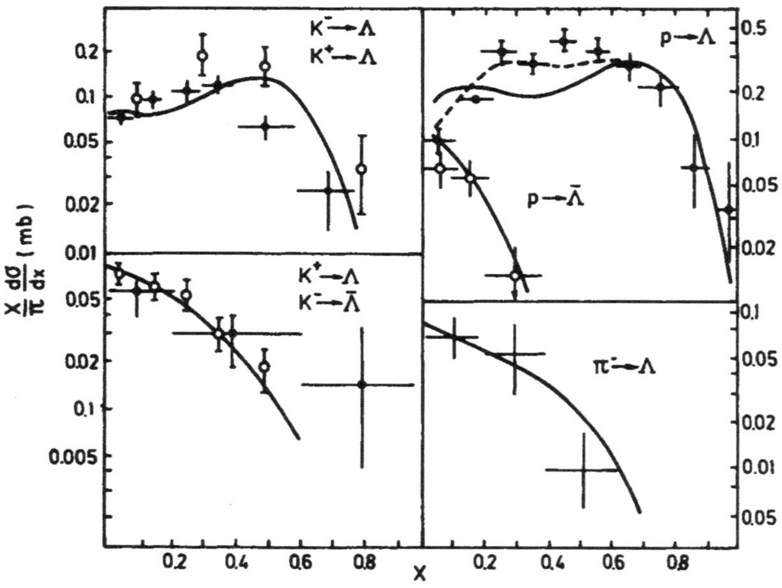

Fig. 7.17 Secondary Λ, $\bar{\Lambda}$-hyperon spectra in pp, πp, and Kp collisions. Solid lines are quark model calculation with $\xi_2 = 1/12$ in Eq. (7.2); dashed ones, the same with $\xi_2 = 0.215$. Full list of references for experimental data is given in Ref. 35.

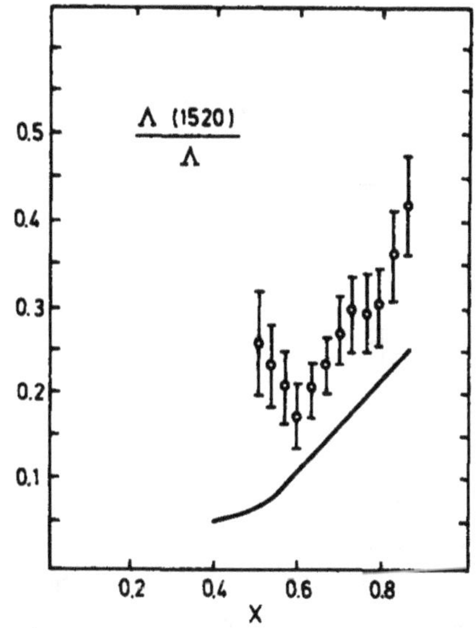

Fig. 7.18 Λ(1520) to Λ spectrum ratio at $p_T = 0.65$ GeV/c[39]. The curve is the quark model prediction, which corresponds to the P-wave baryon fraction, determined by the Eqs. (7.41), (7.42).

7.6 Multiple Rescattering and Scale Breaking at Superhigh Energies

7.6.1 *Shadow corrections*

The impulse approximation is a typical first approximation in the description of hadron-hadron collisions in the framework of the additive quark model. The next approximations have to account for shadow corrections which appear due to the multiple quark interactions. Such corrections were considered for elastic pp scatterings in Sec. 6.5. Here we shall discuss the modifications in the calculated inclusive spectra when the inelastic interactions of two or three initial quarks is taken into account.

Shadow corrections diminish the number of quark-spectators and the corresponding multiplicity of the spectator hadrons. As was already considered, in the process of Fig. 7.19a there are two spectators

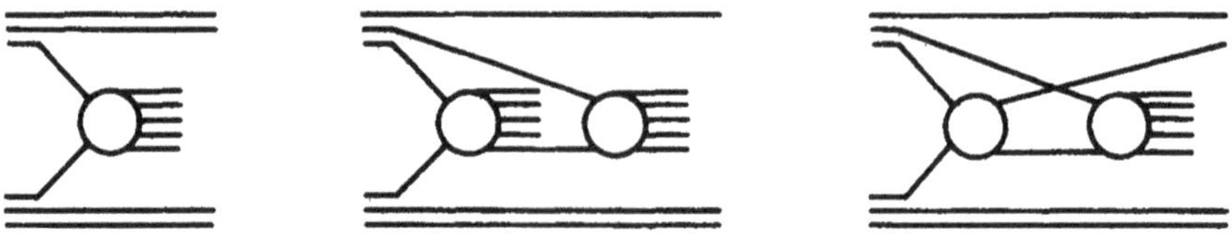

Fig. 7.19 Quark model diagrams with the interaction of one (a) and two (b, c) fast quarks of the projectile nucleon.

which may produce one secondary baryon B_{ij} with $x \sim 2/3$ or a pair of particles (two of B_i, B_j, M_i, M_j) with $x \sim 1/3$. In the process of Fig. 7.19b there is only one quark-spectator, but instead there are two tagged quarks and the number of sea quarks is also doubled. The diagram like Fig. 7.19c, where one quark interacts inelastically, while the other one undergoes elastic scattering, are also to be considered. The usual value of the momentum transfer in qq scattering is comparatively large in the hadron scale, it is of the order of r_q^{-1}. Thus the pair of remaining dressed quarks in Fig. 7.19c hardly can be joined in one secondary baryon B_{ij}. As a result, when the probabilities of the transitions "quark-hadron" and "diquark-hadron" are fixed, shadow corrections diminish the multiplicity of B_{ij} and increase the multiplicities of the tagged and sea particles. In other words, shadow corrections effectively renormalize the values of $\tilde{\xi}_i$ in Eq. (7.1).

Let $\langle n_q \rangle$ be the average number of the inelastically interacting quarks. Then the sum rules for $\tilde{\xi}_i$ in (7.1) including shadow corrections will be:

$$\tilde{\xi}_4 + \tilde{\xi}_5 = \langle n_q \rangle \qquad \text{(the conservation of the average number of inelastically interacted quarks)}$$

$$2\tilde{\xi}_1 + 2\tilde{\xi}_2 + 2\tilde{\xi}_3 = 3 - \langle n_q \rangle \qquad \text{(the conservation of the average number of the remaining dressed quarks)}$$

$$\tilde{\xi}_1 + 2\tilde{\xi}_2 + \tilde{\xi}_4 = 1 \qquad \text{(baryon number conservation)} \tag{7.74}$$

Moreover, there is one more relation

$$\tilde{\xi}_1 = \alpha\sigma_{react}(pp \to d_{qq}x)/\sigma_{react}(pp) \qquad 0 \le \alpha \le 1 \tag{7.75}$$

where α is the probability that the diquark in Figs. 19a, c turns into one baryon B_{ij}. Here $\sigma_{react}(pp \to d_{qq}X)$ is the inclusive cross section of "diquark-spectator" production (i.e. a pair of quarks with small relative momentum); it was calculated in Sec. 4.7. $\sigma_{react}(pp) = \sigma_{inel}(pp) - \sigma_{DD}(pp \to pX)$ is the value of the "true inelastic" pp collision cross section, which is caused by the inelastic interaction of at least one initial dressed quark.

The value of $\langle n_q \rangle$ may be written as

$$\langle n_q \rangle = 3\sigma_{react}(qp)/\sigma_{react}(pp) \tag{7.76}$$

where the value of the quark-proton cross section $\sigma_{react}(qp)$ is defined similarly to $\sigma_{react}(pp)$. Numerical calculations at $E \sim 10^2$ GeV in the multiple rescattering model give $\langle n_q \rangle \simeq 1.1$. If we set $\alpha = 1/2$ in (7.75) and $\tilde{\xi}_4 = 1/3$ in (7.74) (as it would be in the case of independent spectator and tagged quark hadronization), and the cross section $\sigma_{react}(pp \to d_{qq}x)$ is calculated via Eq. (4.100), we obtain[40,41]

$$\tilde{\xi}_1 = 0.39 \quad ; \quad \tilde{\xi}_2 = 0.12 \quad ; \quad \tilde{\xi}_3 = 0.44 \quad . \tag{7.77}$$

These values are rather close to the values 1/2, 1/12, and 5/12, which were obtained from Eq. (7.51) without accounting for shadow corrections. The spectra of secondary protons, Λ-hyperons and Δ^{++} resonances in pp collisions, calculated with the parameter set (7.77), are shown in Fig. 7.20. They differ only slightly from the spectra, calculated in the impulse approximation. This is an argument in favour of the self-consistency of the model.

If the value of α in Eq. (7.75) is chosen so that $\tilde{\xi}_1 = 1/2$, then *both for* $\tilde{\xi}_4 = 1/3$ *and* $\tilde{\xi}_4 = 0.07$ *the values of* $\tilde{\xi}_i$ *(i = 2, 3) and the* forms of the inclusive spectra will be very close to those in the impulse approximation.

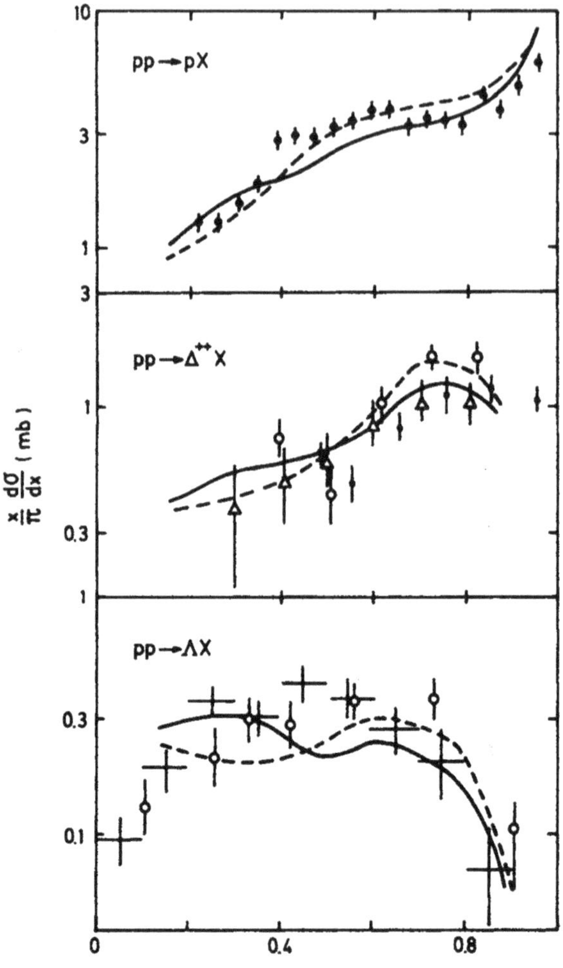

Fig. 7.20 Secondary p (a), Δ^{++} (b) and Λ (c) spectra in pp collisions in the impulse approximation (dashed curves) and with shadow corrections (solid curves), calculated for the value $\sigma_{tot}(pp) = 40$ mb.

7.6.2 *Scale breaking*

Due to the growth of the total pp cross section for $E_{lab} > 10^2$ GeV the role of multiquark interactions of Fig. 7.19b, c type increases and, as a result, the average number of quark-spectators decreases. It leads to the breaking of the Feynman scaling law in the fragmentation region[40,41]. The $\sigma_{tot}(pp)$ dependence of the parameters $\tilde{\xi}_1, \tilde{\xi}_2, \tilde{\xi}_3$ for $\alpha = 1/2$ in (7.75) and $\tilde{\xi}_4 = 1/3$ as an input is presented[41] in Table 7.2 together with the values of the initial energy E_{lab}, which are given only as a guide.

Table 7.2 The $\sigma_{tot}(pp)$ dependence of the coefficients $\tilde{\xi}_1, \tilde{\xi}_2, \tilde{\xi}_3$, and the number of inelastically interacting quarks $\langle n_q \rangle$. IA stands for impulse approximation values, taken as an input.

$\sigma_{tot}(pp)$,mb	$E^{*)}$, GeV	$\tilde{\xi}_1$	$\tilde{\xi}_2$	$\tilde{\xi}_3$	$\langle n_q \rangle$
40	IA	1/2	1/12	5/12	1
40	10^2	0.39	0.125	0.445	1.09
66	10^5	0.345	0.117	0.408	1.25
93	10^7	0.322	0.112	0.384	1.37

*) The estimated initial energy value E is given only as a guide, for all the values $\tilde{\xi}_i$ and $\langle n_q \rangle$ depend only on $\sigma_{tot}(pp)$.

Strictly speaking, the calculations performed are justified only in the fragmentation region. The reason is that some scale breaking may occur in the elementary quark-quark collision, e.g. due to Regge cut contributions (see Sec. 2.3). It is possible that such contributions are responsible for the growth of the central plateau in the energy interval 10^2-10^5 GeV[42]. However, some estimates, based on the analysis of hadron-nucleus collisions (see the next chapter) show that Regge cut contributions in quark-quark interactions are negligible at $x > 0.1$. Thus, our calculations of scale breaking in the fragmentation *region* $x > 0.1$ *seem to be reliable.*

In order to calculate the inclusive spectra of secondaries at superhigh energies it is necessary to know the probability of diffraction dissociation Δ in (7.2) as well. In the simplest case we assumed it to be energy independent.

The calculated values of scale breaking for different secondaries in pp collisions are presented in Fig. 7.21. Here the curves stand for the ratios of the inclusive spectra at $\sigma_{tot}(pp) = 66$ mb (full curves) and 93 mb (dashed curves) to that at $\sigma_{tot}(pp) = 40$ mb. At small x all spectra increase with the energy growth due to the growth of the average number of destroyed quarks $\langle n_q \rangle$; at $x \gtrsim 0.2$ the spectra decrease due to the decrease of the average number of the spectators.

The predicted values of the second momentum of x-distribution for the secondary $\pi^{\pm}$, π^0, $K^{\pm}$, η mesons and p, $\bar{p}$, n, Λ baryons

$$\langle x^2 \rangle_h = \int dx \frac{x^2}{\sigma_{in\ell}(pp)} \cdot \frac{d\sigma(pp \to hX)}{dx} \qquad (7.78)$$

are presented in Table 7.3 for $\sigma_{tot}(pp) = 40$, 66, and 93 mb. The value of $\langle x^2 \rangle_h$ characterizes the softening of the corresponding spectrum.

Table 7.3 The calculated values of $\langle x^2 \rangle$ for the secondaries in pp collisions. All $\langle x^2 \rangle$ in the Table are multiplied by 10^3.

$\sigma_{tot}(pp)$	Secondary					hadron		
	p	n	Λ	π^0	η	$\pi^{\pm}$	$K^{\pm}$	$\bar{p}$
40	246	51	9	23	7	45	4	1
66	239	47	8	23	7	44	4	1
93	233	44	8	22	7	43	4	1

Besides shadow corrections of the Glauber type, some other effects may lead to the scale breaking in hadron-hadron collisions at super-high energies. For example, in Ref. 43 the following mechanism was considered. In the quasinuclear quark model every dressed quark is

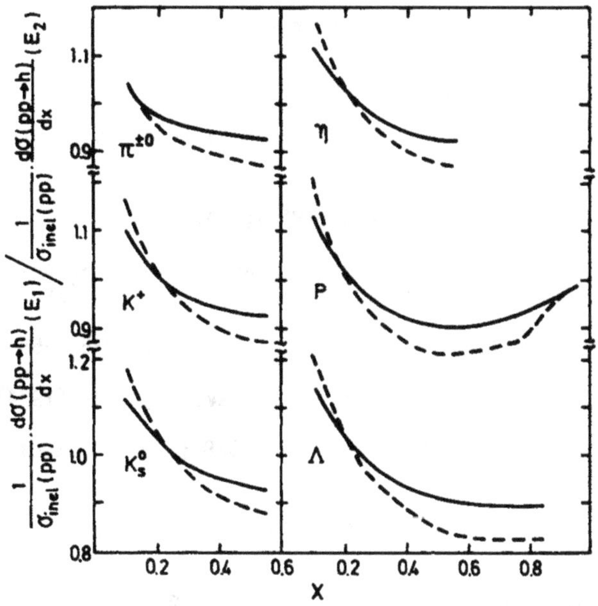

Fig. 7.21 The predictions for scale breaking in pp collisions. The curves are the calculated ratios of the inclusive spectra at $\sigma_{tot}(pp) =$ 66 mb (solid lines) and 93 mb (dashed) lines to that at $\sigma_{tot}(pp) = 40$ mb.

assumed to be a cloud of QCD quarks and gluons. The size of this cloud grows with the energy increase as $\alpha_p' \ln(\frac{s}{s_0})$. Thus at superhigh energies these clouds will probably overlap inside the hadron. In this case any inelastic interaction of one of the dressed quarks will destroy all the quarks of the colliding hadrons into partons, and there will be no quark-spectators at all. The corresponding scale breaking in the fragmentation region would be very significant: the spectra in pp collisions would probably be similar to those in e^+e^- annihilation. The natural parameter for such an effect is the ratio of the radii r_q^2/R_h^2, while the Glauber-type corrections are proportional to $\sigma(qq)/R_h^2$. Thus non-Glauber type effects may dominate at superhigh energies if the dressed quark size r_q^2 grows faster than the cross section $\sigma(qq)$.

References

1. H. Satz, Phys. Lett. 25B, 220 (1967).
2. I. Ya. Pomeranchuk, E.L. Feinberg, Doklady Ak. Nauk SSSR 93, 439 (1953) (in Russian).
3. E.L. Feinberg. ZhETF 50 202 (1966); Sov. Phys. JETP 23, 132 (1966).
4. V.V. Anisovich, M.N. Kobrinsky, J. Nyiri, Yu.M. Shabelski Yad. Fiz. 38, 425 (1983); Sov. J. Nucl. Phys. 38, 254 (1983).
5. V.V. Anisovich and V.M. Shekhter, Nucl. Phys. B55, 455 (1973).
6. J.D. Bjorken and G.E. Farrar, Phys. Rev. D9, 1449 (1974).
7. V.V. Anisovich, Yad. Fiz. 28, 761 (1978); Sov. J. Nucl. Phys. 28, 390 (1978).
8. V.V. Anisovich, M.N. Kobrinsky and J. Nyiri, Phys. Lett. B102, 357 (1981).
9. H. Kichimi et al., Lett. Nuovo Cim. 24, 129 (1979).
10. C. Cochet et al., Nucl. Phys. B155, 339 (1979).
11. V.M. Shekhter and L.M. Shcheglova, Yad. Fiz. 27, 1070 (1978); Sov. J. Nucl. Phys. 27, 567 (1978).
12. V.V. Anisovich, M.N. Kobrinsky and J. Nyiri, Preprint JINR P2-82-294 (1982) (in Russian).
13. M.I. Strikman and L.L. Frankfurt, Yad. Fiz. 32, 220 (1980); Sov. J. Nucl. Phys. 32, 113 (1980).
14. V.N. Guman and V.M. Shekhter, Nucl. Phys. B99, 523 (1975).
15. M.A. Voloshin, Yu.P. Nikitin and P.I. Porfirov, Yad. Fiz. 35, 1006, 1259 (1982); Sov. J. Nucl. Phys. 35, 586 (1982).
16. G.S. Abrams et al., Phys. Rev. 44, 10 (1980).
17. V.V. Anisovich and M.N. Kobrinsky, Yad. Fiz. 36, 972 (1982); Sov. J. Nucl. Phys. 36, 569 (1982).
18. P.K. Malhotra and R. Orava FERMILAB-PUB-82/79 (1982).
19. A. Wroblewski, Preprint VTL Pub-83 (1982).
20. M.W. Coles et al., Preprint SLAC-PUB-2916 (1982).

21. J.G. Rushbrooke, Preprint CERN-EP/84-34 (1984).

22. M. Basile et al., Phys. Lett. 92B, 367 (1980); 95B, 311 (1980).

23. D. Brick et al., Phys. Lett. 103B, 241 (1981).

24. V.G. Grishin et al., "Comparison of meson jet behaviour in π^-p collisions at 40 GeV/c with the data on e^+e^- annihilation.", Preprint JINR P1-82-754, Dubna 1982 (in Russian).

25. F.Z. Takagi, Phys. Rev. C13, 301 (1982).

26. V.G. Grishin, in Proceedings of the XVIII Int. Conf. on High Energy Physics, Tbilisi, p. A2-6 (Dubna 1977).

27. V.G. Grishin, Uspekhi Fiz. Nauk 127, 51 (1979); Sov. Phys. Uspekhi, 22, 1 (1979).

28. G. Jancso et al., Nucl. Phys. B124, 1 (1977).

29. R.P. Feynman, Photon-hadron Interactions, W.A. Benjamin, N.Y., 1972.

30. V.A. Matveev, R.M. Muradyan and A.N. Tavkhelidze, Lett. Nuovo Cim. 7, 719 (1973).

31. S.J. Brodsky and G.R. Farrar, Phys. Rev. Lett. 31, 1153 (1973).

32. V.V. Anisovich, M.N. Kobrinsky, P.E. Volkovitski, Z. Phys. C19, 221 (1983).

33. Y.J. Aubert et al., Phys. Lett. 103B, 338 (1981).

34. R. Brandelik et al., Preprint DESY 82-046 (1982).

35. V.V. Anisovich, M.N. Kobrinsky, J. Nyiri, Preprint LNPI-982 (1984).

36. C.P. Ward et al., Nucl. Phys. B153, 299 (1974).

37. D. Brick et al., Phys. Rev. D21, 632 (1980).

38. D. Drijard et al., Preprint CERN/EP 83-5 (1982).

39. G.J. Bobbink et al., Nucl. Phys. B217, 11 (1983).

40. V.V. Anisovich, V.M. Braun, YU.M. Shabelski, Yad. Fiz. 36, 1556 (1982); Sov. J. Nucl. Phys. 36, 904 (1982).

41. V.V. Anisovich, V.M. Braun, Yu.M. Shabelski, Yad. Fiz. 39, 932 (1984); Sov. J. Nucl. Phys. 39 (to be published).

42. A.B. Kaidalov, K.A. Ter-Martirosyan, Phys. Lett. 117B, 247 (1982).

43. V.V. Anisovich, V.M. Shekhter, Yad. Fiz. 28, 1079 (1978); Sov. J. Nucl. Phys. 28, 554 (1978).

Chapter 8

HADRON-NUCLEUS COLLISIONS

Soft hadron-nucleus collisions present an excellent field of application for the quasinuclear quark model, because they allow us to discover all the possibilities of the model. On the other hand, these processes provide us with a nice test of the basic assumptions of the quark model, as well as of the assumptions about hadron structure. The reason is that the calculations of the soft hadron-nucleus collision properties in the quasinuclear quark model are essentially unambiguous and do not involve any fitting procedure. We have already seen that the assumption of dressed quarks is very fruitful, it gives a lot of relations which describe the experimental data reasonably well. However, in every case some free parameters were introduced, their values being determined from the fit. Such a situation takes place, e.g. for the multiparticle production processes in hadron-hadron collisions. However, when hadron-nucleus collisions are considered, it appears that all necessary characteristics are already determined, and no additional free parameters are required.

This explains our heightened interest in hadron-nucleus collisions. The required calculation technique was described in detail in Chap. 4; the results of the first two sections of the present chapter are rather evident and may be understood without it.

The calculations which are considered in Secs. 8.3, 8.4 are

essentially based on this technique; however we have tried to present the results obtained in such a form that they should be clear to one who has not studied Chap. 4.

8.1 Spectator Mechanism and its Consequences

There are processes which allow one to observe in a relatively clear way the consequences of the spectator mechanism, i.e. to prove the hypothesis which is crucial from the point of view of the hadron structure. These are the multiple production processes of secondary particles on nuclei. They enable us to test the hadron structure because of the well-known fact that fast secondary hadrons do not multiply by possible repeated scatterings with the nuclear matter. Such rescatterings are obviously impossible, since, as was shown in Chap. 7, the formation time of sufficiently fast secondaries is so large that they can be produced only outside the nucleus, after the constituents have passed through. Consequently, while for a scattering process on a nucleon target the main contribution is given by the impulse approximation, in the case of a (not very light) target nucleus the interaction probabilities of one, two or three quarks of the projectile are of the same order of magnitude. These probabilities can be calculated in a relatively simple way as functions of the atomic weight A of the target nucleus.

Accepting the hadron picture with spatially separated quarks one assumes that each constituent quark interacts independently with nuclear matter. The nucleus plays the role of a filter which allows us to select interaction processes with different number of quarks belonging to the initial hadron. Thus the spectator mechanism of the fragmentational particle production may be tested in detail. Similarly the role of the dressed quarks in the formation of the central region spectra may be revealed.

8.1.1 *A-dependence of the inclusive spectra in the fragmentation region*

The accepted hadron picture suggests the assumption that the projectile dressed quarks interact in the nuclear matter independently. The inelastic collision of a quark with the target nucleon can be characterized by a quite definite cross section

$$\sigma_{\text{ine}\ell}(qN) \simeq \frac{1}{3}\,\sigma_{\text{ine}\ell}(pp) \simeq \frac{1}{2}\,\sigma_{\text{ine}\ell}(\pi p) \simeq 10\text{mb} \tag{8.1}$$

and each quark carries away about one-third of the nucleon momentum. These facts make it possible to predict the A-dependence of the production of different secondary particles. The comparison of such predictions with experiment provides us with a good opportunity to probe the quasi-nuclear hadron structure. As was mentioned already, in hadron-hadron collisions only one pair of constituent quarks takes part in the interaction (Fig. 8.1a); the other constituents remain spectators. The

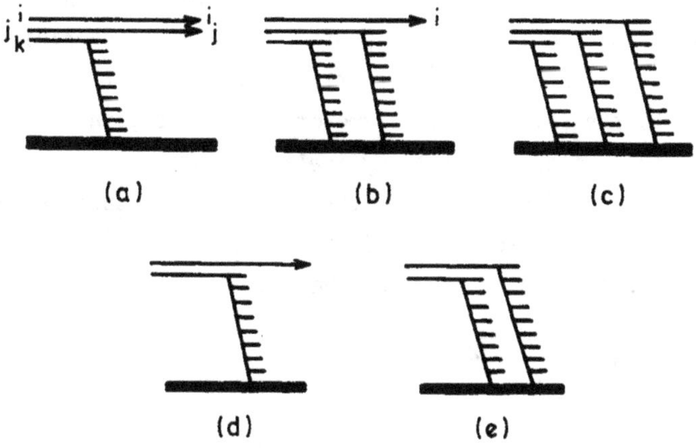

Fig. 8.1 Inelastic baryon-target (a, b, c) and meson-nucleus (d, e) collisions in the quasinuclear quark model.

probability of a proton-proton collision in which two (or three) quarks of a projectile interact (Figs. 8.1b and 8.1c) is not large (< 20%). The situation changes, when a collision of a nucleon with a nucleus (especially with a heavy one) is considered[1-4]. Since the target is large, the probabilities for the interactions of two or three quarks of the incident proton are not small anymore: while proceeding through nuclear matter, all the constituents can interact. The assumption of a definite value (8.1) for the quark-nucleon cross section enables us to calculate the probabilities $V_1^p(A)$, $V_2^p(A)$ and $V_3^p(A)$ of the three possible processes shown in Fig. 8.1a,b,c. $V_1^p(A)$ is the interaction probability of one constituent quark of the proton with the nuclues A; $V_2^p(A)$ and $V_3^p(A)$, respectively, stand for the probabilities when two or all three constituents interact.

The processes of high energy inelastic collisions with heavy nuclei $A \gtrsim 10$ were considered in detail in Sec. 4.4 with the help of a diagram technique. Here we will present simple macroscopic considerations, which give the same final expressions for the values $V_i^h(A)$ and $\sigma_{prod}(hA)$.

Let the impact parameter of the projectile proton in pA collision be $\vec{r}_T$. Neglecting the size of the proton compared to the radius of the nucleus, each constituent (dressed quark) of the incident proton can be characterized by the same impact parameter $\vec{r}_T$. If so, the probability to go through the nucleus without any inelastic interactions is $\exp[-\sigma_{ine\ell}(qN)T(\vec{r}_T)]$ for such a constituent; the corresponding probability of an inelastic interaction is $1 - \exp[-\sigma_{ine\ell}(qN)T(\vec{r}_T)]$. The function $T(\vec{r}_T)$ is expressed in terms of the nuclear density $\rho(r)$:

$$T(\vec{r}_T) = A\int_{-\infty}^{+\infty} dz\rho(\sqrt{\vec{r}_T^{\,2} + z^2}) \tag{8.2}$$

Integrating over the impact parameter $\vec{r}_T$ and taking into account that in both Figs. 8.1a and 8.1b there are three possibilities for quark interactions, we obtain the probabilities[4)]

$$V_1^p(A) = \frac{3}{\sigma_{prod}(pA)}\int d^2r_T e^{-2\sigma_{ine\ell}(qN)T(\vec{r}_T)}\left[1 - e^{-\sigma_{ine\ell}(qN)T(\vec{r}_T)}\right]$$

$$V_2^p(A) = \frac{3}{\sigma_{prod}(pA)}\int d^2r_T e^{-\sigma_{ine\ell}(qN)T(\vec{r}_T)}\left[1 - e^{-\sigma_{ine\ell}(qN)T(\vec{r}_T)}\right]^2$$

$$V_3^p(A) = \frac{1}{\sigma_{prod}(pA)}\int d^2r_T\left[1 - e^{-\sigma_{ine\ell}(qN)T(\vec{r}_T)}\right]^3 \tag{8.3}$$

where

$$\sigma_{prod}(pA) = \int d^2r_T[1 - e^{-3\sigma_{ine\ell}(qN)T(\vec{r}_T)}] \tag{8.4}$$

has the meaning of the inelastic pA cross section with the production of at least one fast secondary and is obtained from the condition $V_1^p + V_2^p + V_3^p = 1$.

The Fermi parametrization is used to describe the distribution of the nuclear matter

$$\rho(r) = \frac{\rho_0}{1 + \exp[(r - c_1)/c_2]} , \quad \int d^3r\rho(r) = 1 \quad . \tag{8.5}$$

The parameters c_1 and c_2 can be the best obtained from the data on elastic electron and hadron scattering on nuclei. The value c_1 is proportional to $A^{1/3}$: $c_1 = r_0A^{1/3}$; while c_2 is roughly the same for all nuclei, except the lightest ones. The values of r_0 and c_2 vary somewhat for different nuclei. For the required accuracy one can, however, neglect these slight differences and take

$$c_1 = 1.15\ A^{1/3}\ \text{fm}, \quad c_2 = 0.51\ \text{fm} \quad . \tag{8.6}$$

In Fig. 8.2 the cross sections $\sigma_{prod}(hA)$ calculated with the

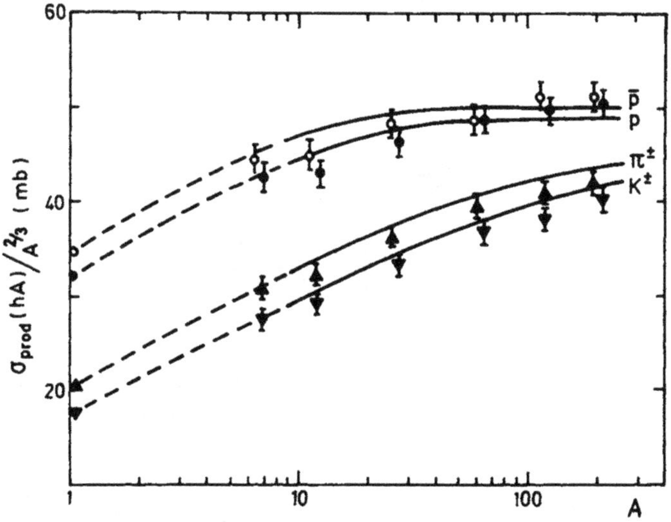

Fig. 8.2 A-dependences of hadron-nucleus production cross-sections.

above parameters are compared with the experimental data[5] at the energy of 200 GeV. (The values $\sigma_{\text{inel}}(hp)$ are taken from Ref. 6.) For the sake of convenience both the calculated and the experimental quantities are divided by $A^{2/3}$. As can be seen, the agreement is quite good. The cross sections $\sigma_{\text{prod}}(pA)$ are proportional to $A^{2/3}$ for $A \gtrsim 30$; for the cross sections $\sigma_{\text{prod}}(\pi A)$ and $\sigma_{\text{prod}}(KA)$ the increase is somewhat faster.

The calculated values of the probabilities V_1^p, V_2^p and V_3^p as functions of A are shown in Fig. 8.3a. Due to the expectations for light nuclei the one-quark-interaction process with the probability V_1^p domonates. However even for Be the probability of the process of

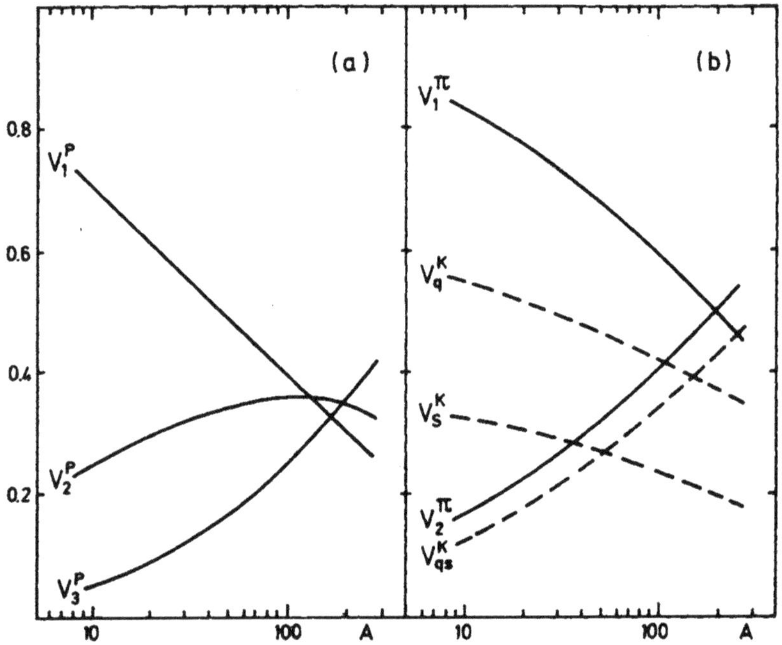

Fig. 8.3 The probability of absorbing a different number of incident quarks in hadron-nucleus interactions versus A. (a) The probabilities to absorb one, two or three quarks in a pA collision. (b) The quark absorption probabilities for πA (solid lines) and KA (dashed lines) interactions.

Fig. 8.1b, with two interacting quarks, is not small (~ 25%). When $A > 30$ the probability of the process of Fig. 8.1a, with two spectators, decreases roughly as $A^{-1/3}$. For $A > 100$ the probabilities of all the processes shown in Fig. 8.1 a,b,c are of the same order (~ 1/3).

The interaction probabilities of one or two incident quarks in the case of pion-nucleus collisions are calculated similarly (see Fig. 8.1 d,e):

$$V_1^{\pi}(A) = \frac{2}{\sigma_{prod}(\pi A)}\int d^2_{r_T}\, e^{-\sigma_{ine\ell}(qN)T(\vec{r}_T)}\left[1 - e^{-\sigma_{ine\ell}(qN)T(\vec{r}_T)}\right]$$

$$V_2^{\pi}(A) = \frac{1}{\sigma_{prod}(\pi A)}\int d^2 r_T\left[1 - e^{-\sigma_{ine\ell}(qN)T(\vec{r}_T)}\right]^2 \tag{8.7}$$

where

$$\sigma_{prod}(\pi A) = \int d^2 r_T[1 - e^{-2\sigma_{ine\ell}(qN)T(\vec{r}_T)}] \ . \tag{8.8}$$

The probabilities $V_1^{\pi}(A)$ and $V_2^{\pi}(A)$ are shown in Fig. 8.3b.

For the kaon-nucleus interactions the different behaviours of the strange quark and the nonstrange ones in the inelastic interaction have to be taken into account. This modifies somewhat the expressions (8.7) and (8.8):

$$V_s^K(A) = \frac{1}{\sigma_{prod}(KA)}\int d^2 r_T e^{-\sigma_{ine\ell}(qN)T(\vec{r}_T)}\left[1 - e^{-\sigma_{ine\ell}(sN)T(\vec{r}_T)}\right]$$

$$V_q^K(A) = \frac{1}{\sigma_{prod}(A)}\int d^2 r_T e^{-\sigma_{ine\ell}(sN)T(\vec{r}_T)}\left[1 - e^{-\sigma_{ine\ell}(qN)T(\vec{r}_T)}\right]$$

$$V_{qs}^K(A) = \frac{1}{\sigma_{prod}(KA)}\int d^2 r_T\left[1 - e^{-\sigma_{ine\ell}(sN)T(\vec{r}_T)}\right]\left[1 - e^{-\sigma_{ine\ell}(qN)T(\vec{r}_T)}\right] \tag{8.9}$$

where

$$\sigma_{prod}(KA) = \int d^2 r_T\left[1 - e^{-[\sigma_{ine\ell}(qN) + \sigma_{ine\ell}(sN)]T(\vec{r}_T)}\right] \tag{8.10}$$

whereas the cross section $\sigma_{ine\ell}(sN)$ is equal to

$$\sigma_{ine\ell}(sN) = \sigma_{ine\ell}(KN) - \sigma_{ine\ell}(qN) \simeq 6.5 \text{ mb} \quad . \tag{8.11}$$

The calculated values of $V_s^K(A)$, $V_q^K(A)$, $V_{qs}^K(A)$ are presented in Fig. 8.3b.

The numerical values of probabilities V_i for several nuclei in the cases of proton, pion and kaon beams are given in Table 8.1. For not very light nuclei these probabilities can be approximated by the form $V_i(A) \sim A^\alpha$. The results of such an approximation are given in Table 8.2.

Table 8.1 The probabilities of inelastic interactions of different numbers of quarks with the nucleus for proton (V_1^p, V_2^p, V_3^p), pion (V_1^π, V_2^π) and kaon (V_q^K, V_s^K, V_{qs}^K) beams.

	Nucleus						
	^1H	^9Be	^{12}C	$^{27}A\ell$	^{63}Cu	^{108}Ag	^{207}Pb
V_1^p	1	0.76	0.72	0.60	0.47	0.39	0.30
V_2^p	0	0.21	0.24	0.31	0.36	0.37	0.36
V_3^p	0	0.03	0.04	0.09	0.17	0.24	0.34
V_1^π	1	0.86	0.84	0.76	0.65	0.58	0.49
V_2^π	0	0.14	0.16	0.24	0.35	0.42	0.51
V_q^K	0.6	0.55	0.54	0.50	0.46	0.42	0.37
V_s^K	0.4	0.34	0.33	0.30	0.26	0.23	0.19
V_{qs}^K	0	0.11	0.13	0.20	0.28	0.35	0.44

As was mentioned in the beginning, hadron-nucleus interactions provide us with a tool to verify the nucleus-like hadron structure. Indeed, the latter leads to the quark-fragmentation mechanism, which implies that a large momentum fraction of the incident hadron remains with the constituent quark-spectators. The production of a fast secondary baryon

Table 8.2 Power approximation for A-dependences of the probabilities $V_i^h(A) = CA^{\alpha}$ for $30 \leq A \leq 240$.

Value	Approximation
V_1^p	$1.7\ A^{-0.32}$
$V_1^p + \frac{4}{5} V_2^p$	$1.45\ A^{-0.17}$
V_1^{π}	$1.5\ A^{-0.21}$
V_q	$0.82\ A^{-0.15}$
V_s	$0.6\ A^{-0.21}$

with $x = 2/3$ in the fragmentation reaction $p \to$ baryon B is due to two non-interacting quark-spectators picking up the newly created quark from the sea with a relatively small momentum, i.e. to the process which is shown in Fig. 8.1a (it is highly improbable that such a baryon can be formed by one spectator with an anomalously large momentum $x = 2/3$ and by two sea quarks). The production process of the $x = 2/3$ baryon is the same in both the proton-nucleus and the proton-proton interactions. The upper vertices in Fig. 8.1a coincide for pA and pp collisions, so they cancel in the ratio of the production probabilities. This means that the unknown dynamics in the formation process of a concrete baryon disappears and there remains only the probability ratio of the interaction shown in Fig. 8.1a on nuclear (V_1^p) and on proton ($V_1^p \simeq 1$) targets[4). In other words

$$\frac{\dfrac{1}{\sigma_{prod}(pA)} \cdot \dfrac{d^3\sigma}{d^3p}(pA \to B)}{\dfrac{1}{\sigma_{in\ell}(pp)} \cdot \dfrac{d^3\sigma}{d^3p}(pp \to B)} = V_1^p(A) \qquad (x \sim \tfrac{2}{3}) \ . \qquad (8.12)$$

This equation means that the ratio of the experimentally observable quantities at the left-hand side does not depend on the sort of the secondary baryon B, on the initial energy or on the x-value near $x = 2/3$. It can depend only on A, and this shows — in a quite definite way, corresponding to Eq. (8.3) and Fig. 8.4 — how well this prediction is fulfilled. Here for the left-hand side of Eq. (8.12) the experimental ratio of the production probabilities of secondary protons with $x = 0.7$ in pA and pp collisions[7] at 100 GeV/c is drawn.

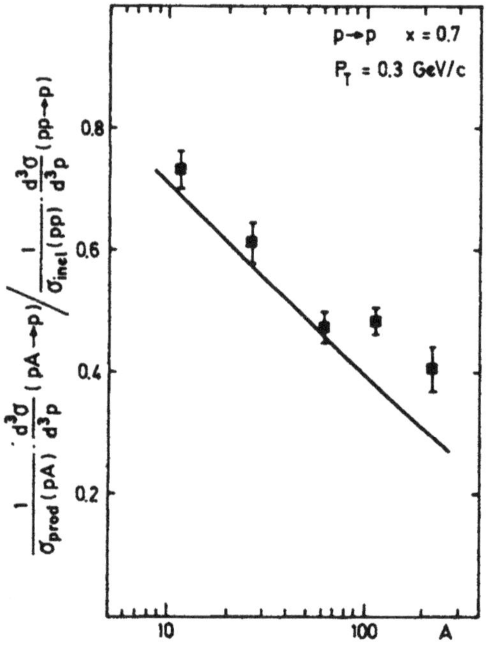

Fig. 8.4 Experimental check of the relation (8.12).
The solid line is the quark model prediction.
Data at 100 GeV/c are taken from Ref. 7.

For pion-nucleus and pion-nucleon collisions the production of fast secondaries with $x \sim 1/2$ is defined by the same process, in which one quark of the incident pion interacts (Fig. 8.1d). Because of this fact the ratio of the multiplicities for the secondary fragmentational particles in πA and πp collisions must not depend on x in the $x \sim 1/2$ region and is equal to

$$\frac{\dfrac{1}{\sigma_{prod}(\pi A)} \cdot \dfrac{d^3\sigma}{d^3p}(\pi A \to h)}{\dfrac{1}{\sigma_{in e\ell}(\pi p)} \cdot \dfrac{d^3\sigma}{d^3p}(\pi p \to h)} = V_1^{\pi}(A) \qquad (x \sim \tfrac{1}{2}) \tag{8.13}$$

As can be seen from Fig. 8.5, this relation is in good agreement with the experimental data for different secondary particles ($\pi^{\pm}$, K^{+}, p).

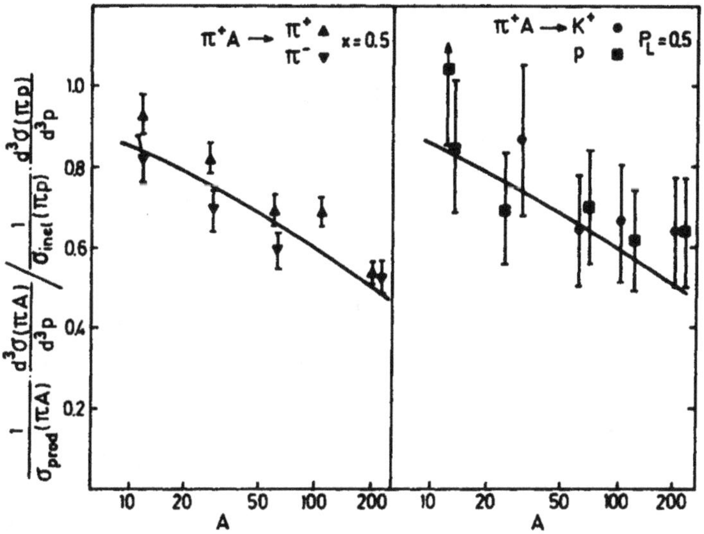

Fig. 8.5 Experimental check of the relation (8.13). Solid lines are the quark model predictions. Data at 100 GeV/c are taken from Ref. 7.

The production of mesons with $x \simeq 1/3$ in a proton fragmentation process $p \to M$ can be considered in a similar way. A secondary meson is formed by a quark-spectator ($x \simeq 1/3$) picking up a newly-produced antiquark ($x \ll 1$). Hence such a meson can occur either in a process with two spectators such as Fig. 8.1a (where in principle even two fast mesons can be produced) or in a process with one spectator (Fig. 8.1b). The contributions of these processes depend on the probabilities of one or two spectators to be transferred into secondary mesons. By the quark-statistical rules the probability of the process of Fig. 8.1a is larger by a factor 5/4 than that of Fig. 8.1b. Therefore we have

$$\frac{\dfrac{1}{\sigma_{prod}(pA)}\cdot\dfrac{d^3\sigma}{d^3p}(pA\to M)}{\dfrac{1}{\sigma_{inel}(pp)}\cdot\dfrac{d^3\sigma}{d^3p}(pp\to M)} = V_1^p(A)+\frac{4}{5}V_2^p(A) \qquad (x\sim\tfrac{1}{3}) \tag{8.14}$$

In Fig. 8.6 we compare the values $V_1^p + \frac{4}{5}V_2^p$ calculated according to Eq. (8.3) with the data on the fragmentation processes[7] $pA\to\pi^{\pm}X$.

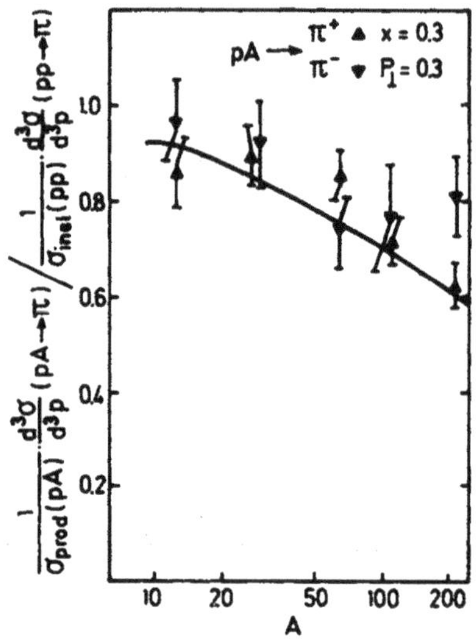

Fig. 8.6 Experimental check of the relation (8.14). The solid line is the quark model prediction. Data at 100 GeV/c are taken from Ref. 7.

The agreement between the prediction and the experiment means, presumably, that indeed, the secondary hadrons are formed later than the constituents' passage through the nucleus. It is interesting to see what happens if one supposes the contrary, i.e. that the secondaries are produced immediately when the projectile and the nucleus collide. Since the secondary hadron contains one more quark than the number of spectators, the cross sections of its absorption in the nuclear matter is larger. Consequently, in this case the production of fragmentational particles must be decreased. In Fig. 8.7 the secondary meson multiplicities in the $p \to M$ fragmentation process are presented. The solid line corresponds to the calculation with the assumption that secondary mesons are formed outside the nucleus; the dotted line represents the multiplicities calculated using the contrary assumption. The experimental data on the reactions $pA \to \pi^- X$ and $pA \to K^+ X$ at 19.2 GeV/c[8] are in a good agreement with the solid line and contradict the dashed one.

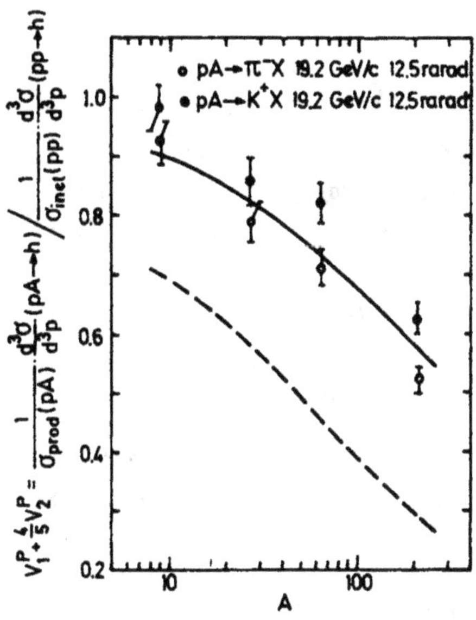

Fig. 8.7 Secondary meson multiplicity ratios for meson formation just at the point of inelastic interaction (dashed line) and outside the nucleus (solid line). Data at 19.2 GeV/c are taken from Ref. 8.

Writing down the formation time of secondaries with momentum p in the form

$$\tau = \frac{p}{\mu^2} \tag{8.15}$$

we obtain the experimental restriction for the parameter $\mu^2 \lesssim 0.4$ GeV^2 in agreement with the estimates of Chap. 6.

8.1.2 *Multiplicity ratio of pA and πA collisions in the central region of the spectrum*

The decrease of the number of secondary hadrons in the fragmentation region of the projectile has to be followed by the increase of the multiplicity in the central region. As is known, the hadrons in the central region are formed by sea quarks and antiquarks belonging to ladders which take their origin from interactions of the projectile and the target (Fig. 8.1). The ladders which belong to different projectile quarks are spatially separated, so that quarks from the different ladders cannot join each other. Such a picture becomes invalid for the slow parts of the ladders, e.g. due to cascade multiplication[9]. So let us consider only the high-energy part of the central region, where nuclear cascades are negligible. Then the density of secondaries in the processes where two constituent quarks are absorbed (Fig. 8.1b, 8.1e) will be twice as large, and for the process of Fig. 8.1c (three quarks absorbed) three times as large as those of the processes of Figs. 8.1a, 8.1d, where only one quark of the projectile interacts.

Let us define the value of the charged particle multiplicity in the central region of the collision, initiated by the hadron of the sort h with the nucleus A:

$$n_{hA} = \frac{1}{\sigma_{prod}(hA)}\, d\sigma/dy \quad . \tag{8.16}$$

The values of $n_{\pi A}$ and n_{pA} may be expressed in the quark model in terms of the charge multiplicity[3,10] in quark-nucleus collision n_{qA}:

$$R\left(\frac{pA}{qA}\right) = \frac{n_{pA}}{n_{qA}} = \sum_{k=1}^{3} kV_k^p(A) = \frac{3}{\sigma_{prod}(pA)} \cdot \int d^2r_T\left[1 - e^{-\sigma_{in\ell}(qN)T(\vec{r}_T)}\right]$$

$$R\left(\frac{\pi A}{qA}\right) = \frac{n_{\pi A}}{n_{qA}} = \sum_{k=1}^{2} kV_{k(A)}^{\pi} = \frac{2}{\sigma_{prod}(\pi A)} \cdot \int d^2r_T\left[1 - e^{-\sigma_{in\ell}(qN)T(\vec{r}_T)}\right] \tag{8.17}$$

The ratio of the secondary multiplicities in πA and pA collisions[10] does not depend on n_{qA}:

$$R(\frac{\pi A}{pA}) = \frac{n_{\pi A}}{n_{pA}} = \frac{2\sigma_{prod}(pA)}{3\sigma_{prod}(\pi A)} \tag{8.18}$$

For a hypothetical superheavy nucleus $A\to\infty$, $\sigma_{prod}(pA) \simeq \sigma_{prod}(\pi A)$, then[1])

$$R(\pi A/pA) \to 2/3 \tag{8.19}$$

The relation (8.18) is compared with the experimental data in Fig. 8.8.

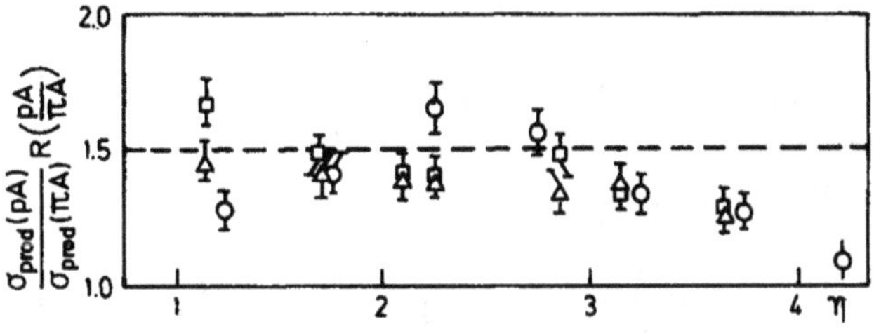

Fig. 8.8 Experimental check of the relation (8.18). The data correspond to an incident energy of 200 GeV.

8.2 Inclusive Spectra in Hadron-Nucleus Collisions

Previously we saw that in the case of a nuclear target the proportion of spectators and of the interacting quarks depends on the atomic weight of the target nucleus: the average number of the quark-spectators decreases with the increase of A. Consequently, a decrease in the multiplicity of secondary particles in the fragemantation region is also expected. Using the calculated values of these probabilities, as well as the derived inclusive spectra in hadron-hadron scattering, we can obtain the spectra in the case of nuclear targets. Let us first investigate the case of an interacting baryon consisting of three dressed quarks i, j, k. One, two or all three quarks interact inside the nucleus with the probabilities V_1^B, V_2^B, and V_3^B, respectively. In the process of Fig. 8.1a the secondaries of different types are produced with the same multiplicities as those in a hadron-hadron collision (see Chap. 7). In the processes of Figs. 8.1b, 8.1c two or three quarks interact, each producing a ladder (jet) of quarks.

Recalling the assumption that these ladders are spatially separated in their fast parts, it is natural to consider independent hadronization of the corresponding tagged quarks according to the rules of quark statistics (see Eq. (7.50)). As a result we get, for a nucleon beam:

$$
\begin{aligned}
N_{ijk} &\overset{A}{\to} \Delta_A N_{ijk} && (x \gtrsim 0.9)\\
&+ (1-\Delta_A)\xi_1 V_1^p B_{ij} && (x \sim \tfrac{2}{3})\\
&+ (1-\Delta_A)[\xi_2 V_1^p (B_i + B_j) + \tfrac{1}{3} V_2^p B_i] && (x \sim \tfrac{1}{3})\\
&+ (1-\Delta_A)[(1-\xi_1-\xi_2) V_1^p (M_i + M_j) + \tfrac{2}{3} V_2^p M_i] && (x \sim \tfrac{1}{3})\\
&+ (1-\Delta_A)[(1-\xi_1-2\xi_2) V_1^p B_k + \tfrac{1}{3} V_2^p (B_j + B_k) + \tfrac{1}{3} V_3^p (B_i + B_j + B_k)] && (x \sim \tfrac{1}{6})\\
&+ (1-\Delta_A)[(\xi_1+2\xi_2) V_1^p M_k + \tfrac{2}{3} V_2^p (M_j + M_k) + \tfrac{2}{3} V_3^p (M_i + M_j + M_k)] && (x \sim \tfrac{1}{6})\\
&+ (1-\Delta_A)(V_1^p + 2V_2^p + 3V_3^p) N(s)(d\cdot M + B + \bar{B}) \quad . && (x \lesssim 0.1)
\end{aligned}
$$

(8.20)

For a pion beam (in which case, as stated earlier, the absorption of two quarks i and j is possible) one obtains, analogously:

$$
\begin{aligned}
\pi_{ij} &\longrightarrow \delta_A \pi_{ij} && (x \gtrsim 0.9)\\
&+ (1-\delta_A)[\eta V_1^\pi (\tfrac{1}{2}B_i + \tfrac{1}{2}\bar{B}_j) + (1-\eta) V_1^\pi (\tfrac{1}{2}M_i + \tfrac{1}{2}M_j)] && (x \sim \tfrac{1}{2})\\
&+ (1-\delta_A)[\eta (V_1^\pi + 2V_2^\pi)(\tfrac{1}{2}B_i + \tfrac{1}{2}\bar{B}_j) + (1-\eta)(V_1^\pi + 2V_2^\pi)(\tfrac{1}{2}M_i + \tfrac{1}{2}M_j)] && (x \sim \tfrac{1}{4})\\
&+ (1-\delta_A)[V_1^\pi + 2V_2^\pi] N(s)(d\cdot M + B + \bar{B}) \quad . && (x \lesssim 0.15)
\end{aligned}
$$

(8.21)

The consideration of a kaon beam leads to somewhat more complicated formulas because of the different cross sections of strange and non-strange quarks.

The inclusive spectra of fragmentational secondary particles formed in hadron-nucleus collisions can be calculated in the same way as in the case of hadron-nucleon interactions using expression (7.73) of Chap. 7.

The only difference is that the coefficients a_i^{h2} have to be obtained using (8.20) and (8.21) of the present chapter instead of (7.3).

Different contributions to the inclusive spectra in the reaction $pCu \to pX$ are shown in Fig. 8.9. In Fig. 8.9a the contributions of

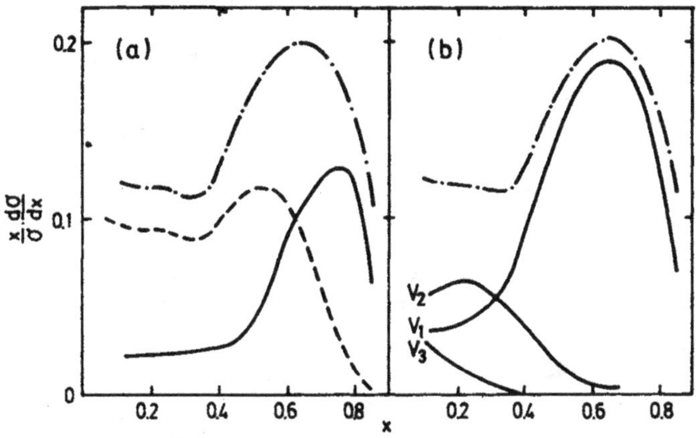

Fig. 8.9 (a) Calculated contributions of direct protons (solid line) and resonance decay products (dashed line) for the reaction $pCu \to pX$. The summed contribution is shown by dash-dotted line. (b) Contributions of inelastic interaction of one (V_1), two (V_2) and three (V_3) quarks of the projectile proton.

direct protons (solid line) and resonance decay products (dashed line) are shown. In Fig. 8.9b the contributions of inelastic interactions of one, two and three quarks of the incident proton are given. The share due to diffraction dissociation, which is also taken into account, is not shown separately in Fig. 8.9.

Let us compare the predictions of Ref. 11 with experimental data. Remember that for hadron-nucleus collision processes, high energy data for not very light nuclei exist either at fixed values of the momentum transfer or without distinguishing between different sorts of secondaries. At the same time, the calculations give x-distributions integrated over p_T. Still, the comparison makes sense, because the cross section ratios of secondary productions in hadron-nucleus and hadron-hydrogen collisions at $p_T \sim \langle p_T \rangle$ have to be almost the same as the ratios of integral distributions.

The results of the calculations[11] are presented in Fig. 8.10

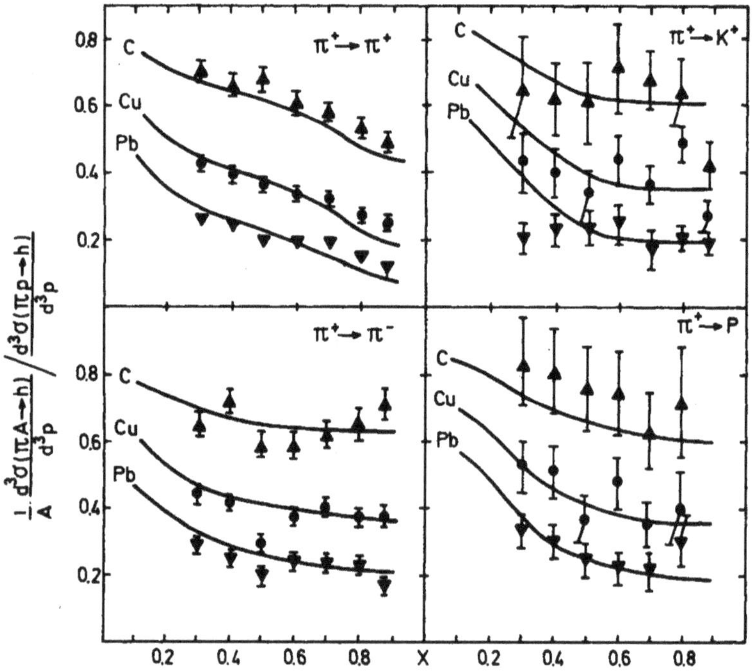

Fig. 8.10 The ratios of inclusive spectra for π^+A and π^+p collisions at 100 GeV/c, p_T = 0.3 GeV/c.

together with the experimental data[7] on pion production in π^+A collisions at 100 GeV/c. In Fig. 8.11 the comparison of predictions with experiment shows sufficiently good agreement for the reactions[7] $pA \to pX$ and $pA \to \pi^{\pm}X$ at 100 GeV/c.

For high initial energies experimental data on particle production in the central region of hadron-nucleus interactions are obtained mainly without identifying the secondaries. The usual variable here is the quasirapidity $\eta = -\ln \mathrm{tg}(\theta_{lab}/2)$. The Feynman variable x can be transformed to η if one assumes that the average transverse momentum

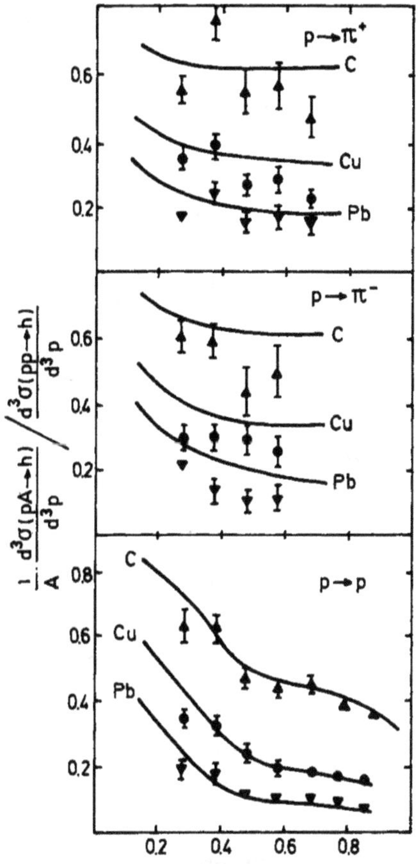

Fig. 8.11 The ratios of inclusive spectra for pA and pp collisions at 100 GeV/c, $p_T = 0.3$ GeV/c.

$\langle p_T \rangle$ of the secondary particle does not depend on x:

$$\eta \simeq \ell n \frac{2xp_0}{\langle p_T \rangle} ,$$

where p_0 is the momentum of the incident hadron in the lab system.

Calculations are compared with experimental data on the production of positively and negatively charged particles in pA collisions[12] at 200 GeV/c. It is obvious that in the region where nuclear effects like cascades are essential the approach described cannot be applied. As can be seen from Fig. 8.12 these nuclear effects play an important role only when $x \lesssim 0.05$.

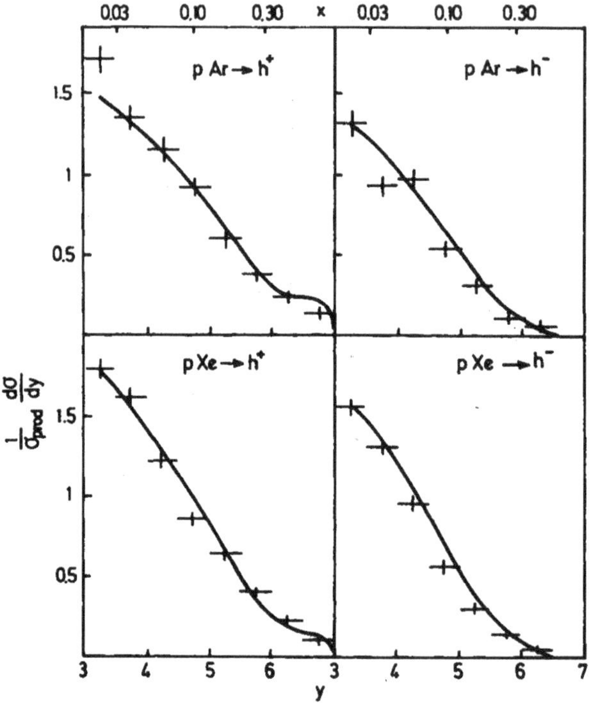

Fig. 8.12 Rapidity distributions of positive and negative secondaries in pA collisions at 200 GeV/c. Solid lines show the quark model calculations without accounting for any nuclear cascade.

8.3 Inelastic Diffraction Scattering of Hadrons on Nuclei

In this section we investigate reactions of the type

$$h + A \to h + X$$

for the case when the incident hadron h looses only a small fraction of its momentum. In other words, we shall consider hadron spectra in the region $x \gtrsim 0.9$, which will be called the *region of diffraction dissociation* for hadron-nucleus collisions.

Two types of processes might give hadrons h in the region $x \to 1$.

This can be, first of all, the disintegration of a nucleus without the formation of new hadrons (the cross section σ_{dis} of such a process was considered in Sec. 4.5). Besides, a hadron h can appear in the region mentioned as a result of the diffractive dissociation of the target nucleon; in this case new particles (e.g. pions) are produced. In other words, we are interested in two types of hadron-nucleon interactions: these are the elastic scattering $hN \to hN$ and the diffraction dissociation $hN \to hX$. (In Fig. 8.13 a,b quark graphs of these processes are presented for the case when h is a baryon.) Both processes are due to elastic quark-quark scatterings (see Sec. 6.6). As a consequence of the condition of completeness for a three-quark system the sum of the cross sections for the above two processes $\sigma_{scat}(hN \to h)$ is determined by the discontinuities of the amplitudes shown in Fig. 8.13 c,d.

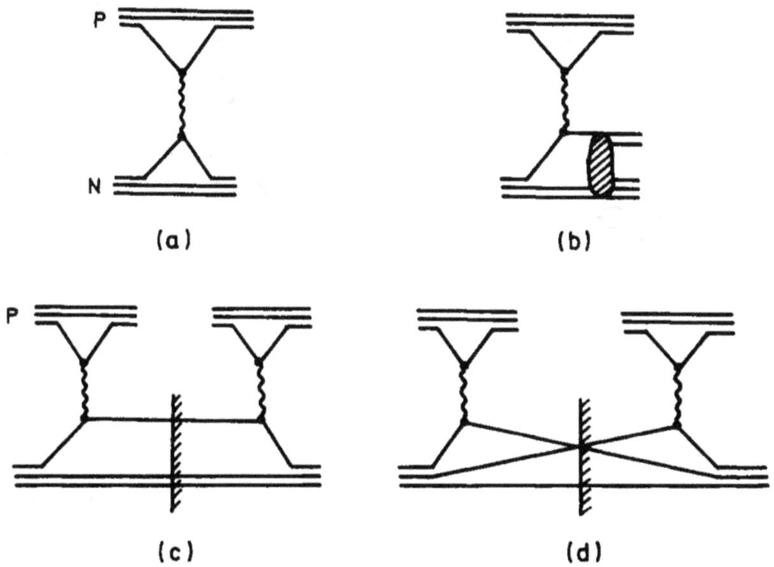

Fig. 8.13 a,b Elastic quark-quark scattering processes resulting in elastic scattering (a) or diffraction dissociation (b) of a target nucleon. (c,d) Discontinuities of the diagrams which determine the cross section $\sigma_{scat}(pN \to p)$.

$$\frac{d\sigma_{scat}(hN \to hX)}{d^2k_T} \simeq n_h^2 F_h(-k^2) \frac{d\sigma_{scat}(qN \to qX)}{d^2k_T} \quad . \tag{8.22}$$

We assume here, that the cross section is integrated over x in the $x \sim 1$ region, and therefore Eq. (8.22) depends only on the transverse component of the momentum transferred $\vec{k}_T$. In expression (8.22) n_h is the number of dressed quarks belonging to the hadron h (i.e. $n_h = 2$ for a meson and $n_h = 3$ for a baryon); F_h is the form factor of the hadron h. The quark-nucleon cross section itself is a fundamental quantity which defines the cross section $h + A \to h + X$ in the diffraction-dissociation region.

Let us consider first the contribution of a single quark-nucleon interaction (Fig. 8.14a) to the cross section of the nucleus diffraction

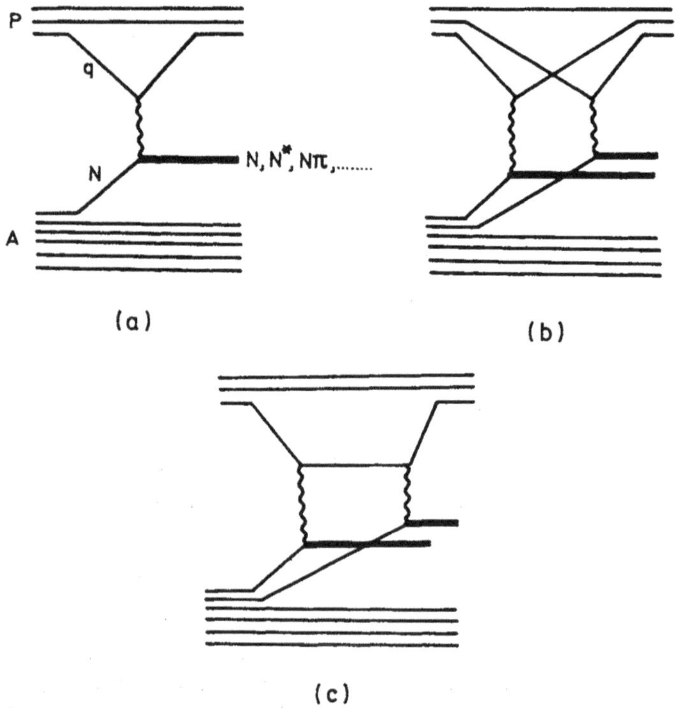

Fig. 8.14 Contributions of (a) single- and (b,c) double-nucleon interactions in the cross section of reaction $pA \to p$ for $x \sim 1$.

dissociation (supposing, as in Eq. (8.22) that the cross section is integrated over x in the region $x \sim 1$). This contribution can be obtained easily by multiplying the cross section (8.22) (which can be taken, for instance, from experiment) by the probability of the hadron (for the sake of definiteness, proton) projectile to interact only once with the nuclear matter:

$$\frac{d\sigma_{DD}^{(1)}(pA \to p)}{d^2k_T} = \frac{d\sigma_{scat}(pN \to p)}{d^2k_T} \int d^2r_T\, T(\vec{r}_T) e^{-\sigma_{tot}(pN)T(\vec{r}_T)} \quad . \qquad (8.23)$$

The factor $\exp[-\sigma_{tot}T]$ accounts for the probability that there are no other interactions between the nucleus and the proton.

Double interactions are defined by graphs 8.14 b,c. The corresponding cross section can be written as

$$\frac{d\sigma_{DD}^{(2)}(pA \to p)}{d^2k_T} = \int d^2k_{1T} d^2k_{2T} \delta^2(\vec{k}_{1T} + \vec{k}_{2T} - \vec{k}_T)[6G_p(\vec{k}_{1T}, \vec{k}_{2T}, 0) + 3F_p(\vec{k}_T^2)]^2$$

$$\times \frac{d\sigma_{scat}(qN \to q)}{d^2k_{1T}} \frac{d\sigma_{scat}(qN \to q)}{d^2k_{2T}} \int d^2r_T \frac{1}{2!} [T(\vec{r}_T)]^2 e^{-\sigma_{tot}(pN)T(\vec{r}_T)} \quad . \qquad (8.24)$$

Here $T(\vec{r}_T)\sigma_{scat}(qN \to q)$ is the probability for q to be scattered in the interaction process with the nucleon. As before, the factor $\exp[-\sigma_{tot}(pN)T]$ means that the quarks of the proton do not take part in any other interactions. The factor 1/2! accounts for the permutations of the quark interactions in the nucleus. The form factor F_p and the difactor G_p of the proton define the probability that the proton did not disintegrate during the collision process. The combinatorial coefficients standing before F_p and G_p correspond to the number of the possible permutations in the upper vertices of the diagrams 8.14b, c. The difactor $G_p(\vec{k}_{1T}, \vec{k}_{2T}, 0)$ can be expressed in *terms of the quark density in the proton in impact parameter space*

$$G_p(\vec{k}_{1T},\vec{k}_{2T},\vec{k}_{3T}) = \int d^2r_{1T} d^2r_{2T} d^2r_{3T} \rho_p^{(t)}(\vec{r}_{1T},\vec{r}_{2T},\vec{r}_{3T}) e^{i(\vec{k}_{1T}\vec{r}_{1T}+\vec{k}_{2T}\vec{r}_{2T}+\vec{k}_{3T}\vec{r}_{3T})} \qquad (8.25)$$

where the density $\rho_p^{(t)}$ is given by

$$\rho_p^{(t)}(\vec{r}_{1T}, \vec{r}_{2T}, \vec{r}_{3T}) = \int dz_1 dz_2 dz_3 |\psi_p(\vec{r}_1, \vec{r}_2, \vec{r}_3)|^2 \quad . \tag{8.26}$$

The proton form factor can be written as $F_p(\vec{k}^2) = G_p(\vec{k}, 0, 0)$. (The definition of densities in impact parameter space was considered in Sec. 4.3; see Eqs. (4.50) and (4.51).)

An expression for $d\sigma_{DD}(pA - p)/d^2k_T$ in which all possible rescatterings of the quarks are taken into account was obtained in Ref. 13 with the help of the method presented in Sec. 4.5. Without going into details, we give here the resulting formula

$$\begin{aligned} \frac{d\sigma_{DD}(pA \to p)}{d^2k_T} &= \frac{1}{(2\pi)^2} \int d^2r_T \; e^{-\sigma_{tot}(pN)T(\vec{r}_T)} \int d^2\Delta \; e^{i\vec{k}_T\vec{\Delta}} \\ &\cdot \prod_{i=1}^{3} d^2r_{iT} \prod_{j=1}^{3} d^2r'_{jT} \cdot \rho_p^{(t)}(\vec{r}_{1T}, \vec{r}_{2T}, \vec{r}_{3T}) \rho_p^{(t)}(\vec{r}\,'_{1T}, \vec{r}\,'_{2T}, \vec{r}\,'_{3T}) \\ &\cdot \left\{-1 + \exp[T(\vec{r}_T) \sum_{i,j=1}^{3} \sigma_q(\vec{\Delta} + \vec{r}_{iT} - \vec{r}\,'_{jT})]\right\} \end{aligned} \tag{8.27}$$

where

$$\sigma_q(\vec{\Delta}) = \int d^2k_T \frac{d\sigma_{scat}(qN \to q)}{d^2k_T} e^{-i\vec{k}_T\vec{\Delta}} \quad . \tag{8.28}$$

The expressions (8.23) and (8.24) for the contributions of the single and double quark rescatterings can be obtained by decomposing (8.27) into a series over σ_q.

The analogy of Eq. (8.27) for the reaction $\pi A \to \pi$ can be derived easily.

In the following, let us investigate the results of the above calculations. The quantity $d\sigma_{scat}(qN \to q)/d^2k_T$ which determines the cross sections was found in Sec. 6.6 (see Eq. (6.85)) as a sum of two exponential terms with phenomenological parameters. The cross sections $d\sigma_{DD}(pA \to p)/dk_T^2$ calculated for the collision processes of protons and pions with the nuclei C, Aℓ, Cu, and Pb are presented in Fig. 8.15. The solid and the dashed lines correspond to the values

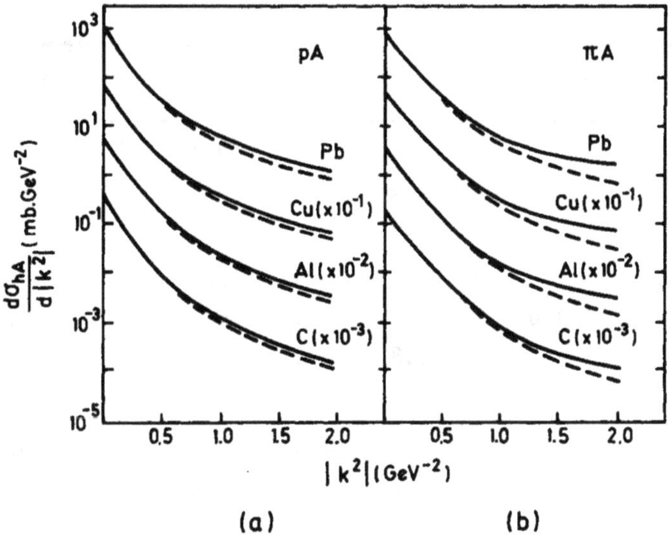

Fig. 8.15 Calculated cross sections for the reactions (a) $pA \to p$ and (b) $\pi A \to \pi$ a function of momentum transfer k^2 for the values $b_q = 1\ GeV^{-2}$ (solid lines) and $2\ GeV^{-2}$ (dashed lines).

$b_q = 1\ GeV^{-2}$ and $b_q = 2\ GeV^{-2}$, respectively. As can be seen, for small momentum transfers the results derived at different values of b_q practically coincide. This is a result of the dominance of the single interaction (Fig. 8.14a). For $|k^2| > 1\ GeV^2$ higher order interactions turn out to be important and therefore the model becomes sensitive to the b_q value. In Table 8.3 the relative contributions of different orders of interactions are demonstrated for $pA\ell$, pPb, $\pi A\ell$, and πPb collisions at $|k^2| = 0, 1, 2\ GeV^2$.

The integrated cross sections $\sigma_{DD}(hA) = \int d^2k_T \cdot \frac{d\sigma_{DD}(hA \to h)}{d^2k_T}$ i.e. the summed cross sections of quasi-elastic nucleus disintegration and diffraction-dissociation processes — are presented in Table 8.4. The main contribution comes here from the region of small momentum transfers; this explains why the change of b_q almost does not influence the value

Table 8.3 The relative contributions of n-fold quark interactions $(d\sigma_{DD}/d|k^2|)_n = D_n$ in the diffraction-dissociation cross section $d\sigma_{DD}(hA)/d|k^2|$.

hA	$\|k^2\|$(GeV^2)	D_1(%)	D_2(%)	D_3(%)	D_4(%)	D_5(%)
pAℓ	0	91.9	7.0	0.9	0.1	-
	1.0	26.0	44.7	20.8	6.7	1.8
	2.0	9.4	38.4	32.0	15.0	5.2
pPb	0	89.8	8.1	1.5	0.4	0.1
	1.0	18.3	37.1	24.8	13.3	6.6
	2.0	5.3	25.5	30.4	23.7	15.2
πAℓ	0	95.1	4.6	0.3	-	-
	1.0	42.3	46.1	10.0	1.4	0.2
	2.0	1.1	72.1	22.1	4.1	0.6
πPb	0	91.6	7.2	1.0	0.2	-
	1.0	26.4	47.0	19.3	5.8	1.5
	2.0	0.5	53.2	31.1	11.7	3.5

Note: D_n are calculated at $b_q = 1.5\ \text{GeV}^{-2}$.

Table 8.4 The sum of the quasielastic nucleus disintegration cross section and that of the diffraction dissociation of one of the target nucleons.

hA	C	Aℓ	Cu	Pb
σ_{DD} (πA), mb	28.2	45.2	67.2	101.5
σ_{DD} (pA), mb	44.9	63.9	84.7	115.1

of the integrated cross sections.

Switching over to multiple rescatterings, the A-dependence of hadron production in the region $x \sim 1$ changes abruptly with increase of $|k^2|$. Figure 8.16a shows the ratio of secondary proton productions in pA and pp collisions

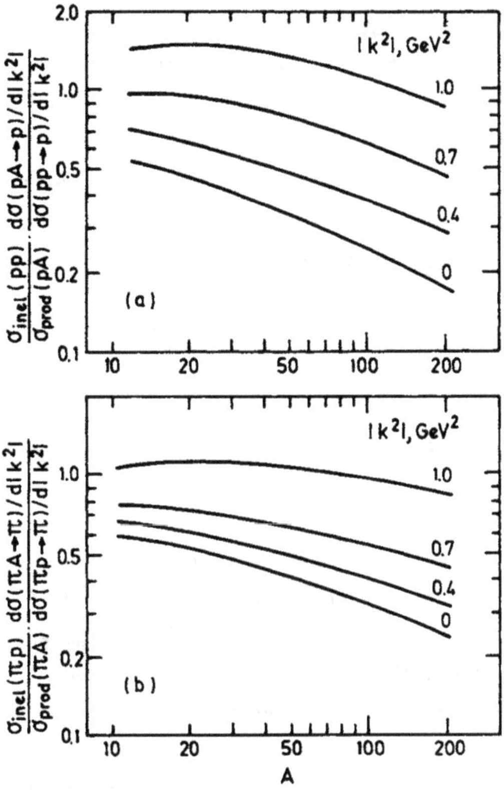

Fig. 8.16 Calculated ratios of (a) proton → proton and (b) pion → pion inclusive cross sections on the nuclear and nucleon targets in the region of $x \sim 1$.

$$R_{p\to p}(A) = \frac{\sigma_{inel}(pp)}{\sigma_{prod}(pA)} \cdot \frac{d\sigma(pA\to p)/d^2k}{d\sigma(pp\to p)/d^2k} \tag{8.29}$$

as functions of A at different $|k^2|$ values. For $|k^2| \leqslant 0.2$ GeV^2 this ratio decreases quickly with the increase of A, reaching the value 0.2 for the Pb nucleus. However, at $|k^2| \sim 1$ GeV^2 already there can be almost no A-dependence observed for the secondary multiplicities.

An analogous ratio for $R_{\pi\to\pi}(A)$ is presented in Fig. 8.16b.

Note, that if a hadron did not consist of two or three dressed quarks, but of a large effective number of constituents, then only a small fraction of these constituents would interact with the nuclei. Such a picture corresponds to the usual model of multiple Glauber-type rescatterings, in the framework of which the form factor of the projectile is written for each interaction. In other words, one assumes that the fast hadron is scattered each time as a whole; the cross section for this case is calculated in Ref. 1. Cross sections corresponding to hadron pictures with and without quasinuclear structure are quite different.

The difference between the models manifests itself in the slopes of differential cross sections $d\sigma_{DD}(hA\to h)/d|k^2|$. In Table 8.5 we list the values of the parameter B_h for

$$\frac{d_\sigma (hA\to h)}{d|k^2|} \sim e^{B_h k^2} , \qquad (8.30)$$

Table 8.5 The slope parameters B_h of the differential cross sections $d\sigma_{DD}(hA-h)/d|k^2| \sim \exp(B_h k^2)$ for nuclear targets at $|k^2| = 1.5\text{-}2\ \mathrm{GeV}^2$ in the quark model (QM) and the multiple scattering model (MSM).

A	B_p (GeV^{-2})		B_π (GeV^{-2})	
	QM	MSM	QM	MSM
C	1.62 1.74	2.47	1.10 1.81	3.05
Aℓ	1.52 1.61	2.42	1.01 1.68	2.85
Cu	1.47 1.54	2.38	0.95 1.61	2.70
Pb	1.46 1.52	2.36	0.93 1.57	2.58

Note: The top numbers correspond to $b_q = 1\ \mathrm{GeV}^{-2}$, the lower ones to $2\ \mathrm{GeV}^{-2}$.

which are obtained in the interval $|k^2| = 1.5\text{-}2\ \mathrm{GeV}^2$. For all versions of the quark model these values are 1.5-2 times less than those for the model of multiple rescattering of a hadron as a whole. Hence, the experimental investigation of the region $x \gtrsim 0.9$-0.95 at large $|k^2|$ and at high energies is of undoubted interest; unfortunately, so far there are no such data yet.

8.4 Scale Breaking in Hadron-Nucleus Interactions

The growth of the total cross section of pp (or $p\bar{p}$) collisions with the energy increase from FNAL ($\sqrt{s} \sim 20$ GeV) to the SPS Collider ($\sqrt{s} = 540$ GeV) must lead to an appreciable breaking of the Feynman scaling in the beam fragmentation region of hadron-nucleus collisions. Indeed, due to the growth of the quark-nucleon inelastic cross section from 10 mb to 18 mb (which corresponds to the increase of $\sigma_{tot}(pp)$ from 40 mb to 62 mb) the value of $V_1^p(A)$ decreases[15] by approximately 20% for light nuclei ($A = 10$-20), and by 30% for heavy nuclei ($A \sim 100$). Thus the secondary baryon production multiplicities at $x \sim 2/3$ in pA collisions decrease by the same factors with respect to those in pp collisions. Even if scaling in pp collisions were unbroken, it would break for fast secondaries produced on the nuclear targets. In fact, in the previous chapter (see Sec. 7.5) it was shown that the growth of the shadow corrections in hadron-hadron collisions with the energy increase also breaks Feynman scaling. This will obviously enhance the effect for nuclear targets.

In order to give numerical predictions, it is necessary to calculate V_1^p, V_2^p, V_1^π, etc. taking into account the possibility for several dressed quarks of the projectile interacting with the same nucleon of the target. In Ref. 16 it was noted that instead of the direct calculations the expressions (7.74), (7.75), (7.76) of the previous chapter may be used, just as in the case of the nucleon-nucleon interaction. For example:

$$V_1^p = \frac{3}{\sigma_{prod}(pA)} \left(\sigma_{prod}(pA) - \sigma_{prod}(d_{qq}A)\right) \quad . \tag{8.31}$$

Here d_{qq} is a diquark; its production cross section was discussed in Chap. 4. When $\sigma_{prod}(d_{qq}A)$ in (8.31) is calculated with the help of Glauber expressions like (8.4), (8.8), the value of $\sigma_{in\ell}(qN)$, obtained with shadow corrections for the nucleon target accounted for, must be used. As a result, all the coefficients V_i^p, V_i^π, V_i^K are renormalized, but the A-dependence of the inclusive spectra remains practically unchanged. The predictions of the scale breaking due to the cross section growth in pA, πA, and KA collisions are shown in Figs. 8.17 and 8.18.

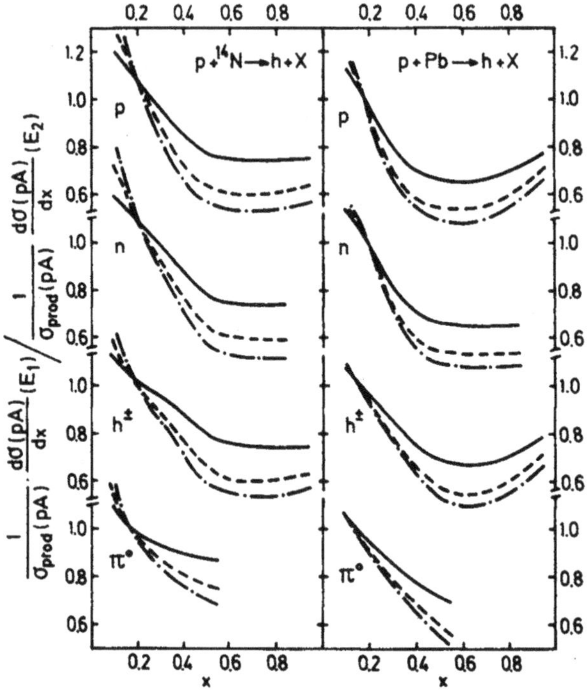

Fig. 8.17 Scaling violation effects in the yields of various secondaries in $p^{14}N$ and pPb collisions. The ratios of the spectra at such energies when $\sigma_{tot}(pp)$ = 66 mb (solid curves), $\sigma_{tot}(pp)$ = 93 mb (dashed curves) and $\sigma_{tot}(pp)$ = 115 mb (dash-dotted curves) to the calculations for energies ~ 10^2 GeV are presented.

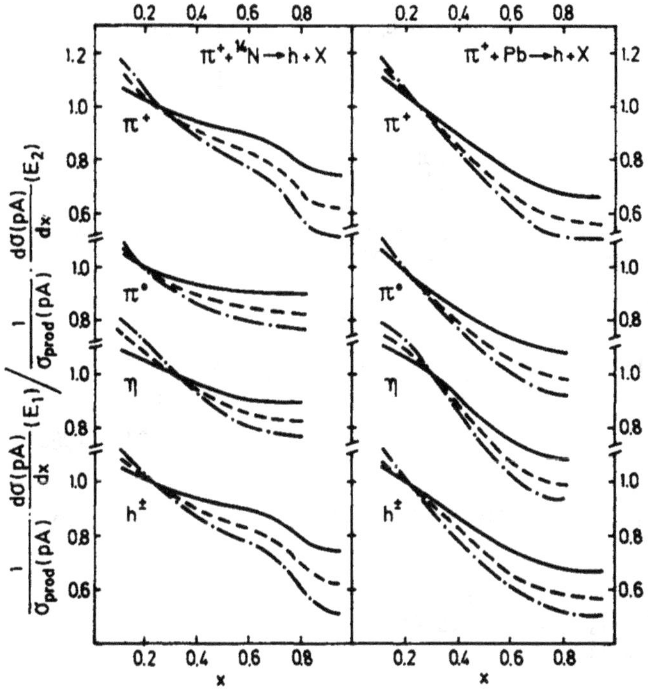

Fig. 8.18 Scaling violation effects in the yields of various secondaries in $\pi^{+14}N$ and π^+Pb collisions at the same energies as those in Fig. 8.17.

References

1. V.V. Anisovich, Phys. Lett. 57B, 87 (1975).

2. N.N. Nikolaev, Phys. Lett. 70B, 95 (1977).

3. A. Bialas, W. Czyz, W. Furmanski, Acta Phys. Pol. B8, 585 (1977).

4. V.V. Anisovich, Yu. M. Shabelski, and V.M. Shekhter, Nucl. Phys. B133, 477 (1978).

5. A.S. Carrol *et al.*, Phys. Lett. 80B, 319 (1979).

6. D.S. Ayres *et al.*, Phys. Rev. D15, 3105 (1977).

7. D.S. Barton *et al.*, Preprint FERMILAB-PUB-82/64-EXP (1982).

8. J.V. Allaby et al., CERN preprint 70-12 (1970).

9. N.N. Nikolaev, Uspekhi Fiz. Nauk. 134, 369 (1981); Sov. Phys. Uspekhi 24, 531 (1981).

10. V.V. Anisovich, F.G. Lepekhin, and Yu. M. Shabelski, Yad. Fiz. 27, 1639 (1978); Sov. J. Nucl. Phys. 27, 861 (1978),

11. V.V. Anisovich, M.N. Kobrinsky, Yu. M. Shabelski, Yad. Fiz. 38, 763 (1983); Sov. J. Nucl. Phys. 38, 455 (1983).

12. C. De Marzo et al., Phys. Rev. D26, 1019 (1982).

13. V.V. Anisovich, V.M. Braun, Yu. M. Shabelski, and V.M. Shekhter, Yad. Fiz. 36, 732 (1982); Sov. J. Nucl. Phys. 36, 428 (1982).

14. V.M. Braun and Yu. M. Shabelski, Yad. Fiz. 37, 379 (1983); Sov. J. Nucl. Phys. 37, 228 (1983).

15. Yu. M. Shabelski, Fizika Elem. Chastitz i Atomogo Yadra, 12, 1070 (1981).

16. V.V. Anisovich, V.M. Braun, and Yu. M. Shabelski, Yad. Fiz. 36, 1556 (1982); 39, 932 (1984); Sov. J. Nucl. Phys. 36, 904 (1982); (to be published).

Chapter 9

RECENT MODEL APPROACHES TO MULTIPLE PRODUCTION PHENOMENA

There exist nowadays a lot of theoretical models, which describe multiple production processes, starting from the quark structure of hadrons. The results obtained from these models seem to be very similar: the agreement achieved with the experimental data is, in most cases, of the same level of accuracy. Since the basic assumptions of the different models often look quite unlike each other, it is interesting to understand, what are the points where they really coincide or differ.

The aim of the present chapter is not to give a detailed description of all the existing models: this could be the subject of an independent book. Nor is it to review these results, for most of the models are still developing, and some new results may be obtained everyday, thus making such a review obsolete. We only want to consider the main ideas of the different approaches and to compare them with each other and with the quasinuclear quark model and rules of quark statistics. Since the latter approach was described in detail here, it is convenient to choose it as our standard reference point for comparison. In doing so, we don't want to outline any advantages or defects of any of the models considered, but rather to find some common meeting ground for them.

For the complete list of papers dealing with these problems, we refer here to a recent review[1] and to review articles and lectures devoted to the concrete models — the latter will be cited in the following.

9.1 The Scheme of Dual Topological Unitarization (DTU)

The picture of multiparticle production which is used in the approach of DTU is essentially based on the ideology of the 1/N expansion. This approach originates from the Refs. 2, 3, where gluon-exchange diagrams of QCD perturbation theory for the elastic scattering were interpreted in terms of Reggeon exchanges (see also Refs. 4-9). In this scheme a Reggeon with non-vacuum quantum numbers is represented by planar-type graphs (Fig. 9.1a), while a Pomeron exchange corresponds to the simplest non-planar (cylinder) graph (Fig. 9.1b). If the number of colours (N_c) and of flavours (N_f) is large, i.e. $N_c \sim N_f \gg 1$, the planar graphs will be dominant. Still, at high energies their contribution is suppressed as $1/\sqrt{s}$ because of the intercept of non-vacuum Reggeons $\alpha_{q\bar{q}}(0) \simeq 1/2$ (see Sec. 2.3). Hence at high energies the main contribution is given by graphs of the type Fig. 9.1b inspite of the small inherent parameter 1/N. Multi-cylindrical graphs corresponding to multi-Pomeron exchanges are corrections to the 1/N expansion. The diagrams of multiparticle production processes are determined by cuts of the graphs Fig. 9.1a, b (or of more complicated graphs in the cases of multi-Pomeron exchanges). Examples for these are shown in Fig. 9.1c, d and Fig. 9.2. Since planar-graph contributions decrease with the increase in s, the main contribution to multiparticle production will be given by diagrams of the sort typified by Fig. 9.2.

Such a multiparticle production picture was discussed by the Orsay group[10,11], by the Saclay group[12,13], and in various other papers[14-16]. Let us consider first of all the contribution of one-Pomeron exchange diagrams of the type Fig. 9.1b.

Quarks in the final state of the cut diagram 9.1b may have suitable colours (as in Fig. 9.1c) or non-suitable (Fig. 9.1d) for hadronization. In the DTU approach it is assumed that in cases like Fig. 9.1d the production of additional $q\bar{q}$ pairs takes place. The new quarks joining each other, lead to hadron formation without any changes in

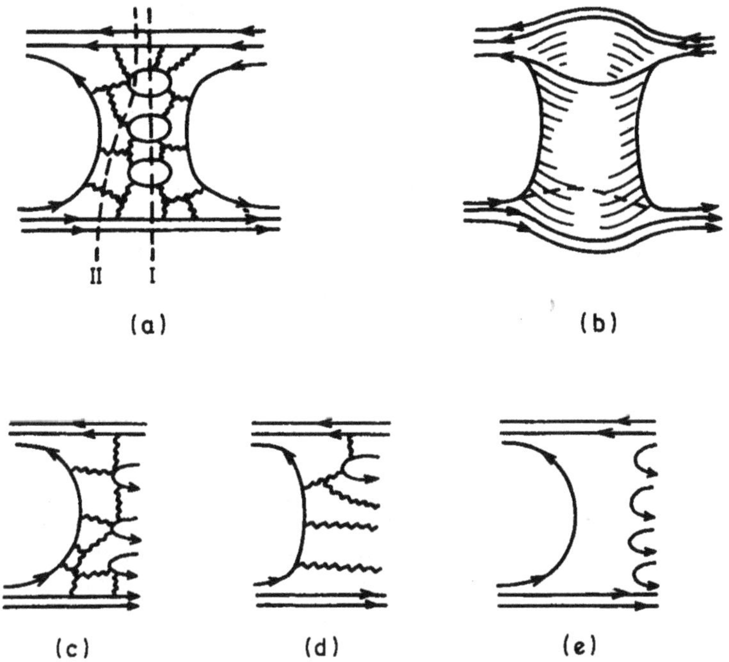

Fig. 9.1 (a) Planar graph in $\bar{p}p$ scattering, corresponding to nonvacuum Reggeon exchange. Gluons and quark-antiquark pairs are placed on a membrane which is stretched between quark lines in the t-channel.

(b) Cylindrical graph in $\bar{p}p$ (or pp) scattering, corresponding to the contribution of one Pomeron. The lines of quark-antiquark loops and gluons are located along the walls of the cylinder.

(c, d) Different types of cuts of the planar graph 9.1a: along the dotted lines I (c) and II (d).

(e) A schematic graph representing diagrams of the type 9.1c, d which supposes the existence of an additional stage in the hadronization process, namely, the stage of soft hadronization and soft colour neutralization (i.e. production of quark-antiquark pairs with suitable quantum numbers for the hadronization).

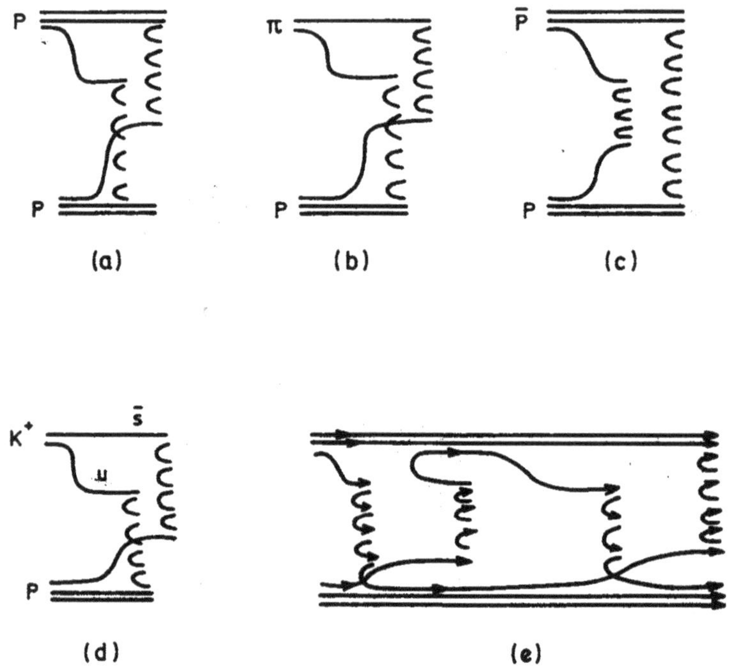

Fig. 9.2 The multiparticle production process in the scheme of DTU.
(a, b, c, d) Cuts of a cylinder in the cases of pp, πp, p̄p and Kp collisions. (In Kp collisions due to (9.5) the slow quark is always the non-strange quark.)
(e) A cut two-cylindric (two-Pomeron) graph in the case of pp collision.

the total probability. The sum of differently cut diagrams of the types Fig. 9.1c, d is therefore drawn in the form of Fig. 9.1e with a chain of quark-antiquark pairs. This means that, as in the case of the quasinuclear quark model, in the DTU scheme one has to assume soft hadronization and soft colour neutralization.

For high energies the main contribution to multiparticle production is thus given by cuts of graphs of the cylinder or multi-cylinder type (Fig. 9.2). The quark chains in these diagrams are "tighted" to the initial quarks or diquarks, so that the chain position on the rapidity axis depends on the initial quark (diquark) distribution over the momentum fraction carried. In one of the versions of the DTU scheme it is assumed (see Refs. 17 and 18), that in the case of a proton (or antiproton) projectile this distribution is

$$F(x_1, x_2, \ldots, x_n) = C_n \frac{1}{\sqrt{x_1}} \cdot \frac{1}{x_2} \cdot \frac{1}{x_3} \cdot \ldots \frac{1}{x_{n-1}} x_n^{3/2} \delta(\Sigma x_i - 1) \quad . \tag{9.1}$$

Here C_n is the normalizing factor; x_1 corresponds to the valence quarks, x_n to the diquark, $x_2, \ldots, x_{n-1}$ to the sea quarks respectively. The behaviours of the quark and the diquark distributions as $x_i \to 0$ correspond in (9.1) to the leading Regge trajectories $x_i^{-\alpha(0)}$ (see Fig. 9.3). For the valence quark this is $\alpha_{q\bar{q}}(0) = 1/2$; for the sea quark $\alpha_{\mathbf{P}}(0) = 1$. The intercept of the Reggeon which corresponds to the diquark exchange (Fig. 9.3c) can be determined from the relation[19] $\alpha_{q\bar{q}}(0) + \alpha_{qq,\bar{q}\bar{q}}(0) = 2\alpha_B(0)$, where $\alpha_B(0) = -1/2$ is the intercept of the baryon (nucleon) trajectory[a]. In the two-chain approximation (Fig. 9.2a) the distributions of the valence quark and the diquark are, respectively:

$$F_q(x) = 0.85\, x^{-1/2}(1-x)^{3/2} \quad , \quad F_{qq}(x) = 0.85\, x^{3/2}(1-x)^{-1/2} \quad . \tag{9.2}$$

The average x values for the valence quark and the diquark are $\bar{x}_q = 0.17$, $\bar{x}_{qq} = 0.83$. In some papers[10,11,20], however, a different parametrization is used for the distributions of valence quarks. These distributions are taken for the quark-partons from data on deep

---

[a] The leading baryon trajectory is in fact the trajectory of the Δ isobar $\alpha_\Delta(0) \simeq 0$ *and not that of the nucleon (see Sec. 2.3). The* application of the nucleon trajectory is justified by the fact that the Regge contributions of the Δ isobar are experimentally suppressed.

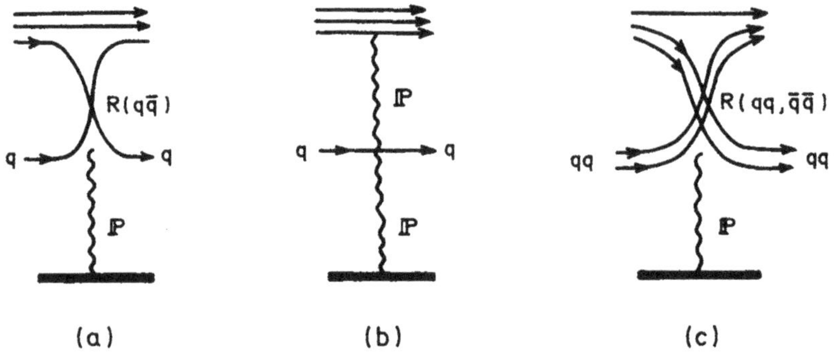

Fig. 9.3 Mueller-Kancheli diagrams, determining in the $x = 0$ region distributions of the a) valence quark, b) sea quark and c) diquark, respectively.

inelastic scatterings, and they are different for the u- and d-quarks: $\bar{x}_u = 0.07$, $\bar{x}_d = 0.03$.

For a pion projectile the distributions of quarks (antiquarks) are determined by considerations analogous to those which lead to (9.1)

$$F(x_1,\ldots,x_n) = C_n \frac{1}{\sqrt{x_1}} \cdot \frac{1}{x_2} \cdot \ldots \cdot \frac{1}{x_{n-1}} \cdot \frac{1}{\sqrt{x_n}} \delta(\Sigma x_i - 1) \quad . \tag{9.3}$$

In the two-chain approximation the distribution of the valence quark (antiquark) in a pion is equal to

$$F_q(x) = 0.75\, x^{-1/2}(1-x)^{-1/2} \quad . \tag{9.4}$$

The distribution of strange and non-strange quarks in a K-meson looks different in the two-chain approximation from the distribution in a pion, since $\alpha_{s\bar{s}}(0) = \alpha_\phi(0) \simeq 0$:

$$F_s(x) = \frac{0.5}{1-x} \quad , \qquad F_q(x) = \frac{0.5}{x} \quad . \tag{9.5}$$

In the scheme discussed a further step is required: it is necessary to define the transition functions $D_{q \to h}(z)$ and $D_{qq \to h}(z)$ of quarks and diquarks into hadrons. The production of a secondary particle can then be written in the following form:

$$\frac{1}{\sigma} \cdot \frac{d\sigma}{dx} = \int_x^1 \frac{dx_1}{x_1} F(x_1)\, D\left(\frac{x}{x_1}\right) \quad . \tag{9.6}$$

In the papers cited the functions $D_{q \to \pi}$ and $D_{q \to K}$ were taken from the Feynman-Field approximation[21], where the hadronization of quark-partons in hard processes was considered. The transition functions for diquark→baryon and diquark→meson (the latter was taken into account in Ref. 20) are obtained from the fit of the experimental data for the inclusive spectra.

Before proceeding, let us discuss an important question: what quarks are considered in the DTU scheme? Looking at the distributions (9.2), (9.4), (9.5), and the corresponding $\bar{x}_q$, $\bar{x}_{qq}$ values we may conclude that these are dressed or constituent quarks: here the quarks carry the whole momentum of the initial hadron, $\bar{x}_q + \bar{x}_{qq} = 1$ (or $\bar{x}_q + \bar{x}_{\bar{q}} = 1$). On the other hand, the transition functions $D_{q \to h}$ of Ref. 21 describe the hadronization of the bare quark-partons. The possibility of applying the approximation[21] for the description of the "dressed quark→hadron" transition seems to be doubtful. The transition of a dressed quark into a hadron has to be different from that of a quark-parton, because the dressed quark is prepared for hadronization to a greater extent. If an attempt is made to relate the distributions (9.2), (9.4), (9.5) with the quark-partons, another question arises: "where are there gluons in DTU?" (see Ref. 13), for the gluons must carry about half of the initial momentum.

Forgetting about the inconsistency mentioned, one can see, first of all, that the picture of hadron production in the two-chain approximation of DTU is rather similar to the impulse approximation of the quasinuclear quark model. To illustrate this, we present in Fig. 9.4 the two-chain graphs 9.2a, b which are redrawn in a manner similar to the graphs of the impulse approximation: two chains are combined into

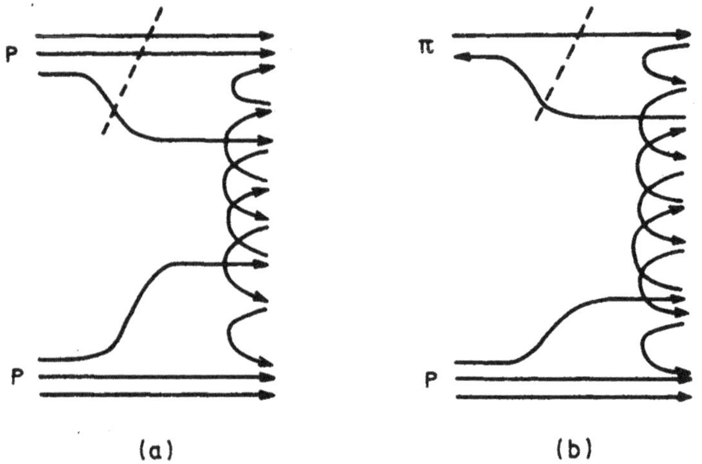

Fig. 9.4 Redrawn graphs of Fig. 9.2a, b: two chains are combined into one. The two-chain graphs became analogous to the impulse-approximation graphs of the additive quark model. The dashed lines show at which stage of the process the distribution of dressed quarks in hadrons is considered by the additive quark model.

one. The analogy with the impulse approximation is rather impressive. Indeed, there is a fast quark (Fig. 9.4b) or a fast diquark (Fig. 9.4a) analogous to a quark-spectator (in a meson beam) or to a pair of quark-spectators (in a baryon beam). There is, further, a slowed valence quark analogous to the tagged quark. The rapidity gap between hadrons produced by fast quarks (or diquarks) and those produced by slow quarks is similar (one-two units) to that between hadrons produced by quark-spectators and by tagged quarks.

Nevertheless the quark distributions (9.2), (9.4), (9.5) in the DTU scheme and those in the quasinuclear quark model (7.7), (7.11), (7.14) are quite different. The reason is that these distributions are considered at different stages of the multiple production process. Indeed, the quasinuclear quark model deals with distributions "before" the interaction (dashed lines in Fig. 9.4), while in the DTU scheme one considers distributions "during" the interaction process: "after" the emission of quark chains, but "before" the hadronization, which is, of course, followed by the production of new quarks and a new re-distribution of the momentum.

An analogy between these two approaches can also be observed in some results obtained. Within the framework of the two-chain approximation the relations of πp and pp cross sections derived in the additive quark model can be easily reproduced in the DTU scheme. Indeed, the number of configurations with slowed quarks is nine (see Fig. 9.4a) in pp collision and six (Fig. 9.4b) in πp collision. Assuming that these configurations are of equal probabilities we come immediately to[11,17] $\sigma_{\text{ine}\ell}(\pi p)/\sigma_{\text{ine}\ell}(pp) = 2/3$. In fact this derivation does not differ from the corresponding one in the additive model, if one assumes that the slowed quark in the DTU scheme plays the role of the tagged (or interacting) quark in the additive model. However, this analogy is by far not complete. For example, if for the kaon beam the distribution (9.5) is applied, the slowed quark in Kp collisions must be almost always the nonstrange one. The considerations which led in the DTU scheme to the ratio $\sigma_{\text{ine}\ell}(\pi p)/\sigma_{\text{ine}\ell}(pp)=2/3$, give in this case $\sigma_{\text{ine}\ell}(Kp)= \frac{1}{2}\sigma_{\text{ine}\ell}(\pi p)$; this, of course, contradicts the experimental data. Thus in some papers (see e.g. Ref. 15) quark distributions in the kaon beam were taken in the form of Eq. (9.4) instead of Eq. (9.5).

Another important point, which is not evident in the DTU scheme, refers to the hadron distribution (9.6): the question is, whether this distribution describes the prompt secondaries (as in quark statistics) or effectively accounts for the resonance decays due to some duality properties of the transition functions $D_{q\to h}(x)$. Most authors of the papers cited consider the latter case; however there are several attempts to account the production and subsequent decays of the resonances in the framework of DTU scheme (see Refs. 22, 23). For instance, in Ref. 23 the production of pseudoscalar, vector, and tensor mesons is considered.

The distribution functions of quarks are taken in Ref. 23 in the form $F_h^q(x) \sim x^A(1-x)^B$, where A and B are free parameters. The transition function $D(z)$ is assumed to be $D(z) = c_1/z + c_2/\sqrt{z}$. The values A and B turn out to be not very different from those used in the standard DTU scheme. The probabilities for pseudoscalar, vector and tensor meson productions are supposed to be

$$P(0^-) : V(1^-) : T(2^+) = 0.29 : 0.55 : 0.16 \quad . \tag{9.7}$$

While the experimental spectra of the resonances produced in πp and Kp collisions are described quite satisfactorily in Ref. 23, the resulting spectra of the stable secondaries sometimes disagree with the data. As a possible reason for these deviations the author points out, that the production of other higher excitations is not considered. According to this, it is interesting to compare (9.7) with the meson-production probabilities in our approach. Within the framework of quark statistics one has to take into account not only the tensor multiplet, but other P-wave mesons, the scalar multiplet 0^+ and the two axial vector multiplets 1^+ too. The production probability of P-wave meson multiplets is supposed to be 30% (see Sec. 7.3). This means that in our approach

$$P(0^-) : V(1^-) : T(2^+) : A(1^+) : A(1^+) : S(0^+)$$

$$= 0.175 : 0.525 : 0.125 : 0.075 : 0.075 : 0.025 \quad . \tag{9.8}$$

It is seen, that the production of vector and tensor mesons are almost the same in quark statistics and in Ref. 23. However, in our approach the prompt production of pseudoscalar mesons is substantially less because of the contribution of axial and scalar mesons.

If in the DTU scheme Eq. (9.6) describes distributions of the observable particles (resonance decay products included), then the particle production in the beam-fragmentation region of meson-nucleon collisions may be obtained without any additional parameters. Examples for DTU-calculated spectra from Refs. 11, 14, 15 are shown in Fig. 9.5 together with the experimental data. The predictions for kaon spectra need some additional remarks. The structure of t-channel exchanges for $\pi^{\pm} \to \pi^{\pm}$ and $K^{\pm} \to K^{\pm}$ transitions are very similar; thus in the frame-

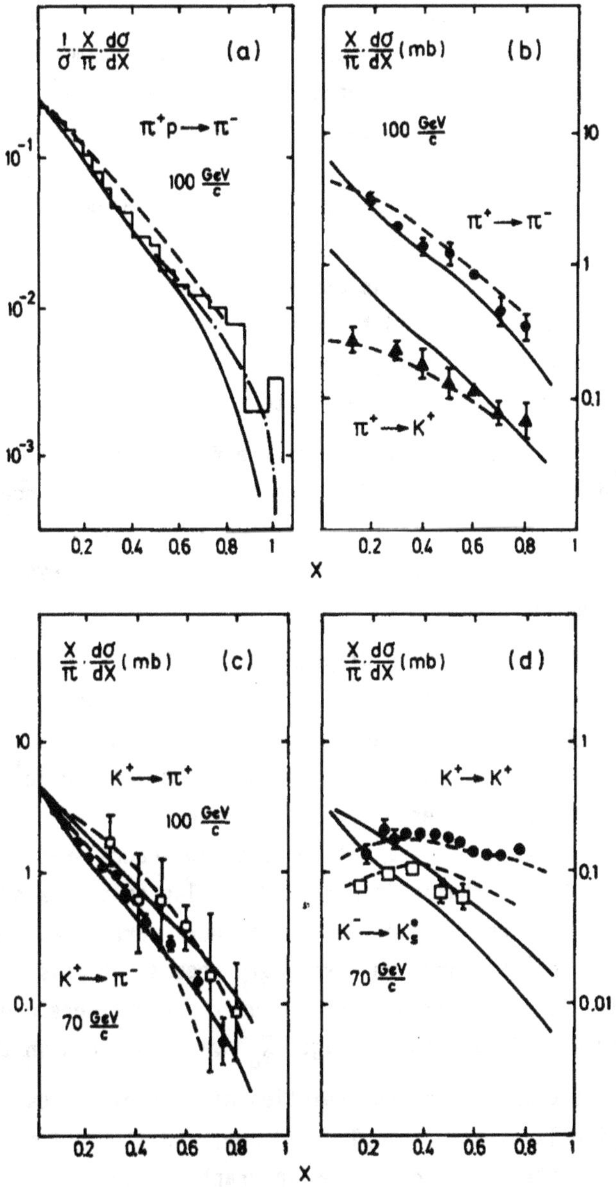

Fig. 9.5 Meson spectra in πp and Kp collisions. The calculated curves are taken from Refs. 11, 14. The dotted curve (a) corresponds to the inclusion of the contribution of diffraction dissociation, the dashed curves stand for AQM predictions.

work of DTU we should expect a similar behaviour of the corresponding spectra. Indeed, in Ref. 15, where, for the kaon beam the distributions of Eq. (9.4) were used, the spectra obtained for $K^{\pm}\to K^{\pm}$ and $K^{\pm}\to K^{0}_{S}$ are similar to $\pi\to\pi$ spectra and contradict the experimental data. The application of the distribution (9.5) does not improve the situation either. Let us remind ourselves that in the additive quark model the difference in kaon and pion spectra is mainly due to resonance decay contributions: the decay pions carry relatively small fractions of the parent resonance momentum and their spectra are appreciably softened. Thus the present understanding of resonance decays through duality arguments is probably inadequate for the secondary kaons.

Particle spectra in proton beams are formed by quark$\to$hadron and diquark$\to$hadron transitions. The effect of baryon resonance decays (e.g. the Δ-isobar) requires the consideration of the diquark$\to$pion transition. This is carried out in Ref. 20, where, however, new parameters are introduced.

Looking at the two-chain diagrams of the Fig. 9.2a-d type, one may conclude that the total multiplicity of the particles produced in these processes is twice as large as that in the one-chain graphs; the latter is assumed to be equivalent to the quark jets in e^+e^- annihilation or other hard processes. This is true, of course, only at asymptotically high energies: at the available accelerator energies like FNAL or ISR the chains in Fig. 9.2 overlap only in a small part of the rapidity interval, for the slowed quark carries away only a small fraction of the momentum. Probably because of this reason, in Refs. 10, 11, 20 the distributions of quark-partons were used instead of (9.2), (9.4), (9.5): the average $\bar{x}_q$ value is less in this case.

In fact the problem of the relation of multiplicities in the central region is a rather complicated one, and it might not be solved by calculating the number of chains in graphs of the Fig. 9.2 type. This is the weak point of the model considered. Indeed, in the case of hadron-hadron collisions quarks belonging to different chains are relatively close to each other, the distances between them being smaller than the radius of confinement. Because of this, quarks of different chains may interact, forming one common jet. Remember that in DTU the presence of a sufficiently strong interaction of the quark-gluon matter at the last stage of hadronization process is assumed (see the transition from graph 9.1d to graph 9.1e). Describing spectra in the framework of the quasinuclear hadron model a different assumption is made, namely, it is supposed that the hadron spectra are equal in the central regions of both the hadron interaction and the e^+e^- annihilation processes (see Secs. 7.1 and 7.2).

The two-chain approximation was considered here in order to clarify the basic ideas of the DTU approach. However, the recent descriptions of the experimental data in this scheme usually take into account the multi-chain graphs of the Fig. 9.2e type. In the multi-chain approximation Eqs. (9.2), (9.4) and (9.5) are no longer valid, since some fraction of the total momentum is carried away by additional quark chains. The way of energy division between the chains cannot be fixed unambiguously in the DTU scheme; in fact, Eqs. (9.1) and (9.3) present one of the possibilities of the energy division. Thus the way of energy division between the quark chains is an implicit free parameter of the models, based on the DTU scheme.

The model, developed in Refs. 14, 24, 25 is essentially based on the consideration of multi-Pomeron graphs with $\alpha_{\mathbf{P}}(0) > 1$. The parameters of the Pomeron are obtained from data on total and elastic cross sections of the pp and $\bar{p}p$ interactions: $\alpha_{\mathbf{P}}(0) = 1 + \Delta$, $\Delta = 0.070 \pm 0.005$ and $\alpha'_{\mathbf{P}}(0) = (0.25 \pm 0.05)\ (\mathrm{GeV}/c)^2$. In the case of $\alpha_{\mathbf{P}}(0) > 1$ the total cross section is determined by the total contribution of graphs with a large number of Pomeron exchanges. Due to[24)]

$$\sigma_{tot} = 8\pi g_1 g_2 e^{\xi\Delta} \sum_{\nu=1}^{\infty} \frac{1}{\nu \cdot \nu!} \left(-\frac{a}{2}\right)^{\nu-1} \tag{9.9}$$

where $a = \dfrac{2Cg_1g_2}{R^2 + \alpha'_{\mathbf{P}}\xi} \exp(\xi\Delta)$, $\xi = \ell n\, s$, *while the vertices which connect* Pomerons and hadrons are assumed to be equal to $g_i \exp(R_i^2 t)$. The parameter of the quasi-eikonal $C = 1 + \sigma_{DD}/\sigma_{e\ell}$ is defined by the diffraction-dissociation cross section; the value $g_1 g_2 = 3.64\ (\mathrm{GeV/c})^{-2}$, $R^2 = 3.56\ (\mathrm{GeV/c})^2$, $C = 1.5$ are taken. As $\xi \to \infty$ the expression (9.9) approximates the Froissart limit $\sigma_{tot} \sim \xi^2$. The same parameters define in the quasi-eikonal approach[26] the cross sections of production of n Pomeron showers $(n \geq 1)$

$$\sigma_n = \frac{8\pi g_1 g_2}{na} e^{\xi\Delta} \left(1 - e^{-a} \sum_{k=0}^{n-1} \frac{a^k}{k!} \right) \tag{9.10}$$

Diagrams with cut Pomerons are interpreted in Refs. 14 and 25 as graphs of the type of Fig. 9.2, quarks (valence quarks and sea quarks) are understood as constituent (dressed) quarks. In the papers of the ITEP group two variants of energy division are considered. First, a continuous emission of showers by the leading baryon is supposed; the latter looses at every step a fraction (x_0) of its momentum $p_1 = (1-x_0)p$, $p_2 = (1-x_0)^2 p, \ldots, p_n = (1-x_0)^n p$. The second version is an even division of energy between the showers: each of them carries a momentum p/n. The distribution of mesons in each of the showers is given by the expression (9.6), where x has to be understood as the momentum fraction of the shower. The distribution function of a quark is defined by an expression similar to (9.2) (the value $\alpha_B(0)$ is taken to be - 0.4). The transition function $D_{q\to h^\pm}$ of a quark into a charged meson is chosen to be of the form

$$D_{q\to h^\pm}(z) = z^{-1}(1-z)^{-1/2 + 2k_\perp^2 \alpha_R'(0)} \tag{9.11}$$

where $\alpha_R'(0) \simeq \alpha_B'(0) \simeq 1\ (\mathrm{GeV/c})^2$. Such a function $D(z)$ reproduces the correct (from the point of view of Reggeon theory) behaviour of the spectra as $x \to 1$ and $x \to 0$. As $x \to 1$, Eq. (9.11) leads to $\frac{1}{\sigma} d\sigma/dx \sim (1-x)^{1-2\alpha_B(k_\perp^2)}$ which corresponds to the three-Reggeon asymptotics BBP. (At this point the approach considered differs from Refs. 11, 12, where this condition is not fulfilled.) In the $x \to 0$ region the contribution of one shower to $\frac{1}{\sigma} d\sigma/dx$ behaves like $1/x - 1.6/\sqrt{x}$, i.e.

it is determined by both the two-Pomeron Mueller-Kancheli graph (see Sec. 2.3) and diagrams RP (P'P included). Note, that the second term enters here with a minus sign: the necessity of the introduction of such a term for the description of experimental data was pointed out in Ref. 27. The ITEP model gives a good description for charged-particle spectra in the central region (see Fig. 9.6a). In the fragmentation region the model predicts a slight decrease of the spectra of charged mesons with the increase of energy (Fig. 9.6b). The value of the scale breaking is here, however, model dependent: it is determined by the way energy divides between the showers.

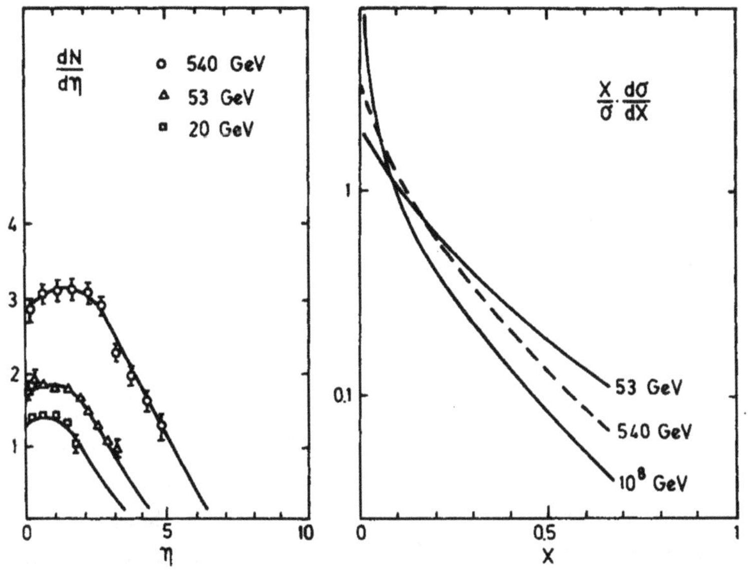

Fig. 9.6 (a) Spectra of charged particles as functions of the quasi-rapidity η obtained at the ISR and the SPS collider and calculations of the ITEP group (the first version of energy division between showers[14]).
(b) Scale breaking at large x values for the same spectra.

If the energy is high enough and the hadron density in the central region of an arbitrary chain does not depend on the number of the produced chains (i.e. on the number of cut cylinders), then due to the Abramovsky-Gribov-Kancheli rules (see Sec. 2.3), the inclusive cross section has to be equal in the $x \to 0$ region to

$$x \frac{d\sigma}{dx}(x \to 0) = \sigma_{\mathbf{P}} \quad , \quad \sigma_{\mathbf{P}} = 8\pi g_1 g_2 e^{\xi\Delta} \quad . \tag{9.12}$$

A non-diffractive inelastic cross section which is connected with the production of at least one chain is, due to (9.10),

$$\sigma_{abs} = \sum_{n=1}^{\infty} \sigma_n = f(a)\sigma_{\mathbf{P}} \tag{9.13}$$

where

$$f(a) = \sum_{\nu=1}^{\infty} \frac{1}{\nu \cdot \nu!} (-a)^{\nu-1} = \frac{1}{a} \int_0^a \frac{dx}{x} (1 - e^{-x}) \tag{9.14}$$

If so, the relative height of the plateau, i.e. the density of the hadrons formed in the central region is equal to

$$F(0) = \frac{x}{\sigma_{abs}} \cdot \frac{d\sigma}{dx} = \frac{F_0}{f(a)} \tag{9.15}$$

With the values of the parameters taken from Ref. 14 the quantity $F(0)$ increases by about 10% if the energy rises from ISR values to the SPS collider energy. Experimentally $F(0)$ increases by 50%, and this is well described by the DTU model[14]. The reason for this seeming contradiction is the fact that these energies are not asymptotic ones yet from the point of view of the model: the density of hadrons produced by one chain equals[14]

$$F_1(x) = \frac{F_0}{2} \int_{x_+}^{1} dx_1 u(x_1) D\left(\frac{x_+}{x_1}\right) \quad , \quad x_+ = \frac{1}{2}\left(x + \sqrt{x^2 + \frac{m_T^2}{s}}\right) \quad . \tag{9.16}$$

As $x \to 0$ the function $D(x_+/x_1)$ turns out to be less than unity, and with the increase of the energy the quantity $F_1(x)$ approaches slowly its asymptotic limit $F_0/2$: $F_1(x)$ does not reach it at ISR and even at Collider energies. Hence, the calculated increase of spectra with the energy in the central region is connected with the choice of $D(x_+/x)$ and not with the growth of the contribution from Reggeon cuts.

The total cross section of all diffractive processes (including elastic scatterings) is

$$\sigma_{dif} = \sigma_{tot} - \sigma_{abs} = \sigma_{\mathbf{P}} \left[f(\tfrac{a}{2}) - f(a) \right] \tag{9.17}$$

The limit of "black" interaction corresponds to $\sigma_{dif} = \frac{1}{2}\sigma_{tot}$; here

$$f(a/2) = 2f(a) \quad . \tag{9.18}$$

From (9.14) it is seen, that in the quasi-eikonal scheme this situation occurs as $a \to \infty$ (since $f(a \to \infty) \sim \frac{\ell n\, a}{a}$)). Using parameters from Ref. 14, at FNAL-ISR energies $\sigma_{dif} \simeq 10$ mb. (From this about 3 mb is the cross section of inelastic diffraction dissociation.) Experimentally σ_{dif} is about 15 mb. The deviation is at first sight not large, but in the quasi-eikonal approach $\sigma_{dif} = 10$ mb corresponds to $a = 3$, while $\sigma_{dif} = 15$ mb requires a parameter $a = 20$. In the last case the effective number of Regge cuts which have to be calculated is a few dozens. (For the description of the diffraction-dissociation cross section it is, probably, necessary to take into account the so-called enhanced graphs. This, however, might lead to changes in the structure of the model.) Thus, the possibility to take into account multi-chain graphs of the DTU scheme in a simple quasi-eikonal approach is closely connected with some additional problems in the description of high energy scattering processes. We think, that the approach of the ITEP group to multiparticle production processes and that developed in the framework of the quasi-nuclear quark model are in many respects complementary. To study the spectra the ITEP model starts from the Regge behaviour at $x \to 0$ and $x \to 1$, while the intermediate x forms that region to which one interpolates; the way of interpolation here is an important element for success. On the other hand, the quasinuclear approach starts with the region of intermediate x, while the region $x \to 1$ cannot be described at all. In the model of the ITEP group so far the spectra of the stable hadrons are considered — here the problem is how to extend this scheme to the production of baryon and meson resonances.

The DTU scheme may be applied to hadron-nucleus collisions[16,17,28]. This is based on the formalism of the multi-Reggeon exchange for hadron-nucleus amplitudes[29-34]. In multiparticle production on nuclei the multi-chain graphs are relevant: their contribution is here magnified by the combinatorial factor which takes into account the possibilities for joining, by chains, different nucleons of the nuclei. The weight of the individual multi-chain graphs is determined by the number of different Pomeron cuts and can be found with the help of the rules formulated in Refs. 24, 29-34. Obviously, the multi-chain graphs lead to an increase of the multiparticle production in the central region. Nuclear screening in the fragmentation region of the nucleus is a consequence of the consideration of energy conservation: the production of additional chains requires energy and therefore, with the increase of the number of chains the multiplicity decreases in the region of large x $(x > 0.2)$.

The region $x \to 1$ in the DTU scheme is dominated by the two-chain graphs, which are suppressed by the combinatorial factor in hadron-nucleus collisions. Thus the spectra calculated for large x values usually underestimate the data (see e.g. Ref. 17).

The region of target fragmentation in hA collisions needs some special consideration of absorption and cascade effects; in terms of the DTU scheme it is equivalent to the consideration of the enhanced Regge diagrams. The problems that arise were discussed in Refs. 16 and 35.

9.2 The Recombination Model

In the recombination model the hadron production is considered as a two-stage process[36]; the first one is the production of quarks as a result of an interaction, the second one, their transition into hadrons. This model has been developing for quite a long time already. One of its realizations is the approach of quark statistics. We will give here a brief description of two directions, namely, the recombination model of quark-partons into hadrons and the recombination model of valons.

The first question which has to be asked is: which quarks recombine into hadrons. In Refs. 37-39 it was assumed, that the hadrons were formed by dressed (constituent) quarks, while Refs. 40-54 deal with the recombination of quark-partons. Let us consider in detail the latter approach.

The idea of the recombination of quark-partons into hadrons was stimulated by the observation of the similarity of pion spectra in pp collisions to the structure functions of valence quark-partons in protons[44]:

$$\frac{d\sigma}{dx}(pp\to\pi^+X)\sim u(x) \quad , \quad \frac{d\sigma}{dx}(pp\to\pi^-X)\sim d(x) \tag{9.19}$$

These relations led to the idea that the spectra of secondary pions were formed by fast valence quark-partons and relatively slow sea antiquarks joining each other. On this basis in Ref. 45 the following formula was suggested for the cross section of meson production

$$\frac{x}{\sigma}\frac{d\sigma}{dx} = \int \frac{dx_1}{x_1}\cdot\frac{dx_2}{x_2}\, F(x_1, x_2)R(x_1, x_2)\delta(x_1+x_2-x) \tag{9.20}$$

Here $F(x_1, x_2)/(x_1x_2)$ is the probability of finding a quark with the momentum fraction x_1 and an antiquark with x_2 in the initial hadron; $R(x_1, x_2)$ is the recombination function of these quarks into a meson. Formula (9.20) itself is not very informative if the aim is to calculate the x-dependence of the inclusive cross section, since it contains two unknown functions F and R. In Ref. 45 an additional assumption was made; due to this the distribution function $F(x_1, x_2)$ of the quark-partons is almost equal to the product of the distributions of the quark and of the antiquark which are known from the data on deep inelastic scatterings at not very large momentum transfers (see, e.g. Ref. 55):

$$F(x_1, x_2) = F_q(x_1)F_{\bar{q}}(x_2)\rho(x, x_1, x_2) \quad . \tag{9.21}$$

The "phase space" factor $\rho(x, x_1, x_2)$ which is here a function not given a priori, forces the inclusive cross section (9.20) to vanish at $x = 1$. In Ref. 45 $\rho(x, x_1, x_2) \sim (1 - x)$ was taken; such a behaviour corresponds to the description of the reaction $pp \to \pi X$ by a triple-Reggeon diagram $R_\Delta R_\Delta \mathbf{P}$ (see Sec. 2.3) with Δ-pole exchange. (Remember, that the intercept of the Δ trajectory is $\alpha_\Delta(0) \simeq 0.$) For the recombination function the following simple form is assumed:

$$R(x_1, x_2, x) = \alpha \frac{x_1 x_2}{x} \tag{9.22}$$

where α is a normalization constant. One can prove, that in the limits $x \to 0$ and $x \to 1$ the expressions (9.21) and (9.22) correspond to the leading Reggeon exchanges in the reaction $pp \to \pi X$.

In the framework of such an approach it was possible to give quite a good description for the ratios of meson inclusive spectra (π^+/π^-, π^-/K^-, K^+/K^-, π^+/K^+) in pp collisions[46]. To describe fragmentational spectra in pion or kaon beams, the structure functions of these mesons (F_q^π and F_q^K) have to be known. F_q^π and F_q^K are introduced as functions which have to be fitted — they are determined from experiment. Figure 9.7a illustrates the result of such a procedure[54]. In Fig. 9.7b the structure function obtained for the valence quark in a pion is compared to the functions which are estimated from data on $\mu^+\mu^-$ pair production.

Let us now briefly consider some weak points of the quark-parton recombination model.

First of all, the role of gluons is not clear in this model. The

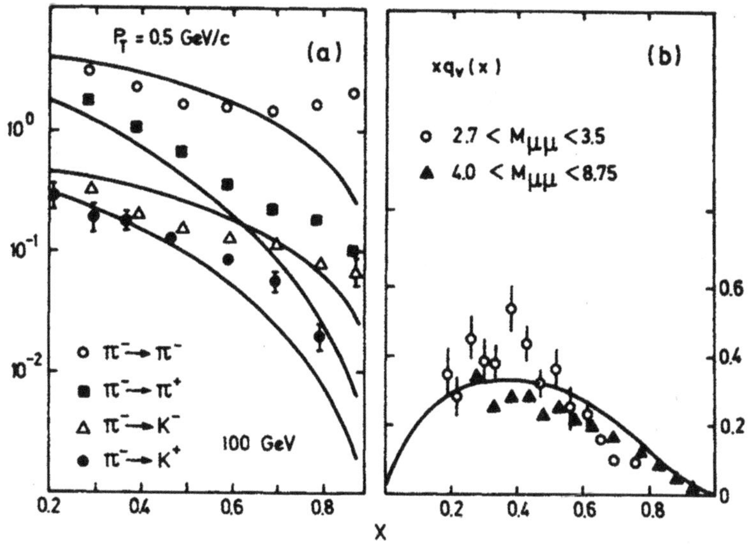

Fig. 9.7 Fits to the single-meson inclusive production in the recombination model[46] (a). A comparison of the valence quark structure function in a pion, used in the recombination model, with the data on dimuon production in pion collisions.

gluons carry away about half of the hadron momenta, and they must participate in the hadron formation process. This is most evident when the spectra of secondary baryons are considered in the framework of the recombination model. Since the spectra of secondary baryons are rather hard (the average x values being about 0.5 - 0.7), they cannot be formed by the recombination of quark-partons only. In order to describe the spectra of secondary baryons usually diquarks are introduced, and it is assumed, that they carry a relatively large momentum fraction of the initial protons, i.e. they practically contain the gluons. This assumption goes beyond the logic of the simple recombination model — it shows an analogy with the additive quark model.

Another confusing point was mentioned by De Grand during the discussion of Hwa's talk[56]. Let us imagine that the density of sea quarks and antiquarks is increased, say, by the factor of 2. Then one would obviously expect that the density of sea mesons has also to be doubled while according to Eq. (9.21) it increases by the factor 4.

Finally, one can ask, which distributions of the quark-partons should be introduced into Eq. (9.21)? It is well known, that the distributions of partons depend on Q^2. Because of that, the choice of Q^2 is essential, if one intends to give a quantitative description of the spectra in a wide x range.

An attempt to include gluons into the framework of recombination model is made in the valon model of Hwa[57-59]. In this model, the recombination of quark-partons into hadrons is a two-stage process: in the first step quark-partons emit and absorb gluons in the process of evolution of the quark-gluon cloud and become "valons"; then these valons recombine into hadrons according to Eq. (9.20).

Up to this point the model is very similar to the quasinuclear quark model and valons may be understood as dressed quarks. However, deeper scrutiny reveals some serious differences.

First of all, the valon model does not assume any additive structure of hadron interaction: there are no valon-spectators in hadron-hadron collisions; instead, all initial valons are destroyed into partons.

Next, there are different estimates of the valon's size. In Ref. 58 the ratio r_v^2/R_h^2 was estimated from the analysis of deep inelastic collisions and appeared to be about 1/5, which is in agreement with the size of the dressed quark; however, in Ref. 59 the same ratio was estimated to be 0.8 - 0.9 (see the discussion in Sec. 5.4). The x-distributions of the valons inside the mesons are also very different from those of the dressed quarks (see Fig. 9.8).

The distribution of the secondary valons (which enter Eq. (9.20)) does not coincide with that in Fig. 9.8: there are no spectators in the valon model. These distributions are to be calculated from quark-parton distributions via the evolution equations. It is possible, however, that the re-distribution of the valon momenta in the final

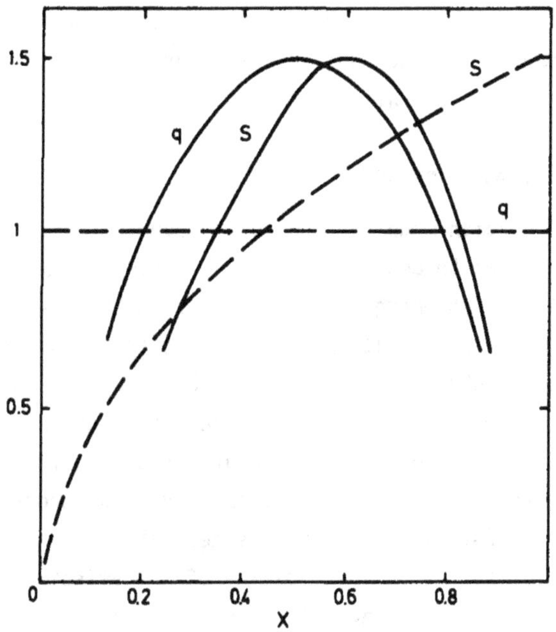

Fig. 9.8 The distributions of the quark q in a pion and of the strange quark s in a kaon in the quasinuclear quark model (solid lines) and in the valon model (dashed lines) respectively.

state would effectively reproduce the distributions of the quark-spectators and tagged quark of the quasinuclear approach.

The role of resonance decays is not clear either in the valon model. It is assumed, that these contributions to the observable spectra are effectively accounted for by Eq. (9.20) due to some duality properties of the recombination function. In this case the observable spectra of all the secondary mesons of the same quark content would have the same x-dependence (only the value of α in Eq. (9.22) is different for different particles), which seems to contradict the data. In the quark statistics approach only prompt particles have similar spectra, and resonance decays destroy this similarity.

It seems to us that the model of valon recombination is not yet definitely established, and thus further changes are still possible in some important points of this model.

9.3 The Lund Fragmentation Model

The fragmentation model of the Lund group (see Ref. 60 and references given therein) tries to describe the whole picture of multiple production phenomena in high energy collisions: soft hadron-hadron collisions, quark and gluon-jet production in hard collisions, heavy flavour production, high p_T phenomena, etc. It is based on the model of relativistic string interaction, in which particle production is interpreted as a stochastic process of the breakup of the string (the colour flux tube) into pieces (Fig. 9.9). Such a fragmentation process was considered e.g. in Refs. 21, 61 and may be described by the simple integral equation. Let $f(z)$ be the probability of the emission of the particle (e.g. meson), which carries away the momentum fraction z. Then the final distribution of these particles, $D(z)$, is given by the solution of

$$D(z) = f(z) + \int_z^1 \frac{dz'}{z'} f(1 - z')D(z/z') \quad . \tag{9.23}$$

The simplest case $f(z) = 1$ is known as the simple or naive Lund model[62]; from Eq. (9.23) we have $D(z) = 1/z$, which corresponds to the constant number of produced particle per rapidity interval $dy = dz/z$ (central plateau). The more famous "standard Lund" model starts from $f(z) = (1+c)(1-z)^c$. In this case as $z \to 0$, $D(z) \simeq (1+c)/z$. Here c is a model parameter, which may be roughly estimated by the perturbative diagram of gluon emission; c appears to be flavour dependent: $c_u = c_d = 0.5$, $c_s = 0.35$, $c_c = 0.15$. The "standard Lund" version of the model is often presented as a set of computational programs of Monte-Carlo event generation[63].

In the recent version of "symmetric Lund" model[64] $f(z)$ is taken in a more complicated form. In the case, when flavours i and j are

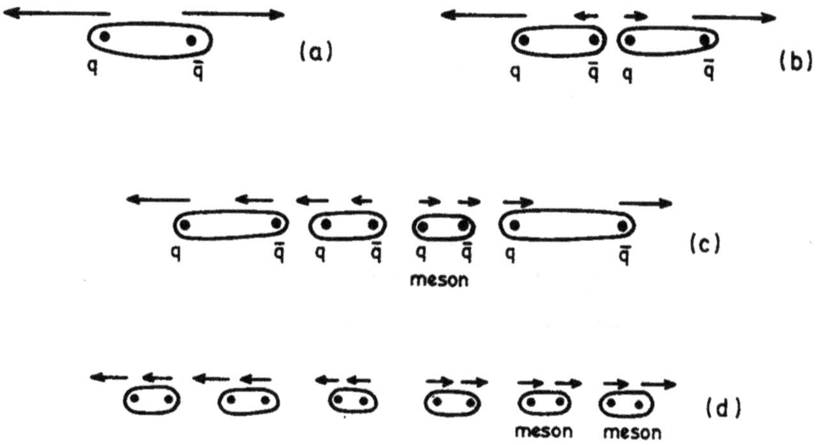

Fig. 9.9 An example for the production of hadrons in the quark jet of the Lund model. Between the quark-antiquark pair a colour force field is stretched (a) which breaks into pieces because of the production of new $q\bar{q}$ pair (b), mesons are produced, first slow ones (c), later fast ones (d).

tighted to the ends of the string it has the form:

$$f_{ij}(z) = N_{ij} \frac{1}{z} z^{a_i} \left(\frac{1-z}{z}\right)^{a_j} \exp\left(-\frac{bm_T^2}{z}\right) \quad . \tag{9.24}$$

Here N_{ij} is a normalization constant, m_T is the transverse mass of the produced hadron; b, a_i and a_j are free parameters of the model. This form of $f(z)$ is rather similar to that used in the DTU approach (see e.g. Eq. (9.11)); however, in the Lund model the values of the parameters a_i, a_j are not determined by the intercepts of the corresponding Reggeons, but used to fit the experimental data. Since for the most part, the results obtained in the Lund model refer to the standard Lund, our following considerations will deal with this version of the model.

Meson production in the Lund model is considered as a tunneling process of the massive $q\bar{q}$ pair production in the uniform linear potential when the string is broken at some given point. Thus, the distribution of the produced quarks over the transverse momentum k_{qT} is given by the function

$$\exp(-m_{qT}^2/\sigma^2) \quad , \tag{9.25}$$

where $m_{qT}^2 = m_q^2 + k_{qT}^2$, σ^2, the value which characterizes the string tension, $\sigma \sim 400 - 500$ MeV. The produced quarks are considered in the Lund model as dressed quarks with masses

$$m_u = m_d = 325 \text{ MeV} \quad , \quad m_s = 500 \text{ MeV} \quad , \quad m_c = 1600 \text{ MeV}$$

Thus, the probability of the strange quark pair $s\bar{s}$ production is suppressed by the factor

$$\lambda = \exp\left(-2\,\frac{m_s^2 - m_q^2}{\sigma^2}\right) \simeq 0.2 - 0.3 \quad , \tag{9.26}$$

which is quite similar to the quark statistics value.

The colour tube may be broken by the production of a diquark-antidiquark pair ($\bar{3}_c + 3_c$ colour state); in this case baryons are produced. The relation of the probabilities for diquark-antidiquark and quark-antiquark pair productions in the Lund model is

$$p(qq)/p(q) \simeq 0.065 - 0.075 \tag{9.27}$$

It is assumed that the distribution of the diquarks is governed by the same expressions as those for quarks. Since diquarks of different flavours and spins have different masses, their production probabilities in the Lund model are different too:

$$p(ud_0) : p(ud_1) : p(uu) : p(us_0) : p(us_1) : p(ss)$$

$$= 1 : 0.35 : 0.35 : 0.06 : 0.02 : 0.007 \quad . \tag{9.28}$$

The subscripts in (9.28) stand for the total spin of the diquark; all vector diquarks have an extra factor 3 from counting the different spin states. The authors of Ref. 60 note, however, that the numbers in the right-hand side of (9.28) are very uncertain.

Since the produced quarks (and diquarks) are dressed ones, soft gluons with small transverse momenta are already accounted for by quark production. The production and fragmentation of high-p_T gluons is considered in the model independently. In spite of very different underlying ideas, there are many similarities in the Lund model and the quasinuclear quark model with quark statistics.

First of all, all particle distributions, defined by Eqs. (9.23), (9.24), (9.25) and so on, are considered as prompt distributions: the resonance decays are to be accounted in both models. In the Lund model the production of the lowest meson 35-plet ($J^P = 0^-, 1^-$) and of the baryon 56-plet is taken into account, for P-wave hadron production is estimated in this model to be negligible. The production ratio of the pseudoscalar and vector mesons is assumed to be a free parameter. While in the earliest version of the model (simple Lund), similarly to quark statistics, $P : V = 1 : 3$, was assumed, in standard Lund the ratio $P : V = 1 : 1$ is taken instead. It is justified on the one hand by the consideration of the tunneling process of $q\bar{q}$ pair production with the inclusion of spin-spin interaction (it gives some suppression for the vector $q\bar{q}$ states); and on the other hand, by the consideration of the data on ρ^0 production in the quark jets of deep inelastic μp collisions[65]. These indicate that in the region $z \to 1$ (where only prompt particles contribute) ρ^0/π^0 production ratio tends to unity. However, as we have considered in the Sec. 7.5, particle production in the region $z \to 1$ is in fact not a typical multiple production process, and thus all the data in this region should be handled with care.

Next, multiparticle production in hadron-hadron collisions is considered in both approaches in a similar manner. In the Lund model the process of inelastic hadron-hadron collision proceeds through the colour exchange between a pair of quarks (one from each of the colliding hadrons). As a result a colour flux tube is stretched between these (interacting) quarks, which loose a large part of their initial momenta; the momenta of the rest quarks are changed less significantly. Thus, the interacting quarks are analogous to the tagged quarks of the quasinuclear quark model, while the leading quark (or antiquark) of the meson beam to the quark-spectator (see Fig. 9.10). In the baryon-fragmentation process there is one more fast quark — the junction quark in the terminology of the Lund model. In such a picture the x-distributions of the secondaries in the central region of hadron-hadron collisions and in the quark jets of hard processes are identical, as in the quasinuclear quark model. However the distributions of the fragmentational particles (i.e. those which contain the leading and junction quarks) are taken in the Lund model in a form just to fit the experimental data.

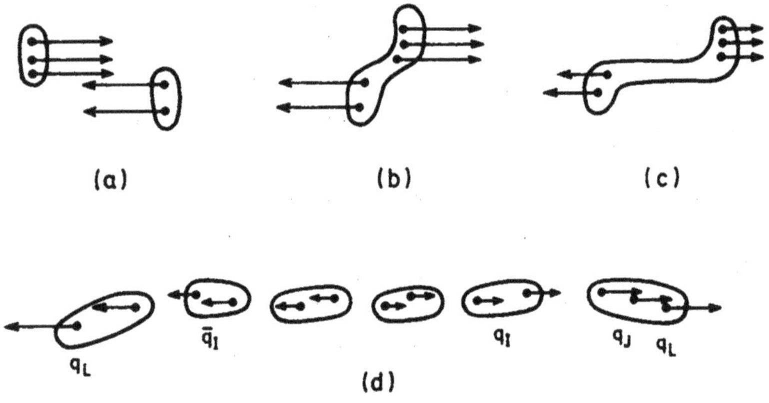

Fig. 9.10 An example of hadron production in meson-nucleon collision. The distribution of hadrons in the central region is similar to the distribution of hadrons in a quark jet, while the distribution of fragmentational hadrons is determined by the distribution of the valence quark after the interaction.

The content of the secondaries in hadron-hadron collisions in the Lund model is also similar to that in the quark statistics in many points. For example, the relative probability of baryon production in the meson-fragmentation region is determined by the probability of the diquark production (9.27), which is very close to the value of η in

Eq. (7.3). In the baryon-fragmentation process two fast quarks (the leading quark q_L and the junction quark q_J) are joined into one hadron $q_L q_J q$ with the probability of about 50%, which is also similar to the value of ξ_1 in Eq. (7.2). Octet/decouplet ratio for these baryons is also chosen just as in quark statistics. However, the relative production probabilities of the baryons, containing sea diquarks, are different in the Lund model and in the quark statistics, for in the former sea diquarks are produced with the probabilities (9.28), which differ from the quark statistical relations.

One may note, that baryon production and fragmentation are considered in the Lund model rather differently: the baryons produced are assumed to be in a quark-diquark configuration, while in the fragmentation processes baryons are assumed to consist of three separated quarks. It is an open question, whether such pictures are compatible with each other.

A comparison of the results of the Lund model with the data on multiparticle production in hadron collisions was presented in the talk of De Wolf[66)] at the Volendam Conference. Let us consider a few data from this talk. Table 9.1 shows average multiplicities of K^{*+} and K^{*0} in $K^0 p$ collisions and the corresponding predictions of the Lund model and of the quark statistics. In Fig. 9.11 the ρ^0 and K_S^0 production cross sections are shown together with the predictions of both of the models. The agreement is in all cases quite satisfactory; however, the Lund model predicts slightly higher values for vector meson production probabilities. This may seem to be surprising, since in the Lund model the ration $V : P$ is taken to be $1 : 1$ instead of the $3 : 1$ ratio used in quark statistics. The reason is that in the Lund model the P-wave meson contributions are neglected; in Sec. 7.5 we have already demonstrated the effect of these contributions for the predicted ρ/π production ratio. We will illustrate it here once more by a simple, though rather artificial example.

Let us assume that S- and P-wave mesons are produced in the ratio $7 : 3$ (as was estimated in the analysis of the average multiplicities; see Sec. 7.4) and that the quark statistical relation $3 : 1$ holds for

Table 9.1 The average numbers of $K^{*0,+}(890)$ and ρ^0 per inelastic K^+p collision compared with the Lund-model (LM) and quark-statistics (QS) predictions.

P_{lab}	Particle	Experiment	LM	QS
	$K^{*+}(890)$	0.23 ± 0.01	0.40	0.34
32 GeV/c	$K^{*0}(890)$	0.21 ± 0.03	0.16	0.19
	ρ^0	0.22 ± 0.02	0.36	0.25
	$K^{*+}(890)$	0.25 ± 0.02	0.41	0.37
70 GeV/c	$K^{*0}(890)$	0.25 ± 0.02	0.21	0.22
	ρ^0	0.26 ± 0.02	0.47	0.35

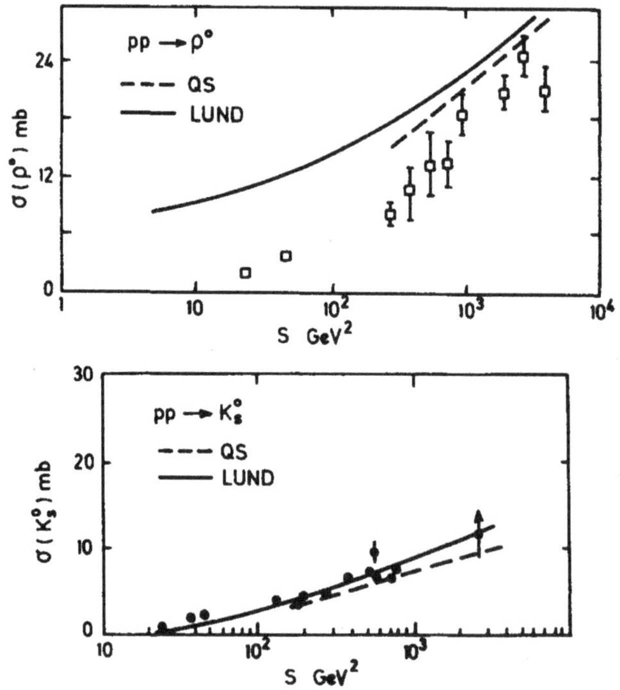

Fig. 9.11 Energy dependence of ρ^0 and K_S^0 production in pp collisions, compared to Lund model and quark statistics.

the $V:P$ production ratio. Then the relative fraction of the prompt vector meson will be $0.7 \times 3/4 = 0.52$, while that for pseudoscalars would be $0.7 \times 1/4 = 0.18$. Further, let us assume that every P-wave meson decays into two pseudoscalars (this is only a rough approximation); if in a model the P-wave meson contribution is neglected, these decay products must be interpreted as "prompt" particles. Thus, if we now try to extract $V:P$ ratio from the experimental data without taking account of the P-wave mesons, we have to estimate the prompt pseudoscalar fraction to be $P = 0.18 + 0.6 = 0.78$ and the ratio will be $V:P = 0.52 : 0.78 = 1 : 1.5$ instead of the input $3:1$.

The most serious differences between the predictions of the Lund model and those of the quark statistics occur in the baryon production in the quark jets. The quark statistics predicts for the tagged hadrons in the jet, initiated by the quark q_i, the average multiplicities (see Eq. (7.48a))

$$< n_{M_i} > = 2/3 \quad , \qquad < n_{B_i} > = 1/3 \quad ,$$

while in the Lund model they are determined by the diquark/quark production ratio of Eq. (9.27)

$$\begin{aligned} < n_{M_i} > &\simeq 0.93 - 0.94 \\ < n_{B_i} > &\simeq 0.06 - 0.07 \quad . \end{aligned} \tag{9.29}$$

The predicted meson/baryon ratios in the central region are also different, but the difference is less than that for the tagged particles.

The values of $R(h) = \sigma(h)/\sigma(\mu^+\mu^-)$ predicted from Eq. (9.29) for the c-quark-initiated jets of e^+e^- annihilation are presented in Table 9.2 together with the corresponding quark statistical predictions. There is a serious disagreement between the data for[67] $R(\Lambda_c^+ + \bar{\Lambda}_{\bar{c}}^-)$ and the predictions of the Lund model. However, these data are obtained not by direct measurements, but from the jump of the baryon number passing the threshold of charm production. Further measurements on

Table 9.2 The values of $R(h) = \sigma(h)/\sigma_{\mu\mu}$ for charmed particle yields in e^+e^- annihilation and the results of LM and QS calculations.

R(h)	Experiment	LM	QS
$R(D^+ + D^-)$	0.75 ± 0.31 (Ref. 68)	0.56	0.40
$R(D^0 + \bar{D}^0)$	1.2 ± 0.4 (Ref. 68)	1.60	1.14
$R(\Lambda_c^+ + \Lambda_c^-)$	0.53 ± 0.13 (Ref. 67)	0.18	0.62

baryon production in quark jets, and especially direct measurements on the production of charmed baryons seem therefore to be rather important.

References

1. K. Fialkowski and W. Kittel, "Parton models of low momentum transfer processes", Preprint NIKHEP Nijmegen, HEN 82-226 (1982).
2. P. Low, Phys. Rev. D12, 163 (1975).
3. S. Nussinov, Phys. Rev. Lett. 34, 1286 (1975).
4. G. t'Hooft, Nucl. Phys. B72, 461 (1974).
5. G. Veneziano, Nucl. Phys. B74, 365 (1974).
6. G. Veneziano, Nucl. Phys. B117, 519 (1976).
7. Chan Hong Mo et al., Nucl. Phys. B86, 470 (1975).
8. Chan Hong Mo et al., Nucl. Phys. B92, 13 (1975).
9. G. Chew and C. Rosenzweig, Phys. Rep. 41C, 263 (1978).
10. A. Capella et al., Phys. Lett. 81B, 68 (1979).
11. A. Capella et al., Zeit, f. Phys. C3, 329 (1980).
12. G. Cohen-Tannoudji et al., Phys. Rev. D21, 2689 (1980).
13. G. Cohen-Tannoudji et al., Phys. Rev. D19, 3397 (1979).

14. K.A. Ter-Martirosyan, "Spectra and multiplicities of hadrons produced at high energies by breaking quark-gluon Pomeron", Preprint ITEP-134, Moscow (1982).

15. H. Minakata, Phys. Rev. D20, 1656 (1979).

16. W.Q. Chao et al., Phys. Rev. Lett. 44, 518 (1980).

17. A. Capella, "Dual Fragmentation Models of Soft Hadron-Hadron and Hadron-Nucleus Interactions", Lecture given at the Europhysics Study Conf., Erice 1981, Preprint LPTPE 81/10 (1981).

18. A. Capella and J. Tran Than Van, Preprint LPTPE 82/8 (1982).

19. A.B. Kaidalov, Zeit. f. Phys. C12, 63 (1982).

20. A. Capella, U. Sukhatme and J. Tran Than Van, Phys. Lett. 119B, 220 (1982).

21. R. Field and R. Feynman, Nucl. Phys. B136, 1 (1978).

22. M. Fukugita et al., Phys. Rev. D19, 187 (1979).

23. M. Uehara, Prog. Theor. Phys. 66, 1697 (1981).

24. M.S. Dubovikov and K.A. Ter-Martirosyan, Nucl. Phys. B124, 163 (1977).

25. A.B. Kaidalov, Preprint ITEP-50, Moscow 1982.

26. K.A. Ter-Martirosyan, Phys. Lett. 44B, 377 (1973).

27. M.N. Kobrinsky, A.K. Likhoded, and A.N. Tolstenkov. Yad. Fiz. 20, 775 (1974); Sov. J. Nucl. Phys. 20, 414 (1974).

28. A. Capella and J. Tran Than Van, Phys. Lett. 93B, 146 (1980).

29. Yu. M. Shabelski, "Multiparticle production in nucleon-nucleus collisions at high energies", Preprint LNPI-248, Leningrad (1976).

30. A. Capella and A.B. Kaidalov, Nucl. Phys. B111, 477 (1976).

31. R.T. Cutler and D.R. Snider, Phys. Rev. D13, 1509 (1976).

32. J. Weis, Preprint CERN-TH. 2197 (1976).

33. A. Capella and A. Krzywicki, Phys. Rev. D18, 3357 (1978).

34. Yu. M. Shabelski, Nucl. Phys. B132, 491 (1978).

35. V.R. Zoller and N.N. Nikolaev, Yad. Fiz. 36, 918 (1982); Sov. J. Nucl. Phys. 36, 538 (1982).

36. H. Satz, Phys. Lett. 25B, 220 (1967).

37. V.V. Anisovich and V.M. Shekhter, Nucl. Phys. B55, 455 (1973).

38. J.D. Bjorken and G. Farrar, Phys. Rev. D9, 1449 (1974).

39. V. Cerny, P. Lichard and J. Pisut, Phys. Rev. D16, 2822 (1977); D17, 4052 (1979).

40. L. Van Hove and S. Pokorski, Nucl. Phys. B86, 243 (1975).

41. A.N. Tolstenkov, Preprint IFVE 76-51, (in Russian) Serpukhov (1976).

42. V.G. Kartvelishvili, A.K. Likhoded and G.P. Pronko, Pis'ma v ZhETF 23, 664 (1976); JETP Letters 23, 609 (1976).

43. L. Van Hove, Acta Phys. Pol. B7, 339 (1976).

44. W. Ochs, Nucl. Phys. B118, 397 (1977).

45. K.P. Das and R.C. Hwa, Phys. Lett. 68B, 459 (1977).

46. D.W. Duke and F.N. Taylor, Phys. Rev. D17, 1788 (1977).

47. J. Ranft, Phys. Rev. D18, 1491 (1978).

48. M. Teper, Rutherford Lab. Rep. RL78-022 (1978).

49. E. Lehman *et al.*, Phys. Rev. D18, 3353 (1978).

50. T. De Grand and H. Miettinen, Phys. Rev. Lett. 40, 612 (1978).

51. F. Takagi, Tohoku Univ. Rep. TU-79-194 (1979).

52. L. Van Hove, Acta Phys. Austriaca. Suppl. XXI, 621 (1979).

53. E. Takasugi and X. Tata, Phys. Rev. D23, 2573 (1981).

54. E. Takasugi and X. Tata, Phys. Rev. D26, 120 (1982).

55. J. Kuti and V.P. Weiskopf, Phys. Rev. D4, 3419 (1971).

56. R.C. Hwa, in *Proceedings of the XII International Symposium on Multiparticle Dynamics*, Notre Dame (1981).

57. R.C. Hwa, Phys. Rev. D22, 759 (1980).

58. R.C. Hwa, Preprint OITS-159 (1981).

59. R.C. Hwa, Preprint OITS-194 (1982); Preprint OITS-206 (1983).

60. B. Andersson, G. Gustafson, G. Ingelman, and T. Sjöstrand, Phys. Rep. 97C, 31 (1983).

61. A. Krzywicki and B. Peterson, Phys. Rev. D6, 924 (1972).

62. B. Andersson, G. Gustafson, and C. Peterson, Nucl. Phys. B135, 273 (1978); Zeit f. Physik. C1, 105 (1979).

63. T. Sjöstrand, "The Lund Monte Carlo for Jet Fragmentation". Preprint LUTP 82-3 (1982): "The Lund Monte Carlo for e^+e^- Jet Physics", Preprint LUTP 82-7 (1982).

64. B. Andersson, G. Gustafson, and B. Söderberg, Lund preprint LUTP 83-2 (1983).

65. EMC Collaboration: J.J. Aubert et al., Phys. Lett. 133B, 370 (1983).

66. E.A. DeWolf, "Inclusive Resonance and Particle Production", in Multiparticle Dynamics 1982, eds. W. Kittel, W. Metzger and A. Stergiou (World Scientific, Singapore, 1983).

67. G.S. Abrams et al., Phys. Rev. Lett. 44, 10 (1980).

68. M.W. Coles et al., Preprint SLAC-PUB-2916 (1982).

Appendix A

NOTATIONS, NORMALIZATION CONDITIONS AND ALL THAT

There are different ways of writing the metric tensor in 4 dimensions, γ-matrices, amplitudes and so on. We present here our choice for these. Besides, some useful relations are given for reference.

A.1 Metric

We use the metric tensor

$$g_{\mu\nu} = \text{diag}\,(1, -1, -1, -1) \quad . \tag{A.1}$$

Covariant and contravariant vectors are not distinguished; we adopt the notation

$$A_\mu B_\mu = A_0 B_0 - \vec{A}\cdot\vec{B} = A_0 B_0 - A_1 B_1 - A_2 B_2 - A_3 B_3 \quad . \tag{A.2}$$

Summation over doubled subscripts is assumed wherever the opposite is not specified.

A.2 SU(N) Groups

The fundamental representation space for SU(N) group is formed by N-component spinors Ψ (columns of N complex numbers or field operators). The transformations of the fundamental representation

$$\Psi \rightarrow \Psi' = S\Psi \tag{A.3}$$

are realized by $N \times N$ complex matrices which obey unitarity and unimodularity conditions:

$$SS^{+} = I \quad , \qquad \det S = 1 \quad . \tag{A.4}$$

Every matrix S has N^2-1 real independent parameters ω_a $(a = 1,2,\ldots,N^2-1)$ and may be presented in the form:

$$S = \exp(i\omega_a t_a) \quad , \tag{A.5}$$

where $\vec{t} = (t_1, t_2, \ldots, t_{N^2-1})$ is a fixed set of (N^2-1) $N \times N$ matrices. Due to (A.4) t_a are hermitian and traceless:

$$t_a^{+} = t_a \quad , \qquad \mathrm{Tr}(t_a) = 0 \quad . \tag{A.6}$$

t_a matrices are the generators of the fundamental representation of the group SU(N). They are normalized according to the condition

$$\mathrm{Tr}(t_a t_b) = \frac{1}{2}\,\delta_{ab} \quad . \tag{A.7}$$

Every Hermitian traceless $N \times N$ matrix may be presented as a linear superposition of t_a. The commutator of two t_a matrices is an anti-hermitian traceless matrix; thus

$$[t_a, t_b] = i\,f_{abc}t_c \quad . \tag{A.8}$$

The structure constants f_{abc} are real and totally antisymmetric.

The matrices $\vec{t}$ obey Fierz' identities:

$$I_{\alpha\beta}I_{\gamma\delta} = \frac{1}{N}\,I_{\alpha\delta}I_{\gamma\beta} + 2\vec{t}_{\alpha\delta}\vec{t}_{\gamma\beta}$$

$$\vec{t}_{\alpha\beta}\vec{t}_{\gamma\delta} = (1/2 - 1/2N^2)I_{\alpha\delta}I_{\gamma\beta} - \frac{1}{N}\,\vec{t}_{\alpha\delta}\vec{t}_{\gamma\beta} \quad , \tag{A.9}$$

where $I_{\alpha\beta}$ is a unit $N \times N$ matrix. For the simplest groups the generators t_a and structure constants f_{abc} are given here explicitly:

SU(2)

$$\vec{t} = \frac{1}{2}\vec{\sigma} \tag{A.10}$$

where $\vec{\sigma}$ are Pauli matrices

$$\sigma_1 = \begin{pmatrix} 0 & 1 \\ 1 & 0 \end{pmatrix} \quad , \quad \sigma_2 = \begin{pmatrix} 0 & -i \\ i & 0 \end{pmatrix} \quad ,$$

$$\sigma_3 = \begin{pmatrix} 1 & 0 \\ 0 & -1 \end{pmatrix} \quad , \tag{A.11}$$

The structure constants form a unit totally antisymmetric tensor ε_{abc}:

$$f_{abc} = \varepsilon_{abc} \quad , \quad \varepsilon_{123} = 1 \quad . \tag{A.12}$$

SU(3)

$$\vec{t} = \frac{1}{2}\vec{\lambda} \tag{A.13}$$

where $\vec{\lambda}$ are the Gell-Mann matrices

$$\lambda_1 = \begin{pmatrix} 0 & 1 & 0 \\ 1 & 0 & 0 \\ 0 & 0 & 0 \end{pmatrix} \qquad \lambda_2 = \begin{pmatrix} 0 & -i & 0 \\ i & 0 & 0 \\ 0 & 0 & 0 \end{pmatrix} \qquad \lambda_3 = \begin{pmatrix} 1 & 0 & 0 \\ 0 & -1 & 0 \\ 0 & 0 & 0 \end{pmatrix}$$

$$\lambda_4 = \begin{pmatrix} 0 & 0 & 1 \\ 0 & 0 & 0 \\ 1 & 0 & 0 \end{pmatrix} \qquad \lambda_5 = \begin{pmatrix} 0 & 0 & -i \\ 0 & 0 & 0 \\ i & 0 & 0 \end{pmatrix} \qquad \lambda_6 = \begin{pmatrix} 0 & 0 & 0 \\ 0 & 0 & 1 \\ 0 & 1 & 0 \end{pmatrix}$$

$$\lambda_7 = \begin{pmatrix} 0 & 0 & 0 \\ 0 & 0 & -i \\ 0 & i & 0 \end{pmatrix} \qquad \lambda_8 = \frac{1}{\sqrt{3}} \begin{pmatrix} 1 & 0 & 0 \\ 0 & 1 & 0 \\ 0 & 0 & -2 \end{pmatrix} . \tag{A.14}$$

The nonzero independent coefficients f_{abc} are

$$f_{123} = 1 \quad , \quad f_{458} = f_{678} = \sqrt{3}/2$$

$$f_{147} = f_{516} = f_{246} = f_{257} = f_{345} = f_{637} = 1/2 \quad . \tag{A.15}$$

A.3 γ Matrices and 4-Component Spinors

We use the following γ-matrices:

$$\gamma_\mu = (\gamma_0, \vec{\gamma})$$

$$\gamma_0 = \begin{pmatrix} 0 & 1\!\!1 \\ 1\!\!1 & 0 \end{pmatrix} \quad , \quad \vec{\gamma} = \begin{pmatrix} 0 & \vec{\sigma} \\ -\vec{\sigma} & 0 \end{pmatrix} \quad , \tag{A.16}$$

where $1\!\!1$ is a unit 2×2 matrix.

$$\gamma_5 = -i\gamma_0\gamma_1\gamma_2\gamma_3 = -\begin{pmatrix} 0 & 1\!\!1 \\ 1\!\!1 & 0 \end{pmatrix} \quad . \tag{A.17}$$

The commutation relations for γ-matrices are

$$\gamma_\mu\gamma_\nu + \gamma_\nu\gamma_\mu = 2I\delta_{\mu\nu} \quad , \quad \gamma_\mu\gamma_5 + \gamma_5\gamma_\mu = 0 \quad . \tag{A.18}$$

The four-component spinors

$$u^j(p) = \begin{pmatrix} u_1^j(p) \\ u_2^j(p) \\ u_3^j(p) \\ u_4^j(p) \end{pmatrix} \tag{A.19}$$

obey the Dirac equation

$$(\hat{p} - mI)u^j(p) = 0 \quad , \quad \text{where} \quad \hat{p} \equiv p_\mu\gamma_\mu \quad . \tag{A.20}$$

They are normalized according to the condition

$$\sum_{\alpha=1}^{4} u_\alpha^{j*}(p)u_\alpha^k(p) = \bar{u}^j(p)\gamma_0 u^k(p) = 2p_0\delta_{jk} \quad , \tag{A.21}$$

where $\bar{u}^j(p) = u^{j+}(p)\gamma_0 = (u^{j*}(p))^T$, i.e. $\bar{u}(p)$ is a row, while $u(p)$ is a column.

The spinors obey also the completeness condition

$$\sum_{j=1,2} u_\beta^j(p)\bar{u}_\alpha^j(p) = (\hat{p}+m)_{\beta\alpha} \quad , \quad \sum_{j=3,4} u_\beta^j(p)\bar{u}_\alpha^j(p) = (\hat{p}+m)_{\beta\alpha} \quad . \tag{A.22}$$

The explicit form of spinors with definite helicity (spin projection on the momentum direction) is:

$$u^{\lambda=\frac{1}{2}}(p) = \begin{pmatrix} \sqrt{p_0+m} \begin{pmatrix} e^{-i\varphi/2}\cdot\cos\frac{\theta}{2} \\ e^{i\varphi/2}\cdot\sin\frac{\theta}{2} \end{pmatrix} \\ \frac{\vec{p}\cdot\vec{\sigma}}{\sqrt{p_0+m}} \begin{pmatrix} e^{-i\varphi/2}\cdot\cos\frac{\theta}{2} \\ e^{i\varphi/2}\cdot\sin\frac{\theta}{2} \end{pmatrix} \end{pmatrix}$$

$$u^{\lambda=-\frac{1}{2}}(p) = \begin{pmatrix} \sqrt{p_0+m} \begin{pmatrix} -e^{-i\varphi/2}\cdot\sin\frac{\theta}{2} \\ e^{i\varphi/2}\cdot\cos\frac{\theta}{2} \end{pmatrix} \\ \frac{\vec{p}\cdot\vec{\sigma}}{\sqrt{p_0+m}} \begin{pmatrix} -e^{-i\varphi/2}\cdot\sin\frac{\theta}{2} \\ e^{i\varphi/2}\cdot\cos\frac{\theta}{2} \end{pmatrix} \end{pmatrix} \cdot \tag{A.23}$$

where $\vec{p}/|\vec{p}| = \{\cos\varphi\sin\theta, \sin\varphi\sin\theta, \cos\theta\}$.

A.4 Amplitude Normalization Conditions

We use the amplitudes A, which are connected with the S-matrix by

$$S = 1 + i(2\pi)^4\delta^4(\Sigma p_{in} - \Sigma p_{out})A \quad , \tag{A.24}$$

where Σp_{in} and Σp_{out} are the total incoming and outgoing particle momenta, respectively. We account for the factors of particle identity directly in the amplitudes. It allows one to write phase space integrals (see Appendix B) in the same form for different and identical particles. Thus, if amplitudes are constructed according to the standard Feynman rules, additional factors enter for groups of identical particles. These are $\prod_i \frac{1}{\sqrt{n_i!}}$ for bosons and $\prod_i \frac{(-1)^{P_i}}{\sqrt{n_i!}}$ for fermions. Here n_i is the number of the identical particles of the i-th sort, $P_i = 0$ or 1 is the parity of the permutation for fermions.

The connection of such amplitudes with the cross sections measured is given in the Appendix B.

Appendix B

CROSS SECTIONS AND AMPLITUDE DISCONTINUITIES

B.1 Exclusive and Inclusive Cross Sections

In the normalization adopted for amplitudes (see Appendix A) the differential cross section of the process $1+2 \to N$ particles (Fig. B.1a is:

$$d\sigma_{2\to N} = \frac{1}{J} |A_{2\to N}|^2 d\Phi_N \tag{B.1}$$

where $J = 4\sqrt{(p_1p_2)^2 - m_1^2m_2^2}$ is the invariant flux factor; p_1, p_2 and m_1, m_2 are 4-momenta and masses of the initial particles. The phase space element for N particles is

$$d\Phi_N = (2\pi)^4\delta^4(p_{in} - \sum_{\ell=1}^{N} k_\ell) \prod_{n=1}^{N} \frac{d^4k_n}{(2\pi)^3} (k_{n0})\delta(k_n^2 - m_n^2) \tag{B.2}$$

where $p_{in} = p_1 + p_2$. The amplitude $A_{2\to N}$ depends on momenta and spins of the incoming and outgoing particles. If the colliding particles are unpolarized, the differential cross section (B.1) should be averaged over their spin projections:

$$\frac{1}{(2j_1+1)(2j_2+2)} \sum_{\mu_1,\mu_2} \quad . \tag{B.3}$$

If the polarization properties of the outgoing particles are not measured, the cross section should be summed over their spin projections:

$$\sum_{\nu_1,\nu_2,\ldots,\nu_N} \qquad . \tag{B.4}$$

The cross section (B.1) integrated over the whole phase space and summarized over all spin projections gives the total exclusive cross section of the given channel:

$$\sigma_N = \sum_{\nu_1,\ldots,\nu_N} \int d\sigma_{2\to N} \qquad . \tag{B.5}$$

A particular case of the Eq. (B.5) is the elastic cross section (Fig. B.1b). The differential elastic cross section at fixed energy of the collision (or fixed $s = (p_1 + p_2)^2$) is a function of two scattering angles. If particles are spinless the elastic amplitude is spin-independent, the cross section depends on one scattering angle or on the associated variable, e.g. $t = (p_1 - p_1)^2 \le 0$:

$$\frac{d\sigma_{e\ell}}{d(-t)} = \frac{d\sigma_{e\ell}}{d|t|} = \frac{1}{J}\int |A_{2\to 2}|^2 d\Phi_2 \delta(t - (p_1 - p_1')^2) \qquad . \tag{B.6}$$

The total elastic cross section is

$$\sigma_{e\ell}(s) = \int_{t_{min}}^{0} dt \, \frac{d\sigma_{e\ell}(s, t)}{d(-t)} \tag{B.7}$$

where $t_{min} = -\frac{1}{s}[s - (m_1 + m_2)^2] \cdot [s - (m_1 - m_2)^2]$.

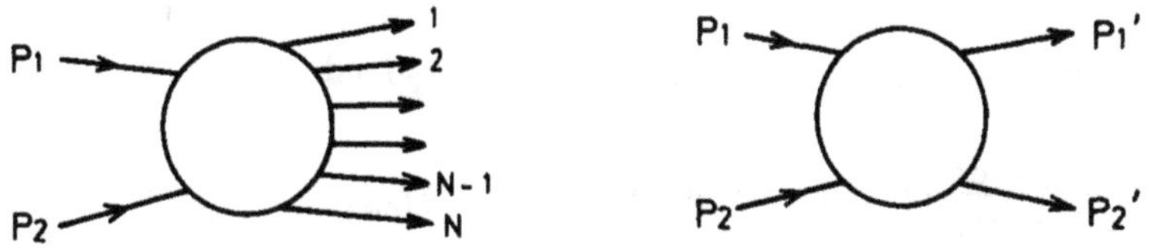

(a) (b)

Fig. B.1 (a) Diagrams for N-particle production process $(2\to N)$ and
(b) elastic scattering $(2\to 2)$.

The sum of all possible exclusive cross sections (B.5) is the total cross section

$$\sigma_{tot} = \sum_N \sigma_N \quad . \tag{B.8}$$

The total inelastic cross section is defined as

$$\sigma_{inel} = \sigma_{tot} - \sigma_{el} \quad . \tag{B.9}$$

If only one secondary particle of the definite sort h is registrated in the experiment, then the inclusive cross section

$$1 + 2 \to h + X \tag{B.10}$$

is measured. The differential inclusive cross section of the production of the secondary h is the sum of different exclusive cross sections:

$$\frac{d\sigma}{d^3k_h}(1+2 \to h+X) = \sum_i n_{ih} \frac{d\sigma_i}{d^3k_h} \tag{B.11}$$

where the sum runs over all opened channels of the collision $(1+2)$ at a fixed energy; n_{ih} is the number of secondaries of the sort h in i-th channel, while $d\sigma_i/d^3k_h$ is defined as

$$(2\pi)^3 2k_{h0} \frac{d\sigma_i}{d^3k_h} = \frac{1}{J}\int |A_{2\to N_i}|^2 d\Phi_{N_i-1} \tag{B.12}$$

where in the phase space element $d\Phi_{N_i-1}$, $p_{in} = p_1 + p_2 - k_h$ is taken. The inclusive cross section (B.11) is normalized according to

$$\int d^3k_h \frac{d\sigma}{d^3k_h} = \sigma(1+2 \to h+X) = \langle n_h \rangle \sigma_{inel} \tag{B.13}$$

where $\langle n_h \rangle$ is the average number of secondaries of the sort h per inelastic event of the collision $(1+2)$.

Similarly multiparticle inclusive cross sections may be defined, when several particles of fixed sorts are registrated in the final state. For two-particle inclusive reactions:

$$1 + 2 \to h_1 + h_2 + X \tag{B.14}$$

the differential cross section

$$\frac{d\sigma}{d^3k_{h_1} d^3k_{h_2}} (1+2 \to h_1 + h_2 + X) = \sum_i n_{ih_1} n_{ih_2} \cdot \frac{d\sigma_i}{d^3k_{h_1} d^3k_{h_2}} \qquad (h_1 \neq h_2)$$

$$= \sum_i n_{ih_1} (n_{ih_1} - 1) \frac{d\sigma_i}{d^3k_{h_1} d^3k_{h_2}} \qquad (h_1 = h_2)$$

$$\tag{B.15}$$

is normalized according to the condition

$$\int d^3k_{h_1} d^3k_{h_2} \frac{d\sigma}{d^3k_{h_1} d^3k_{h_2}} (1+2 \to h_1 + h_2 + X)$$

$$= \langle n_{h_1} n_{h_2} \rangle \sigma_{ine\ell}(12) \qquad (h_1 \neq h_2)$$

$$= \langle n_{h_1} (n_{h_1} - 1) \rangle \sigma_{ine\ell}(12) \qquad (h_1 = h_2) \quad . \tag{B.16}$$

The difference $\langle n_{h_1} n_{h_2} \rangle - \langle n_{h_1} \rangle \langle n_{h_2} \rangle$ measures the correlation in the production of particles h_1 and h_2; it vanishes if they are produced independently.

B.2 Amplitude Discontinuities and Unitary Condition

Cross sections of the collision processes may be expressed in terms of the amplitude discontinuities at their singular points. Two important examples are the elastic $(2 \to 2)$ and $(3 \to 3)$ amplitudes. The elastic $(2 \to 2)$ amplitude has singularities in the physical region of s, which are connected with two-particle, three-particle, four-particle, etc.

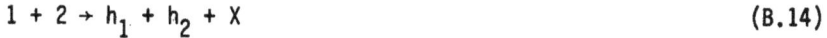

intermediate states (see Fig. B.2). Let us consider e.g. four-particle intermediate states (Fig. B.3a); the amplitude discontinuity at the four-particle threshold singularity is:

$$2i\,\mathrm{disc}_{(4)}A(s,\ldots) = A(s+i0,\ldots) - A(s'-i0,\ldots) \qquad (B.17)$$

The values $s+i0$ and $s-i0$ are shown in Fig. B.2b by the arrows. Hereafter dots will stand for the variables of the amplitude which are not written explicitly. The discontinuity (B.17) is:

$$\mathrm{disc}_{(4)}A(s,\ldots) = \frac{1}{2}\int d\Phi_4 A_{2\to 4}(p_1,p_2,\ldots)A^{+}_{4\to 2}(p_1',p_2',\ldots) \qquad (B.18)$$

Both amplitudes in the integrand in the right-hand side of (B.18) are taken at the same value $(p_1+p_2)^2 = (p_1'+p_2')^2 = s+i0$, i.e., in the physical region. For particles with spin the right-hand side of (B.18) should be summed over spin projections. The calculation of the discontinuities is usually called "the cutting of the diagram"; the right-hand side of the Eq. (B.18) is represented graphically by the diagram of Fig. B.3b.

The sum of all discontinuities is called the total discontinuity

$$\mathrm{disc}\,A(s,\ldots) = \frac{1}{2i}\left[A(s+i0,\ldots) - A(s-i0,\ldots)\right]$$

$$= \sum_{n\geq 2} \mathrm{disc}_{(n)}A(s,\ldots) \qquad (B.19)$$

The values $s+i0$ and $s-i0$ are shown in Fig. B.2a. The total discontinuity of the amplitude is equal to its imaginary part

$$\mathrm{disc}\,A = \mathrm{Im}\,A = \frac{1}{2}\sum_{n\geq 2} d\Phi_N A_{2\to N}(p_1,p_2,\ldots)A^{*}_{2\to N}(p_1',p_2',\ldots) \qquad (B.20)$$

This equality may be obtained directly from the unitarity condition for the S-matrix:

$$SS^{+} = 1 \qquad (B.21)$$

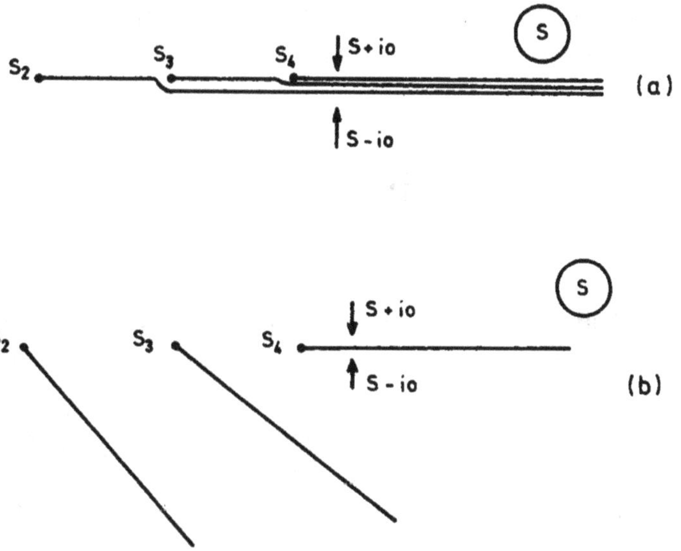

Fig. B.2 Threshold singularities of the elastic amplitude in the complex s plane at $s = s_n = (\sum_{i=1}^{n} m_i')^2$. m_i' are the masses of the particles in the intermediate state;
(a) cuts from the singularities are directed along the real axis;
(b) cuts from the singularities s_2 and s_3 are moved to the lower semiplane.

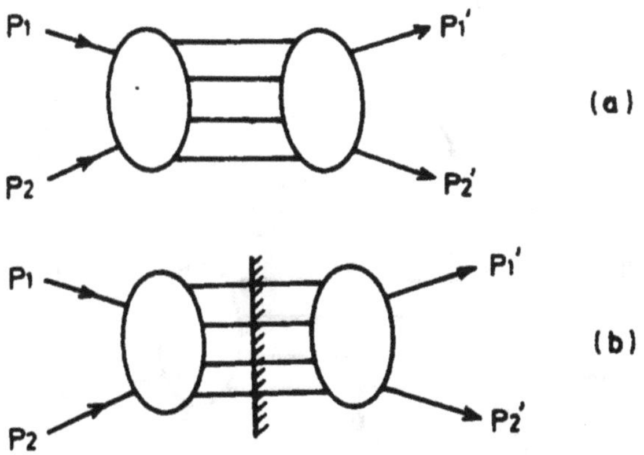

Fig. B.3 (a) Elastic scattering with four-particle intermediate state.
(b) Graphical representation for the discontinuity at the four-particle threshold singularity.

The imaginary part of the elastic amplitude in the forward direction (or at $t=0$) is expressed in terms of the total cross section:

$$\text{Im } A(0) = \frac{1}{2} J \sigma_{tot} \quad . \tag{B.22}$$

For high initial energies $(s \gg m_1^2, m_2^2)$ $J \simeq 2s$ and

$$\text{Im } A(0) \simeq s\sigma_{tot} \tag{B.23}$$

The discontinuities of the $(3\to 3)$ elastic amplitude (Fig. B.4a) are determined similarly to those of the $(2\to 2)$ amplitude. For example, the discontinuity of the $(3\to 3)$ amplitude at the four-particle threshold is defined by the Eq. (B.18) with the replacements $A_{2\to 4} \to A_{3\to 4}$ and $A_{4\to 2} \to A_{4\to 3}$ (see Fig. B.4b). Thus the total discontinuity of $A_{3\to 3}$ at $p_1 = p_1'$, $p_2 = p_2'$ and $k = k'$ (see Fig. B.4a) is expressed in terms of the inclusive cross section of the production of the particle h with momentum k:

$$\frac{2}{J} \text{disc } A_{3\to 3} = (2\pi)^3 2k_0 \frac{d\sigma}{d^3k} (1 + 2 \to h + X) \tag{B.24}$$

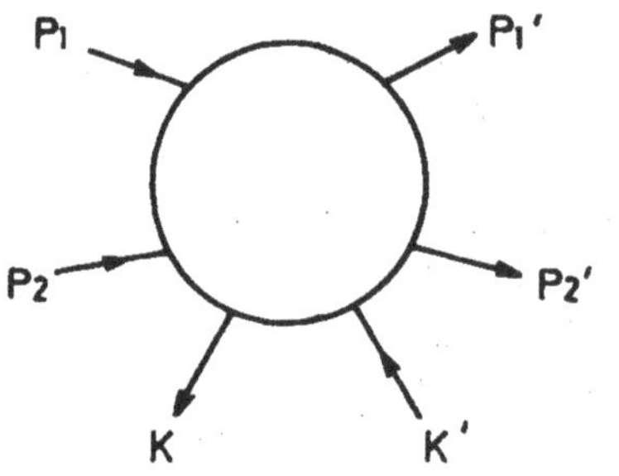

(a)

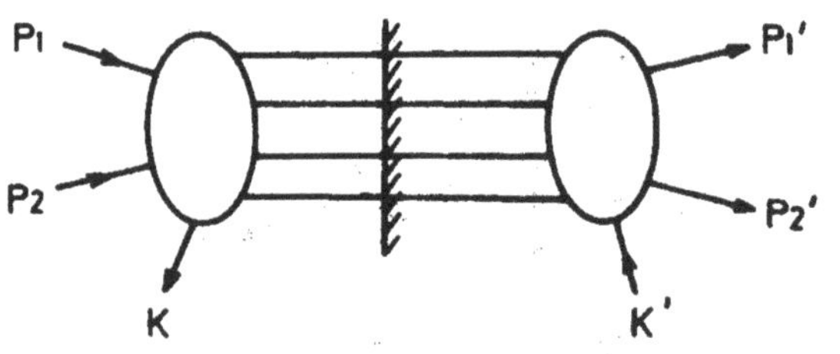

(b)

Fig. B.4 (a) $(3\to 3)$ elastic amplitude and (b) cut $(3\to 3)$ diagram.

The discontinuities of the more complicated amplitudes $(n \to n)$ may be connected with the inclusive cross section of $(n-2)$ particle production in a similar manner.

Appendix C

HADRON WAVE FUNCTIONS

We will present here explicit expressions for the hadron wave functions of S- and P-wave SU(6) multiplets. The radial part of the wave function is omitted everywhere; for the P-wave multiplets the orbital part of the wave function is presented explicitly. In every multiplet only "independent" wave functions are given; the rest may be obtained via spin, isospin or charge conjugation or via permutations $u \to s$, $d \to s$. For instance, $|K^+\rangle = |\pi^+\rangle(\bar{d} \to \bar{s})$; $|\Xi^0\rangle = |p\rangle\,(u \to s,\ d \to u)$; and so on. All the subscripts inside the ket brackets refer to the particle notation, while those outside stand for the J_z value (J is the total angular momentum of the hadron).

C.1 S-Wave Mesons

$$|\pi^+\rangle = \sqrt{\frac{1}{2}}\,|u^{\uparrow}\bar{d}^{\downarrow} - u^{\downarrow}\bar{d}^{\uparrow}\rangle$$

$$|\pi^0\rangle = \frac{1}{2}\,|u^{\uparrow}\bar{u}^{\downarrow} - d^{\uparrow}\bar{d}^{\downarrow} - u^{\downarrow}\bar{u}^{\uparrow} + d^{\downarrow}\bar{d}^{\uparrow}\rangle$$

$$|\eta^1\rangle = \sqrt{\frac{1}{6}}\,|u^{\uparrow}\bar{u}^{\downarrow} + d^{\uparrow}\bar{d}^{\downarrow} + s^{\uparrow}\bar{s}^{\downarrow} - u^{\downarrow}\bar{u}^{\uparrow} - d^{\downarrow}\bar{d}^{\uparrow} - s^{\downarrow}\bar{s}^{\uparrow}\rangle$$

$$|\eta^8\rangle = \frac{1}{2\sqrt{3}}\,|u^{\uparrow}\bar{u}^{\downarrow} + d^{\uparrow}\bar{d}^{\downarrow} - 2s^{\uparrow}\bar{s}^{\downarrow} - u^{\downarrow}\bar{u}^{\uparrow} - d^{\downarrow}\bar{d}^{\uparrow} + 2s^{\downarrow}\bar{s}^{\uparrow}\rangle$$

$$|\rho^+\rangle_1 = |u^{\uparrow}\bar{d}^{\uparrow}\rangle$$

$$|\rho^+>_0 = \sqrt{\frac{1}{2}}\,|u^\uparrow\bar{d}^\downarrow + u^\downarrow\bar{d}^\uparrow>$$

$$|\rho^0>_1 = \sqrt{\frac{1}{2}}\,|u^\uparrow\bar{u}^\uparrow - d^\uparrow\bar{d}^\uparrow>$$

$$|\rho^0>_0 = \frac{1}{2}\,|u^\uparrow\bar{u}^\downarrow - d^\uparrow\bar{d}^\downarrow + u^\downarrow\bar{u}^\uparrow - d^\downarrow\bar{d}^\uparrow>$$

$$|\omega>_1 = \sqrt{\frac{1}{2}}\,|u^\uparrow\bar{u}^\uparrow + d^\uparrow\bar{d}^\uparrow>$$

$$|\omega>_0 = \frac{1}{2}\,|u^\uparrow\bar{u}^\downarrow + d^\uparrow\bar{d}^\downarrow + u^\downarrow\bar{u}^\uparrow + d^\downarrow\bar{d}^\uparrow>$$

$$|\phi>_1 = |s^\uparrow\bar{s}^\uparrow>$$

$$|\phi>_0 = \sqrt{\frac{1}{2}}\,|s^\uparrow\bar{s}^\downarrow + s^\downarrow\bar{s}^\uparrow>$$

We neglected here the possible mixing of $|\omega>$ and $|\phi>$ states.

C.2 P-Wave Mesons

We use the notation $|L_z>$ for the orbital part of the wave function.

$$|A_2^+>_2 = |1>\,|u^\uparrow\bar{d}^\uparrow>$$

$$|A_2^+>_1 = \sqrt{\frac{1}{2}}\,|0>\,|u^\uparrow\bar{d}^\uparrow> + \frac{1}{2}\,|1>\,|u^\uparrow\bar{d}^\downarrow + u^\downarrow\bar{d}^\uparrow>$$

$$|A_2^+>_0 = \sqrt{\frac{1}{3}}\,|1>\,|u^\downarrow\bar{d}^\downarrow> - \sqrt{\frac{1}{6}}\,|0>\,|u^\uparrow\bar{d}^\downarrow + u^\downarrow\bar{d}^\uparrow> + \sqrt{\frac{1}{3}}\,|-1>\,|u^\uparrow\bar{d}^\uparrow>$$

$$|A_2^0>_2 = \sqrt{\frac{1}{2}}\,|1>\,|u^\uparrow\bar{u}^\uparrow - d^\uparrow\bar{d}^\uparrow>$$

$$|A_2^0>_1 = \frac{1}{2\sqrt{2}}\,|1>\,|u^\uparrow\bar{u}^\downarrow - d^\uparrow\bar{d}^\downarrow + u^\downarrow\bar{u}^\uparrow - d^\downarrow\bar{d}^\uparrow> + \frac{1}{2}\,|0>\,|u^\uparrow\bar{u}^\uparrow - d^\uparrow\bar{d}^\uparrow>$$

$$|A_2^0>_0 = \frac{1}{2\sqrt{3}}\,|1>\,|u^\downarrow\bar{u}^\downarrow - d^\downarrow\bar{d}^\downarrow> + \sqrt{\frac{1}{6}}\,|0>\,|u^\uparrow\bar{u}^\downarrow - d^\uparrow\bar{d}^\downarrow + u^\downarrow\bar{u}^\uparrow - d^\downarrow\bar{d}^\uparrow>$$

$$+ \frac{1}{2\sqrt{3}} \,|{-1}\rangle\, |u^{\uparrow}\bar{u}^{\uparrow} - d^{\uparrow}\bar{d}^{\uparrow}\rangle$$

$$|f\rangle_2 = \sqrt{\frac{1}{2}} \,|1\rangle\, |u^{\uparrow}\bar{u}^{\uparrow} + d^{\uparrow}\bar{d}^{\uparrow}\rangle$$

$$|f\rangle_1 = \frac{1}{2\sqrt{2}} \,|1\rangle\, |u^{\uparrow}\bar{u}^{\downarrow} + d^{\uparrow}\bar{d}^{\downarrow} + u^{\downarrow}\bar{u}^{\uparrow} + d^{\downarrow}\bar{d}^{\uparrow}\rangle + \frac{1}{2} \,|0\rangle\, |u^{\uparrow}\bar{u}^{\uparrow} + d^{\uparrow}\bar{d}^{\uparrow}\rangle$$

$$|f\rangle_0 = \frac{1}{2\sqrt{3}} \,|1\rangle\, |u^{\downarrow}\bar{u}^{\downarrow} + d^{\downarrow}\bar{d}^{\downarrow}\rangle + \sqrt{\frac{1}{6}} \,|0\rangle\, |u^{\uparrow}\bar{u}^{\downarrow} + d^{\uparrow}\bar{d}^{\downarrow} + u^{\downarrow}\bar{u}^{\uparrow} + d^{\downarrow}\bar{d}^{\uparrow}\rangle$$

$$+ \frac{1}{2\sqrt{3}} \,|{-1}\rangle\, |u^{\uparrow}\bar{u}^{\uparrow} + d^{\uparrow}\bar{d}^{\uparrow}\rangle$$

$$|f'\rangle_2 = |1\rangle\, |s^{\uparrow}\bar{s}^{\uparrow}\rangle$$

$$|f'\rangle_1 = \frac{1}{2} \,|1\rangle\, |s^{\uparrow}\bar{s}^{\downarrow} + s^{\downarrow}\bar{s}^{\uparrow}\rangle + \sqrt{\frac{1}{2}} \,|0\rangle\, |s^{\uparrow}\bar{s}^{\uparrow}\rangle$$

$$|f'\rangle_0 = \sqrt{\frac{1}{6}} \,|1\rangle\, |s^{\downarrow}\bar{s}^{\downarrow}\rangle + \sqrt{\frac{1}{3}} \,|0\rangle\, |s^{\uparrow}\bar{s}^{\downarrow} + s^{\downarrow}\bar{s}^{\uparrow}\rangle + \sqrt{\frac{1}{6}} \,|{-1}\rangle\, |s^{\uparrow}\bar{s}^{\uparrow}\rangle$$

$$|A_1^+\rangle_1 = \frac{1}{2} \,|1\rangle\, |u^{\uparrow}\bar{d}^{\downarrow} + u^{\downarrow}\bar{d}^{\uparrow}\rangle - \sqrt{\frac{1}{2}} \,|0\rangle\, |u^{\uparrow}\bar{d}^{\uparrow}\rangle$$

$$|A_1^+\rangle_0 = \sqrt{\frac{1}{2}} \,|1\rangle\, |u^{\downarrow}\bar{d}^{\downarrow}\rangle - \sqrt{\frac{1}{2}} \,|{-1}\rangle\, |u^{\uparrow}\bar{d}^{\uparrow}\rangle$$

$$|A_1^0\rangle_1 = \frac{1}{2\sqrt{2}} \,|1\rangle\, |u^{\uparrow}\bar{u}^{\downarrow} - d^{\uparrow}\bar{d}^{\downarrow} + u^{\downarrow}\bar{u}^{\uparrow} - d^{\downarrow}\bar{d}^{\uparrow}\rangle - \frac{1}{2} \,|0\rangle\, |u^{\uparrow}\bar{u}^{\uparrow} - d^{\uparrow}\bar{d}^{\uparrow}\rangle$$

$$|A_1^0\rangle_0 = \frac{1}{2} \,|1\rangle\, |u^{\downarrow}\bar{u}^{\downarrow} - d^{\downarrow}\bar{d}^{\downarrow}\rangle - \frac{1}{2} \,|{-1}\rangle\, |u^{\uparrow}\bar{u}^{\uparrow} - d^{\uparrow}\bar{d}^{\uparrow}\rangle$$

$$|B^+\rangle_1 = \sqrt{\frac{1}{2}} \,|1\rangle\, |u^{\uparrow}\bar{d}^{\downarrow} - u^{\downarrow}\bar{d}^{\uparrow}\rangle$$

$$|B^+\rangle_0 = \sqrt{\frac{1}{2}} \,|0\rangle\, |u^{\uparrow}\bar{d}^{\downarrow} - u^{\downarrow}\bar{d}^{\uparrow}\rangle$$

$$|B^0\rangle_1 = \frac{1}{2} \,|1\rangle\, |u^{\uparrow}\bar{u}^{\downarrow} - d^{\uparrow}\bar{d}^{\downarrow} - u^{\downarrow}\bar{u}^{\uparrow} + d^{\downarrow}\bar{d}^{\uparrow}\rangle$$

$$|B^0\rangle_0 = \frac{1}{2}\,|0\rangle\,|u^{\uparrow}\bar{u}^{\downarrow} - d^{\uparrow}\bar{d}^{\downarrow} - u^{\downarrow}\bar{u}^{\uparrow} + d^{\downarrow}\bar{d}^{\uparrow}\rangle$$

$$|\delta^+\rangle = \sqrt{\frac{1}{3}}\,|1\rangle\,|u^{\downarrow}\bar{d}^{\downarrow}\rangle - \sqrt{\frac{1}{6}}\,|0\rangle\,|u^{\uparrow}\bar{d}^{\downarrow} - u^{\downarrow}\bar{d}^{\uparrow}\rangle + \sqrt{\frac{1}{3}}\,|-1\rangle\,|u^{\uparrow}\bar{d}^{\uparrow}\rangle$$

$$|\delta^0\rangle = \sqrt{\frac{1}{6}}\,|1\rangle\,|u^{\downarrow}\bar{u}^{\downarrow} - d^{\downarrow}\bar{d}^{\downarrow}\rangle - \frac{1}{2\sqrt{3}}\,|0\rangle\,|u^{\uparrow}\bar{u}^{\downarrow} - d^{\uparrow}\bar{d}^{\downarrow} + u^{\downarrow}\bar{u}^{\uparrow} - d^{\downarrow}\bar{d}^{\uparrow}\rangle$$
$$+ \sqrt{\frac{1}{6}}\,|-1\rangle\,|u^{\uparrow}\bar{u}^{\uparrow} - d^{\uparrow}\bar{d}^{\uparrow}\rangle$$

We neglected here the mixing of $|f\rangle$ and $|f'\rangle$ states (the mixing angle is about -8°).

There are 6 isoscalar states with $J^{PG} = 1^{++}$, 1^{+-} and 0^{++}. Since the mixing angles for the observable mesons D(1285) and E(1420) (1^{++} states), ε (1300) and S*(975) (0^{++} states) are not yet well established, and only one candidate H(1190) is observed for 1^{+-} isoscalar states, we present here purely octet and purely singlet wave functions using the notations $|J_1^{PG}\rangle$ and $|J_8^{PG}\rangle$:

$$|1_1^{++}\rangle_1 = \frac{1}{2\sqrt{3}}\,|1\rangle\,|u^{\uparrow}\bar{u}^{\downarrow} + d^{\uparrow}\bar{d}^{\downarrow} + s^{\uparrow}\bar{s}^{\downarrow} + u^{\downarrow}\bar{u}^{\uparrow} + d^{\downarrow}\bar{d}^{\uparrow} + s^{\downarrow}\bar{s}^{\uparrow}\rangle$$
$$- \sqrt{\frac{1}{6}}\,|0\rangle\,|u^{\uparrow}\bar{u}^{\uparrow} + d^{\uparrow}\bar{d}^{\uparrow} + s^{\uparrow}\bar{s}^{\uparrow}\rangle$$

$$|1_1^{++}\rangle_0 = \sqrt{\frac{1}{6}}\,|1\rangle\,|u^{\downarrow}\bar{u}^{\downarrow} + d^{\downarrow}\bar{d}^{\downarrow} + s^{\downarrow}\bar{s}^{\downarrow}\rangle - \sqrt{\frac{1}{6}}\,|-1\rangle\,|u^{\uparrow}\bar{u}^{\uparrow} + d^{\uparrow}\bar{d}^{\uparrow} + s^{\uparrow}\bar{s}^{\uparrow}\rangle$$

$$|1_8^{++}\rangle_1 = \frac{1}{2\sqrt{6}}\,|1\rangle\,|u^{\uparrow}\bar{u}^{\downarrow} + d^{\uparrow}\bar{d}^{\downarrow} - 2s^{\uparrow}\bar{s}^{\downarrow} + u^{\downarrow}\bar{u}^{\uparrow} + d^{\downarrow}\bar{d}^{\uparrow} - 2s^{\downarrow}\bar{s}^{\uparrow}\rangle$$
$$- \frac{1}{2\sqrt{3}}\,|0\rangle\,|u^{\uparrow}\bar{u}^{\uparrow} + d^{\uparrow}\bar{d}^{\uparrow} - 2s^{\uparrow}\bar{s}^{\uparrow}\rangle$$

$$|1_8^{++}\rangle_0 = \frac{1}{2\sqrt{3}}\,|1\rangle\,|u^{\downarrow}\bar{u}^{\downarrow} + d^{\downarrow}\bar{d}^{\downarrow} - 2s^{\downarrow}\bar{s}^{\downarrow}\rangle - \frac{1}{2\sqrt{3}}\,|-1\rangle\,|u^{\uparrow}\bar{u}^{\uparrow} + d^{\uparrow}\bar{d}^{\uparrow} - 2s^{\uparrow}\bar{s}^{\uparrow}\rangle$$

$$|1_1^{+-}\rangle_1 = \sqrt{\frac{1}{6}}\,|1\rangle\,|u^{\uparrow}\bar{u}^{\downarrow} + d^{\uparrow}\bar{d}^{\downarrow} + s^{\uparrow}\bar{s}^{\downarrow} - u^{\downarrow}\bar{u}^{\uparrow} - d^{\downarrow}\bar{d}^{\uparrow} - s^{\downarrow}\bar{s}^{\uparrow}\rangle$$

$$|1_1^{+-}\rangle_0 = \sqrt{\frac{1}{6}}\,|0\rangle\,|u^{\uparrow}\bar{u}^{\downarrow} + d^{\uparrow}\bar{d}^{\downarrow} + s^{\uparrow}\bar{s}^{\downarrow} - u^{\downarrow}\bar{u}^{\uparrow} - d^{\downarrow}\bar{d}^{\uparrow} - s^{\downarrow}\bar{s}^{\uparrow}\rangle$$

$$|1_8^{+-}>_1 = \frac{1}{2\sqrt{3}} |1> |u^\uparrow\bar{u}^\downarrow + d^\uparrow\bar{d}^\downarrow - 2s^\uparrow\bar{s}^\downarrow - u^\downarrow\bar{u}^\uparrow - d^\downarrow\bar{d}^\uparrow + 2s^\downarrow\bar{s}^\uparrow>$$

$$|1_8^{+-}>_0 = \frac{1}{2\sqrt{3}} |0> |u^\uparrow\bar{u}^\downarrow + d^\uparrow\bar{d}^\downarrow - 2s^\uparrow\bar{s}^\downarrow - u^\downarrow\bar{u}^\uparrow - d^\downarrow\bar{d}^\uparrow + 2s^\downarrow\bar{s}^\uparrow>$$

$$|0_1^{++}> = \frac{1}{3} |1> |u^\downarrow\bar{u}^\downarrow + d^\downarrow\bar{d}^\downarrow + s^\downarrow\bar{s}^\downarrow>$$
$$- \frac{1}{3\sqrt{2}} |0> |u^\uparrow\bar{u}^\downarrow + d^\uparrow\bar{d}^\downarrow + s^\uparrow\bar{s}^\downarrow - u^\downarrow\bar{u}^\uparrow - d^\downarrow\bar{d}^\uparrow - s^\downarrow\bar{s}^\uparrow>$$
$$+ \frac{1}{3} |-1> |u^\uparrow\bar{u}^\uparrow + d^\uparrow\bar{d}^\uparrow + s^\uparrow\bar{s}^\uparrow>$$

$$|0_8^{++}> = \frac{1}{3\sqrt{2}} |1> |u^\downarrow\bar{u}^\downarrow + d^\downarrow\bar{d}^\downarrow - 2s^\downarrow\bar{s}^\downarrow>$$
$$- \frac{1}{6} |0> |u^\uparrow\bar{u}^\downarrow + d^\uparrow\bar{d}^\downarrow - 2s^\uparrow\bar{s}^\downarrow + u^\downarrow\bar{u}^\uparrow + d^\downarrow\bar{d}^\uparrow - 2s^\downarrow\bar{s}^\uparrow>$$
$$+ \frac{1}{3\sqrt{2}} |-1> |u^\uparrow\bar{u}^\uparrow + d^\uparrow\bar{d}^\uparrow - 2s^\uparrow\bar{s}^\uparrow>$$

The observable axial strange mesons $Q_1(1280)$ and $Q_2(1400)$ are also mixed states of Q_A and Q_B with a mixing angle of about 50° to 60°.

The calculations of the inclusive spectra and average multiplicities in Chap. 7 require the knowledge of the partial decay modes and the corresponding branching ratios of all the resonant states. For all the S-wave resonances and the most part of the P-wave states these values are taken from the Particle Data Group issues (see e.g. Ref. 1). In addition, the missing probability for f(1270) decay is ascribed to the channel $f \to 4\pi$; only the 2π decay mode is accounted for 0^{++} isoscalar states and only the $\eta\pi$ channel for $\delta(980)$ decay. The 1^+ isoscalar states are assumed to decay into the following channels (see Ref. 2): those which are built of the non-strange quarks decay into $\eta\pi\pi$ and 4π with equal probabilities, while $s\bar{s}$ states decay into $K\bar{K}\pi$.

C.3 S-Wave Baryons

The braces are used for the abbreviated notations of SU(6)-symmetrized expressions:

$$\{q_i q_j q_k\} = \sqrt{\frac{1}{6}}\,(q_i q_j q_k + q_i q_k q_j + q_j q_i q_k + q_j q_k q_i + q_k q_i q_j + q_k q_j q_i) \qquad (i \neq j \neq k)$$

$$= \sqrt{\frac{1}{3}}\,(q_i q_i q_j + q_i q_j q_i + q_j q_i q_i) \qquad (i = k \neq j)$$

$$|p>_{1/2} = \sqrt{\frac{2}{3}}\,|\{u^\uparrow u^\uparrow d^\downarrow\}> - \sqrt{\frac{1}{3}}\,|\{u^\uparrow u^\downarrow d^\uparrow\}>$$

$$|\Lambda>_{1/2} = \sqrt{\frac{1}{2}}\,|\{u^\uparrow d^\downarrow s^\uparrow\}> - \sqrt{\frac{1}{2}}\,|\{u^\downarrow d^\uparrow s^\uparrow\}>$$

$$|\Sigma^0>_{1/2} = \sqrt{\frac{2}{3}}\,|\{u^\uparrow d^\uparrow s^\downarrow\}> - \sqrt{\frac{1}{6}}\,|\{u^\uparrow d^\downarrow s^\uparrow\}> - \sqrt{\frac{1}{6}}\,|\{u^\downarrow d^\uparrow s^\uparrow\}>$$

$$|\Delta^{++}>_{3/2} = |\{u^\uparrow u^\uparrow u^\uparrow\}>$$

$$|\Delta^{++}>_{1/2} = |\{u^\uparrow u^\uparrow u^\downarrow\}>$$

$$|\Delta^{+}>_{3/2} = |\{u^\uparrow u^\uparrow d^\uparrow\}>$$

$$|\Delta^{+}>_{1/2} = \sqrt{\frac{1}{3}}\,|\{u^\uparrow u^\uparrow d^\downarrow\}> + \sqrt{\frac{2}{3}}\,|\{u^\uparrow u^\downarrow d^\uparrow\}>$$

$$|\Sigma^{*0}>_{3/2} = |\{u^\uparrow d^\uparrow s^\uparrow\}>$$

$$|\Sigma^{*0}>_{1/2} = \sqrt{\frac{1}{3}}\,|\{u^\uparrow d^\downarrow s^\uparrow\}> + \sqrt{\frac{1}{3}}\,|\{u^\downarrow d^\uparrow s^\uparrow\}> + \sqrt{\frac{1}{3}}\,|\{u^\uparrow d^\uparrow s^\downarrow\}>$$

C.4 P-Wave Baryons

The notations $|L_z>_\alpha \sim Y_{1L_z}(n_{1,23})$ *and* $|L_z>_\beta \sim Y_{1L_z}(n_{2,3})$ *(see* Chap. 5) are used for the orbital parts of the wave functions. The P-wave baryon states are denoted here as ${}^{2S+1}H_J$ where S is the total quark spin (1/2 or 3/2), J is the total angular momentum (i.e. the hadron spin) and H is the hadron symbol. Λ^1, Λ^8 or Σ^8, Σ^{10} mean singlet, octet or decouplet states respectively.

C.4.1 *Decouplet states*

$$|{}^2\Delta^{++}_{3/2}\rangle_{3/2} = \frac{1}{2\sqrt{3}}\,|1\rangle_\alpha\,|u^\uparrow u^\uparrow u^\downarrow + u^\uparrow u^\downarrow u^\uparrow - 2u^\downarrow u^\uparrow u^\uparrow\rangle + \frac{1}{2}\,|1\rangle_\beta\,|u^\uparrow u^\uparrow u^\downarrow - u^\uparrow u^\downarrow u^\uparrow\rangle$$

$$\begin{aligned}|{}^2\Delta^{++}_{3/2}\rangle_{1/2} = & -\frac{1}{6}\,|1\rangle_\alpha\,|u^\downarrow u^\downarrow u^\uparrow + u^\downarrow u^\uparrow u^\downarrow - 2u^\uparrow u^\downarrow u^\downarrow\rangle \\ & + \frac{1}{3\sqrt{2}}\,|0\rangle_\alpha\,|u^\uparrow u^\uparrow u^\downarrow + u^\uparrow u^\downarrow u^\uparrow - 2u^\downarrow u^\uparrow u^\uparrow\rangle \\ & + \frac{1}{2\sqrt{3}}\,|1\rangle_\beta\,|u^\downarrow u^\uparrow u^\downarrow - u^\downarrow u^\downarrow u^\uparrow\rangle + \sqrt{\frac{1}{6}}\,|0\rangle_\beta\,|u^\uparrow u^\uparrow u^\downarrow - u^\uparrow u^\downarrow u^\uparrow\rangle\end{aligned}$$

$$\begin{aligned}|{}^2\Delta^{++}_{1/2}\rangle_{1/2} = & -\frac{1}{3\sqrt{2}}\,|1\rangle_\alpha\,|u^\downarrow u^\downarrow u^\uparrow + u^\downarrow u^\uparrow u^\downarrow - 2u^\uparrow u^\downarrow u^\downarrow\rangle \\ & - \frac{1}{6}\,|0\rangle_\alpha\,|u^\uparrow u^\uparrow u^\downarrow + u^\uparrow u^\downarrow u^\uparrow - 2u^\downarrow u^\uparrow u^\uparrow\rangle \\ & + \sqrt{\frac{1}{6}}\,|1\rangle_\beta\,|u^\downarrow u^\uparrow u^\downarrow - u^\downarrow u^\downarrow u^\uparrow\rangle \\ & - \frac{1}{2\sqrt{3}}\,|0\rangle_\beta\,|u^\uparrow u^\uparrow u^\downarrow - u^\uparrow u^\downarrow u^\uparrow\rangle\end{aligned}$$

$$\begin{aligned}|{}^2\Delta^{+}_{3/2}\rangle_{3/2} = & \frac{1}{6}\,|1\rangle_\alpha\,|u^\uparrow u^\uparrow d^\downarrow + u^\uparrow u^\downarrow d^\uparrow - 2u^\downarrow u^\uparrow d^\uparrow + u^\uparrow d^\uparrow u^\downarrow + u^\uparrow d^\downarrow u^\uparrow \\ & \quad - 2u^\downarrow d^\uparrow u^\uparrow + d^\uparrow u^\uparrow u^\downarrow + d^\uparrow u^\downarrow u^\uparrow - 2d^\downarrow u^\uparrow u^\uparrow\rangle \\ & + \frac{1}{2\sqrt{3}}\,|1\rangle_\beta\,|u^\uparrow u^\uparrow d^\downarrow - u^\uparrow u^\downarrow d^\uparrow + u^\uparrow d^\uparrow u^\downarrow - u^\uparrow d^\downarrow u^\uparrow + d^\uparrow u^\uparrow u^\downarrow \\ & \quad - d^\uparrow u^\downarrow u^\uparrow\rangle\end{aligned}$$

$$\begin{aligned}|{}^2\Delta^{+}_{3/2}\rangle_{1/2} = & -\frac{1}{6\sqrt{3}}\,|1\rangle_\alpha\,|u^\downarrow u^\downarrow d^\uparrow + u^\downarrow u^\uparrow d^\downarrow - 2u^\uparrow u^\downarrow d^\downarrow + u^\downarrow d^\downarrow u^\uparrow + u^\downarrow d^\uparrow u^\downarrow \\ & \quad - 2u^\uparrow d^\downarrow u^\downarrow + d^\downarrow u^\downarrow u^\uparrow + d^\downarrow u^\uparrow u^\downarrow - 2d^\uparrow u^\downarrow u^\downarrow\rangle \\ & +\end{aligned}$$

$$\begin{aligned}
&+ \frac{1}{3\sqrt{6}} |0\rangle_\alpha \, |u^\uparrow u^\uparrow d^\downarrow + u^\uparrow u^\downarrow d^\uparrow - 2u^\downarrow u^\uparrow d^\uparrow + u^\uparrow d^\uparrow u^\downarrow + u^\uparrow d^\downarrow u^\uparrow \\
&\qquad - 2u^\downarrow d^\uparrow u^\uparrow + d^\uparrow u^\uparrow u^\downarrow + d^\uparrow u^\downarrow u^\uparrow - 2d^\downarrow u^\uparrow u^\uparrow \rangle
\end{aligned}$$

$$\begin{aligned}
&+ \frac{1}{6} |1\rangle_\beta \, |u^\downarrow u^\uparrow d^\downarrow - u^\downarrow u^\downarrow d^\uparrow + u^\downarrow d^\uparrow u^\downarrow - u^\downarrow d^\downarrow u^\uparrow + d^\downarrow u^\uparrow u^\downarrow \\
&\qquad - d^\downarrow u^\downarrow u^\uparrow \rangle
\end{aligned}$$

$$\begin{aligned}
&+ \frac{1}{3\sqrt{2}} |0\rangle_\beta \, |u^\uparrow u^\uparrow d^\downarrow - u^\uparrow u^\downarrow d^\uparrow + u^\uparrow d^\uparrow u^\downarrow - u^\uparrow d^\downarrow u^\uparrow + d^\uparrow u^\uparrow u^\downarrow \\
&\qquad - d^\uparrow u^\downarrow u^\uparrow \rangle
\end{aligned}$$

$$\begin{aligned}
|^2\Delta^+_{1/2}\rangle_{1/2} = &- \frac{1}{3\sqrt{6}} |1\rangle_\alpha \, |u^\downarrow u^\downarrow d^\uparrow + u^\downarrow u^\uparrow d^\downarrow - 2u^\uparrow u^\downarrow d^\downarrow + u^\downarrow d^\downarrow u^\uparrow + u^\downarrow d^\uparrow u^\downarrow \\
&\qquad - 2u^\uparrow d^\downarrow u^\downarrow + d^\downarrow u^\downarrow u^\uparrow + d^\downarrow u^\uparrow u^\downarrow - 2d^\uparrow u^\downarrow u^\downarrow \rangle \\
&- \frac{1}{6\sqrt{3}} |0\rangle_\alpha \, |u^\uparrow u^\uparrow d^\downarrow + u^\uparrow u^\downarrow d^\uparrow - 2u^\downarrow u^\uparrow d^\uparrow + u^\uparrow d^\uparrow u^\downarrow + u^\uparrow d^\downarrow u^\uparrow \\
&\qquad - 2u^\downarrow d^\uparrow u^\uparrow + d^\uparrow u^\uparrow u^\downarrow + d^\uparrow u^\downarrow u^\uparrow - 2d^\downarrow u^\uparrow u^\uparrow \rangle \\
&+ \frac{1}{3\sqrt{2}} |1\rangle_\beta \, |u^\downarrow u^\uparrow d^\downarrow - u^\downarrow u^\downarrow d^\uparrow + u^\downarrow d^\uparrow u^\downarrow - u^\downarrow d^\downarrow u^\uparrow \\
&\qquad + d^\downarrow u^\uparrow u^\downarrow - d^\downarrow u^\downarrow u^\uparrow \rangle \\
&- \frac{1}{6} |0\rangle_\beta \, |u^\uparrow u^\uparrow d^\downarrow - u^\uparrow u^\downarrow d^\uparrow + u^\uparrow d^\uparrow u^\downarrow - u^\uparrow d^\downarrow u^\uparrow + d^\uparrow u^\uparrow u^\downarrow \\
&\qquad - d^\uparrow u^\downarrow u^\uparrow \rangle
\end{aligned}$$

$$\begin{aligned}
|^2\Sigma^{10}_{3/2}\rangle_{3/2} = &\frac{1}{6\sqrt{2}} |1\rangle_\alpha \, |u^\uparrow d^\uparrow s^\downarrow + u^\uparrow d^\downarrow s^\uparrow - 2u^\downarrow d^\uparrow s^\uparrow + u^\uparrow s^\uparrow d^\downarrow + u^\uparrow s^\downarrow d^\uparrow \\
&\qquad - 2u^\downarrow s^\uparrow d^\uparrow + d^\uparrow u^\uparrow s^\downarrow + d^\uparrow u^\downarrow s^\uparrow - 2d^\downarrow u^\uparrow s^\uparrow + d^\uparrow s^\uparrow u^\downarrow \\
&\qquad + d^\uparrow s^\downarrow u^\uparrow - 2d^\downarrow s^\uparrow u^\uparrow + s^\uparrow u^\uparrow d^\downarrow + s^\uparrow u^\downarrow d^\uparrow \\
&\qquad - 2s^\downarrow u^\uparrow d^\uparrow + s^\uparrow d^\uparrow u^\downarrow + s^\uparrow d^\downarrow u^\uparrow - 2s^\downarrow d^\uparrow u^\uparrow \rangle \\
&+ \frac{1}{2\sqrt{6}} |1\rangle_\beta \, |u^\uparrow d^\uparrow s^\downarrow - u^\uparrow d^\downarrow s^\uparrow + u^\uparrow s^\uparrow d^\downarrow - u^\uparrow s^\downarrow d^\uparrow + d^\uparrow u^\uparrow s^\downarrow \\
&\qquad - d^\uparrow u^\downarrow s^\uparrow + d^\uparrow s^\uparrow u^\downarrow - d^\uparrow s^\downarrow u^\uparrow + s^\uparrow u^\uparrow d^\downarrow \\
&\qquad - s^\uparrow u^\downarrow d^\uparrow + s^\uparrow d^\uparrow u^\downarrow - s^\uparrow d^\downarrow u^\uparrow \rangle
\end{aligned}$$

$$
\begin{aligned}
|{}^2\Sigma^{10}_{3/2}>_{1/2} = -\frac{1}{6\sqrt{6}}\,|1>_\alpha\, |&u^\downarrow d^\downarrow s^\uparrow + u^\downarrow d^\uparrow s^\downarrow - 2u^\uparrow d^\downarrow s^\downarrow + u^\downarrow s^\downarrow d^\uparrow + u^\downarrow s^\uparrow d^\downarrow \\
&- 2u^\uparrow s^\downarrow d^\downarrow + d^\downarrow s^\downarrow u^\uparrow + d^\downarrow s^\uparrow u^\downarrow - 2d^\uparrow s^\downarrow u^\downarrow \\
&+ d^\downarrow u^\downarrow s^\uparrow + d^\downarrow u^\uparrow s^\downarrow - 2d^\uparrow u^\downarrow s^\downarrow + s^\downarrow u^\downarrow d^\uparrow \\
&+ s^\downarrow u^\uparrow d^\downarrow - 2s^\uparrow u^\downarrow d^\uparrow + s^\downarrow d^\downarrow u^\uparrow + s^\downarrow d^\uparrow u^\downarrow \\
&- 2s^\uparrow d^\downarrow u^\downarrow >
\end{aligned}
$$

$$
\begin{aligned}
+\frac{1}{6\sqrt{3}}\,|0>_\alpha\, |&u^\uparrow d^\uparrow s^\downarrow + u^\uparrow d^\downarrow s^\uparrow - 2u^\downarrow d^\uparrow s^\uparrow + u^\uparrow s^\uparrow d^\downarrow + u^\uparrow s^\downarrow d^\uparrow \\
&- 2u^\downarrow s^\uparrow d^\uparrow + d^\uparrow u^\uparrow s^\downarrow + d^\uparrow u^\downarrow s^\uparrow - 2d^\downarrow u^\uparrow s^\uparrow \\
&+ d^\uparrow s^\uparrow u^\downarrow + d^\uparrow s^\downarrow u^\uparrow - 2d^\downarrow s^\uparrow u^\uparrow + s^\uparrow u^\uparrow d^\downarrow \\
&+ s^\uparrow u^\downarrow d^\uparrow - 2s^\downarrow u^\uparrow d^\uparrow + s^\uparrow d^\uparrow u^\downarrow + s^\uparrow d^\downarrow u^\uparrow \\
&- 2s^\downarrow d^\uparrow u^\uparrow >
\end{aligned}
$$

$$
\begin{aligned}
+\frac{1}{6\sqrt{2}}\,|1>_\beta\, |&u^\downarrow d^\uparrow s^\downarrow - u^\downarrow d^\downarrow s^\uparrow + u^\downarrow s^\uparrow d^\downarrow - u^\downarrow s^\downarrow d^\uparrow + d^\downarrow u^\uparrow s^\downarrow \\
&- d^\downarrow u^\downarrow s^\uparrow + d^\downarrow s^\uparrow u^\downarrow - d^\downarrow s^\downarrow u^\uparrow + s^\downarrow u^\uparrow d^\downarrow \\
&- s^\downarrow u^\downarrow d^\uparrow + s^\downarrow d^\uparrow u^\downarrow - s^\downarrow d^\downarrow u^\uparrow >
\end{aligned}
$$

$$
\begin{aligned}
+\frac{1}{6}\,|0>_\beta\, |&u^\uparrow d^\uparrow s^\downarrow - u^\uparrow d^\downarrow s^\uparrow + u^\uparrow s^\uparrow d^\downarrow - u^\uparrow s^\downarrow d^\uparrow + d^\uparrow u^\uparrow s^\downarrow \\
&- d^\uparrow u^\downarrow s^\uparrow + d^\uparrow s^\uparrow u^\downarrow - d^\uparrow s^\downarrow u^\uparrow + s^\uparrow u^\uparrow d^\downarrow - s^\uparrow u^\downarrow d^\uparrow \\
&+ s^\uparrow d^\uparrow u^\downarrow - s^\uparrow d^\downarrow u^\uparrow >
\end{aligned}
$$

$$
\begin{aligned}
|{}^2\Sigma^{10}_{1/2}>_{1/2} = -\frac{1}{6\sqrt{3}}\,|1>_\alpha\, |&u^\downarrow d^\downarrow s^\uparrow + u^\downarrow d^\uparrow s^\downarrow - 2u^\uparrow d^\downarrow s^\downarrow + u^\downarrow s^\downarrow d^\uparrow + u^\downarrow s^\uparrow d^\downarrow \\
&- 2u^\uparrow s^\downarrow d^\downarrow + d^\downarrow s^\downarrow u^\uparrow + d^\downarrow s^\uparrow u^\downarrow - 2d^\uparrow s^\downarrow u^\downarrow \\
&+ d^\downarrow u^\downarrow s^\uparrow + d^\downarrow u^\uparrow s^\downarrow - 2d^\uparrow u^\downarrow s^\downarrow + s^\downarrow u^\downarrow d^\uparrow \\
&+ s^\downarrow u^\uparrow d^\downarrow - 2s^\uparrow u^\downarrow d^\downarrow + s^\downarrow d^\downarrow u^\uparrow + s^\downarrow d^\uparrow u^\downarrow \\
&- 2s^\uparrow d^\downarrow u^\downarrow >
\end{aligned}
$$

$+$

$$
\begin{aligned}
- \frac{1}{6\sqrt{6}} \, |0>_{\alpha} \, |u^{\uparrow}d^{\uparrow}s^{\downarrow} &+ u^{\uparrow}d^{\downarrow}s^{\uparrow} - 2u^{\downarrow}d^{\uparrow}s^{\uparrow} + u^{\uparrow}s^{\uparrow}d^{\downarrow} + u^{\uparrow}s^{\downarrow}d^{\uparrow} \\
&- 2u^{\downarrow}s^{\uparrow}d^{\uparrow} + d^{\uparrow}u^{\uparrow}s^{\downarrow} + d^{\uparrow}u^{\downarrow}s^{\uparrow} - 2d^{\downarrow}u^{\uparrow}s^{\uparrow} \\
&+ d^{\uparrow}s^{\uparrow}u^{\downarrow} + d^{\uparrow}s^{\downarrow}u^{\uparrow} - 2d^{\downarrow}s^{\uparrow}u^{\uparrow} + s^{\uparrow}u^{\uparrow}d^{\downarrow} \\
&+ s^{\uparrow}u^{\downarrow}d^{\uparrow} - 2s^{\downarrow}u^{\uparrow}d^{\uparrow} + s^{\uparrow}d^{\uparrow}u^{\downarrow} + s^{\uparrow}d^{\downarrow}u^{\uparrow} \\
&- 2s^{\downarrow}d^{\uparrow}u^{\uparrow} >
\end{aligned}
$$

$$
\begin{aligned}
+ \frac{1}{6} \, |1>_{\beta} \, |u^{\downarrow}d^{\uparrow}s^{\downarrow} &- u^{\downarrow}d^{\downarrow}s^{\uparrow} + u^{\downarrow}s^{\uparrow}d^{\downarrow} - u^{\downarrow}s^{\downarrow}d^{\uparrow} + d^{\downarrow}u^{\uparrow}s^{\downarrow} \\
&- d^{\downarrow}u^{\downarrow}s^{\uparrow} + d^{\downarrow}s^{\uparrow}u^{\downarrow} - d^{\downarrow}s^{\downarrow}u^{\uparrow} + s^{\downarrow}u^{\uparrow}d^{\downarrow} - s^{\downarrow}u^{\downarrow}d^{\uparrow} \\
&+ s^{\downarrow}d^{\uparrow}u^{\downarrow} - s^{\downarrow}d^{\downarrow}u^{\uparrow} >
\end{aligned}
$$

$$
\begin{aligned}
- \frac{1}{6\sqrt{2}} \, |0>_{\beta} \, |u^{\uparrow}d^{\uparrow}s^{\downarrow} &- u^{\uparrow}d^{\downarrow}s^{\uparrow} + u^{\uparrow}s^{\uparrow}d^{\downarrow} - u^{\uparrow}s^{\downarrow}d^{\uparrow} + d^{\uparrow}u^{\uparrow}s^{\downarrow} \\
&- d^{\uparrow}u^{\downarrow}s^{\uparrow} + d^{\uparrow}s^{\uparrow}u^{\downarrow} - d^{\uparrow}s^{\downarrow}u^{\uparrow} + s^{\uparrow}u^{\uparrow}d^{\downarrow} \\
&- s^{\uparrow}u^{\downarrow}d^{\uparrow} + s^{\uparrow}d^{\uparrow}u^{\downarrow} - s^{\uparrow}d^{\downarrow}u^{\uparrow} >
\end{aligned}
$$

C.4.2 *Octet states*

$$
\begin{aligned}
|{}^{4}N^{+}_{5/2}>_{5/2} = \frac{1}{2\sqrt{3}} \, |1>_{\alpha} \, &|u^{\uparrow}u^{\uparrow}d^{\uparrow} + u^{\uparrow}d^{\uparrow}u^{\uparrow} - 2d^{\uparrow}u^{\uparrow}u^{\uparrow} > \\
+ \frac{1}{2} \, |1>_{\beta} \, &|u^{\uparrow}u^{\uparrow}d^{\uparrow} - u^{\uparrow}d^{\uparrow}u^{\uparrow} >
\end{aligned}
$$

$$
\begin{aligned}
|{}^{4}N^{+}_{5/2}>_{3/2} = \frac{1}{2\sqrt{15}} \, |1>_{\alpha} \, &|u^{\uparrow}u^{\uparrow}d^{\downarrow} + u^{\uparrow}u^{\downarrow}d^{\uparrow} + u^{\downarrow}u^{\uparrow}d^{\uparrow} + u^{\uparrow}d^{\uparrow}u^{\downarrow} + u^{\uparrow}d^{\downarrow}u^{\uparrow} \\
&+ u^{\downarrow}d^{\uparrow}u^{\uparrow} - 2d^{\uparrow}u^{\uparrow}u^{\downarrow} - 2d^{\uparrow}u^{\downarrow}u^{\uparrow} - 2d^{\downarrow}u^{\uparrow}u^{\uparrow} > \\
+ \frac{1}{\sqrt{30}} \, |0>_{\alpha} \, &|u^{\uparrow}u^{\uparrow}d^{\uparrow} + u^{\uparrow}d^{\uparrow}u^{\uparrow} - 2d^{\uparrow}u^{\uparrow}u^{\uparrow} > \\
+ \frac{1}{2\sqrt{5}} \, |1>_{\beta} \, &|u^{\uparrow}u^{\uparrow}d^{\downarrow} + u^{\uparrow}u^{\downarrow}d^{\uparrow} + u^{\downarrow}u^{\uparrow}d^{\uparrow} - u^{\uparrow}d^{\uparrow}u^{\downarrow} - u^{\uparrow}d^{\downarrow}u^{\uparrow} \\
&- u^{\downarrow}d^{\uparrow}u^{\uparrow} > \\
+ \frac{1}{\sqrt{10}} \, |0>_{\beta} \, &|u^{\uparrow}u^{\uparrow}d^{\uparrow} - u^{\uparrow}d^{\uparrow}u^{\uparrow} >
\end{aligned}
$$

$$
\begin{aligned}
|{}^4N^+_{5/2}\rangle_{1/2} = {} & \frac{1}{2\sqrt{30}}\,|1\rangle_\alpha\,|u^\uparrow u^\downarrow d^\downarrow + u^\downarrow u^\uparrow d^\downarrow + u^\downarrow u^\downarrow d^\uparrow + u^\uparrow d^\downarrow u^\downarrow + u^\downarrow d^\uparrow u^\downarrow \\
& \quad + u^\downarrow d^\downarrow u^\uparrow - 2d^\uparrow u^\downarrow u^\downarrow - 2d^\downarrow u^\uparrow u^\downarrow - 2d^\downarrow u^\downarrow u^\uparrow\rangle \\
& + \frac{1}{2\sqrt{15}}\,|0\rangle_\alpha\,|u^\uparrow u^\uparrow d^\downarrow + u^\uparrow u^\downarrow d^\uparrow + u^\downarrow u^\uparrow d^\uparrow + u^\uparrow d^\uparrow u^\downarrow + u^\uparrow d^\downarrow u^\uparrow \\
& \quad + u^\downarrow d^\uparrow u^\uparrow - 2d^\uparrow u^\uparrow u^\downarrow - 2d^\uparrow u^\downarrow u^\uparrow - 2d^\downarrow u^\uparrow u^\uparrow\rangle \\
& + \frac{1}{2\sqrt{30}}\,|-1\rangle_\alpha\,|u^\uparrow u^\uparrow d^\uparrow + u^\uparrow d^\uparrow u^\uparrow - 2d^\uparrow u^\uparrow u^\uparrow\rangle \\
& + \frac{1}{2\sqrt{10}}\,|1\rangle_\beta\,|u^\uparrow u^\downarrow d^\downarrow + u^\downarrow u^\uparrow d^\downarrow + u^\downarrow u^\downarrow d^\uparrow - u^\uparrow d^\downarrow u^\downarrow - u^\downarrow d^\uparrow u^\downarrow \\
& \quad - u^\downarrow d^\downarrow u^\uparrow\rangle \\
& + \frac{1}{2\sqrt{5}}\,|0\rangle_\beta\,|u^\downarrow u^\uparrow d^\uparrow + u^\uparrow u^\downarrow d^\uparrow + u^\uparrow u^\uparrow d^\downarrow - u^\downarrow d^\uparrow u^\uparrow - u^\uparrow d^\downarrow u^\uparrow \\
& \quad - u^\uparrow d^\uparrow u^\downarrow\rangle \\
& + \frac{1}{2\sqrt{10}}\,|-1\rangle_\beta\,|u^\uparrow u^\uparrow d^\uparrow - u^\uparrow d^\uparrow u^\uparrow\rangle
\end{aligned}
$$

$$
\begin{aligned}
|{}^4N^+_{3/2}\rangle_{3/2} = {} & -\frac{1}{3\sqrt{10}}\,|1\rangle_\alpha\,|u^\uparrow u^\uparrow d^\downarrow + u^\uparrow u^\downarrow d^\uparrow + u^\downarrow u^\uparrow d^\uparrow + u^\uparrow d^\uparrow u^\downarrow + u^\uparrow d^\downarrow u^\uparrow \\
& \quad + u^\downarrow d^\uparrow u^\uparrow - 2d^\uparrow u^\uparrow u^\downarrow - 2d^\uparrow u^\downarrow u^\uparrow - 2d^\downarrow u^\uparrow u^\uparrow\rangle \\
& + \frac{1}{2\sqrt{5}}\,|0\rangle_\alpha\,|u^\uparrow u^\uparrow d^\uparrow + u^\uparrow d^\uparrow u^\uparrow - 2d^\uparrow u^\uparrow u^\uparrow\rangle \\
& - \frac{1}{\sqrt{30}}\,|1\rangle_\beta\,|u^\uparrow u^\uparrow d^\downarrow + u^\uparrow u^\downarrow d^\uparrow + u^\downarrow u^\uparrow d^\uparrow - u^\uparrow d^\uparrow u^\downarrow - u^\uparrow d^\downarrow u^\uparrow \\
& \quad - u^\downarrow d^\uparrow u^\uparrow\rangle \\
& + \frac{\sqrt{3}}{2\sqrt{5}}\,|0\rangle_\beta\,|u^\uparrow u^\uparrow d^\uparrow - u^\uparrow d^\uparrow u^\uparrow\rangle
\end{aligned}
$$

$$
\begin{aligned}
|{}^4N^+_{3/2}\rangle_{1/2} = {} & -\frac{\sqrt{2}}{3\sqrt{15}}\,|1\rangle_\alpha\,|u^\uparrow u^\downarrow d^\downarrow + u^\downarrow u^\uparrow d^\downarrow + u^\downarrow u^\downarrow d^\uparrow + u^\uparrow d^\downarrow u^\downarrow + u^\downarrow d^\uparrow u^\downarrow \\
& \quad + u^\downarrow d^\downarrow u^\uparrow - 2d^\uparrow u^\downarrow u^\downarrow - 2d^\downarrow u^\uparrow u^\downarrow - 2d^\downarrow u^\downarrow u^\uparrow\rangle \\
& + \frac{1}{6\sqrt{15}}\,|0\rangle_\alpha\,|u^\downarrow u^\uparrow d^\uparrow + u^\uparrow u^\downarrow d^\uparrow + u^\uparrow u^\uparrow d^\downarrow + u^\downarrow d^\uparrow u^\uparrow + u^\uparrow d^\downarrow u^\uparrow \\
& \quad + u^\uparrow d^\uparrow u^\downarrow - 2d^\uparrow u^\uparrow u^\downarrow - 2d^\uparrow u^\downarrow u^\uparrow - 2d^\downarrow u^\uparrow u^\uparrow\rangle \\
& + \frac{1}{\sqrt{30}}\,|-1\rangle_\alpha\,|u^\uparrow u^\uparrow d^\uparrow + u^\uparrow d^\uparrow u^\uparrow - 2d^\uparrow u^\uparrow u^\uparrow\rangle \quad +
\end{aligned}
$$

$$
\begin{aligned}
&- \frac{\sqrt{2}}{3\sqrt{5}} \,|1\rangle_\beta \,|u^{\uparrow}u^{\downarrow}d^{\downarrow} + u^{\downarrow}u^{\uparrow}d^{\downarrow} + u^{\downarrow}u^{\downarrow}d^{\uparrow} - u^{\uparrow}d^{\downarrow}u^{\downarrow} - u^{\downarrow}d^{\uparrow}u^{\downarrow} \\
&\qquad - u^{\downarrow}d^{\downarrow}u^{\uparrow}\rangle \\
&+ \frac{1}{6\sqrt{5}} \,|0\rangle_\beta \,|u^{\uparrow}u^{\uparrow}d^{\downarrow} + u^{\uparrow}u^{\downarrow}d^{\uparrow} + u^{\downarrow}u^{\uparrow}d^{\uparrow} - u^{\uparrow}d^{\uparrow}u^{\downarrow} - u^{\uparrow}d^{\downarrow}u^{\uparrow} \\
&\qquad - u^{\downarrow}d^{\uparrow}u^{\uparrow}\rangle \\
&+ \frac{1}{\sqrt{10}} \,|-1\rangle_\beta \,|u^{\uparrow}u^{\uparrow}d^{\uparrow} - u^{\uparrow}d^{\uparrow}u^{\uparrow}\rangle
\end{aligned}
$$

$$
\begin{aligned}
|^4N^+_{1/2}\rangle_{1/2} = &\frac{1}{6\sqrt{6}} \,|1\rangle_\alpha \,|u^{\uparrow}u^{\downarrow}d^{\downarrow} + u^{\downarrow}u^{\uparrow}d^{\downarrow} + u^{\downarrow}u^{\downarrow}d^{\uparrow} + u^{\uparrow}d^{\downarrow}u^{\downarrow} + u^{\downarrow}d^{\uparrow}u^{\downarrow} \\
&\qquad + u^{\downarrow}d^{\downarrow}u^{\uparrow} - 2d^{\uparrow}u^{\downarrow}u^{\downarrow} - 2d^{\downarrow}u^{\uparrow}u^{\downarrow} - 2d^{\downarrow}u^{\downarrow}u^{\uparrow}\rangle \\
&- \frac{1}{6\sqrt{3}} \,|0\rangle_\alpha \,|u^{\downarrow}u^{\uparrow}d^{\uparrow} + u^{\uparrow}u^{\downarrow}d^{\uparrow} + u^{\uparrow}u^{\uparrow}d^{\downarrow} + u^{\downarrow}d^{\uparrow}u^{\uparrow} + u^{\uparrow}d^{\downarrow}u^{\uparrow} \\
&\qquad + u^{\uparrow}d^{\uparrow}u^{\downarrow} - 2d^{\downarrow}u^{\uparrow}u^{\uparrow} - 2d^{\uparrow}u^{\downarrow}u^{\uparrow} - 2d^{\uparrow}u^{\uparrow}u^{\downarrow}\rangle \\
&+ \frac{1}{2\sqrt{6}} \,|-1\rangle_\alpha \,|u^{\uparrow}u^{\uparrow}d^{\uparrow} + u^{\uparrow}d^{\uparrow}u^{\uparrow} - 2d^{\uparrow}u^{\uparrow}u^{\uparrow}\rangle \\
&+ \frac{1}{6\sqrt{2}} \,|1\rangle_\beta \,|u^{\uparrow}u^{\downarrow}d^{\downarrow} + u^{\downarrow}u^{\uparrow}d^{\downarrow} + u^{\downarrow}u^{\downarrow}d^{\uparrow} - u^{\uparrow}d^{\downarrow}u^{\downarrow} - u^{\downarrow}d^{\uparrow}u^{\downarrow} \\
&\qquad - u^{\downarrow}d^{\downarrow}u^{\uparrow}\rangle \\
&- \frac{1}{6} \,|0\rangle_\beta \,|u^{\downarrow}u^{\uparrow}d^{\uparrow} + u^{\uparrow}u^{\downarrow}d^{\uparrow} + u^{\uparrow}u^{\uparrow}d^{\downarrow} - u^{\downarrow}d^{\uparrow}u^{\uparrow} - u^{\uparrow}d^{\downarrow}u^{\uparrow} \\
&\qquad - u^{\uparrow}d^{\uparrow}u^{\downarrow}\rangle \\
&+ \frac{1}{2\sqrt{2}} \,|-1\rangle_\beta \,|u^{\uparrow}u^{\uparrow}d^{\uparrow} - u^{\uparrow}d^{\uparrow}u^{\uparrow}\rangle
\end{aligned}
$$

$$
\begin{aligned}
|^2N^+_{3/2}\rangle_{3/2} = &\frac{1}{6} \,|1\rangle_\alpha \,|u^{\uparrow}u^{\uparrow}d^{\downarrow} + d^{\uparrow}u^{\downarrow}u^{\uparrow} + u^{\downarrow}d^{\uparrow}u^{\uparrow} + d^{\uparrow}u^{\uparrow}u^{\downarrow} + u^{\uparrow}u^{\uparrow}d^{\uparrow} \\
&\qquad + u^{\uparrow}d^{\downarrow}u^{\uparrow} - 2u^{\uparrow}d^{\uparrow}u^{\downarrow} - 2u^{\uparrow}u^{\downarrow}d^{\uparrow} - 2d^{\downarrow}u^{\uparrow}u^{\uparrow}\rangle \\
&+ \frac{1}{2\sqrt{3}} \,|1\rangle_\beta \,|u^{\uparrow}u^{\uparrow}d^{\downarrow} + d^{\uparrow}u^{\downarrow}u^{\uparrow} + u^{\downarrow}d^{\uparrow}u^{\uparrow} - d^{\uparrow}u^{\uparrow}u^{\downarrow} \\
&\qquad - u^{\uparrow}d^{\downarrow}u^{\uparrow} - u^{\downarrow}u^{\uparrow}d^{\uparrow}\rangle
\end{aligned}
$$

$$
\begin{aligned}
|{}^{2}N^{+}_{3/2}\rangle_{1/2} = &-\frac{1}{6\sqrt{3}}\,|1\rangle_{\alpha}\,|u^{\downarrow}u^{\downarrow}d^{\uparrow} + d^{\downarrow}u^{\downarrow}u^{\uparrow} + u^{\downarrow}d^{\uparrow}u^{\downarrow} + d^{\downarrow}u^{\uparrow}u^{\downarrow} + u^{\uparrow}u^{\downarrow}d^{\downarrow} \\
&\qquad + u^{\uparrow}d^{\downarrow}u^{\downarrow} - 2u^{\downarrow}d^{\downarrow}u^{\uparrow} - 2u^{\downarrow}u^{\uparrow}d^{\downarrow} - 2d^{\uparrow}u^{\downarrow}u^{\downarrow}\rangle \\
&+\frac{1}{3\sqrt{6}}\,|0\rangle_{\alpha}\,|u^{\uparrow}u^{\uparrow}d^{\downarrow} + d^{\uparrow}u^{\uparrow}u^{\downarrow} + u^{\uparrow}d^{\downarrow}u^{\uparrow} + d^{\uparrow}u^{\downarrow}u^{\uparrow} + u^{\downarrow}u^{\uparrow}d^{\uparrow} \\
&\qquad + u^{\downarrow}d^{\uparrow}u^{\uparrow} - 2u^{\uparrow}d^{\uparrow}u^{\downarrow} - 2u^{\uparrow}u^{\downarrow}d^{\uparrow} - 2d^{\downarrow}u^{\uparrow}u^{\uparrow}\rangle \\
&+\frac{1}{6}\,|1\rangle_{\beta}\,|u^{\downarrow}d^{\uparrow}u^{\downarrow} - d^{\downarrow}u^{\uparrow}u^{\downarrow} - u^{\downarrow}u^{\downarrow}d^{\uparrow} + d^{\downarrow}u^{\downarrow}u^{\uparrow} + u^{\uparrow}u^{\downarrow}d^{\downarrow} \\
&\qquad - u^{\uparrow}d^{\downarrow}u^{\downarrow}\rangle \\
&+\frac{1}{3\sqrt{2}}\,|0\rangle_{\beta}\,|u^{\uparrow}u^{\uparrow}d^{\downarrow} + d^{\uparrow}u^{\downarrow}u^{\uparrow} + u^{\downarrow}d^{\uparrow}u^{\uparrow} - d^{\uparrow}u^{\uparrow}u^{\downarrow} - u^{\uparrow}d^{\downarrow}u^{\uparrow} \\
&\qquad - u^{\downarrow}u^{\uparrow}d^{\uparrow}\rangle
\end{aligned}
$$

$$
\begin{aligned}
|{}^{2}N^{+}_{1/2}\rangle_{1/2} = &-\frac{1}{3\sqrt{6}}\,|1\rangle_{\alpha}\,|u^{\downarrow}u^{\downarrow}d^{\uparrow} + d^{\downarrow}u^{\uparrow}u^{\downarrow} + u^{\uparrow}d^{\downarrow}u^{\downarrow} + d^{\downarrow}u^{\downarrow}u^{\uparrow} + u^{\uparrow}u^{\downarrow}d^{\downarrow} \\
&\qquad + u^{\downarrow}d^{\uparrow}u^{\downarrow} - 2u^{\downarrow}d^{\downarrow}u^{\uparrow} - 2u^{\downarrow}u^{\uparrow}d^{\downarrow} - 2d^{\uparrow}u^{\downarrow}u^{\downarrow}\rangle \\
&-\frac{1}{6\sqrt{3}}\,|0\rangle_{\alpha}\,|u^{\uparrow}u^{\uparrow}d^{\downarrow} + d^{\uparrow}u^{\uparrow}u^{\downarrow} + u^{\uparrow}d^{\downarrow}u^{\uparrow} + d^{\uparrow}u^{\downarrow}u^{\uparrow} + u^{\downarrow}u^{\uparrow}d^{\uparrow} \\
&\qquad + u^{\downarrow}d^{\uparrow}u^{\uparrow} - 2u^{\uparrow}d^{\uparrow}u^{\downarrow} - 2u^{\uparrow}u^{\downarrow}d^{\uparrow} - 2d^{\downarrow}u^{\uparrow}u^{\uparrow}\rangle \\
&+\frac{1}{3\sqrt{2}}\,|1\rangle_{\beta}\,|u^{\downarrow}d^{\uparrow}u^{\downarrow} - d^{\downarrow}u^{\uparrow}u^{\downarrow} - u^{\downarrow}u^{\downarrow}d^{\uparrow} + d^{\downarrow}u^{\downarrow}u^{\uparrow} + u^{\uparrow}u^{\downarrow}d^{\downarrow} \\
&\qquad - u^{\uparrow}d^{\downarrow}u^{\downarrow}\rangle \\
&-\frac{1}{6}\,|0\rangle_{\beta}\,|u^{\uparrow}u^{\uparrow}d^{\downarrow} + d^{\uparrow}u^{\downarrow}u^{\uparrow} + u^{\downarrow}d^{\uparrow}u^{\uparrow} - d^{\uparrow}u^{\uparrow}u^{\downarrow} - u^{\uparrow}d^{\downarrow}u^{\uparrow} \\
&\qquad - u^{\downarrow}u^{\uparrow}d^{\uparrow}\rangle
\end{aligned}
$$

$$
\begin{aligned}
|{}^{4}\Sigma^{8}_{5/2}\rangle_{5/2} = &\frac{1}{2\sqrt{6}}\,|1\rangle_{\alpha}\,|u^{\uparrow}d^{\uparrow}s^{\uparrow} + u^{\uparrow}s^{\uparrow}d^{\uparrow} + d^{\uparrow}u^{\uparrow}s^{\uparrow} + d^{\uparrow}s^{\uparrow}u^{\uparrow} - 2s^{\uparrow}u^{\uparrow}d^{\uparrow} \\
&\qquad - 2s^{\uparrow}d^{\uparrow}u^{\uparrow}\rangle \\
&+\frac{1}{2\sqrt{2}}\,|1\rangle_{\beta}\,|u^{\uparrow}d^{\uparrow}s^{\uparrow} + d^{\uparrow}u^{\uparrow}s^{\uparrow} - u^{\uparrow}s^{\uparrow}d^{\uparrow} - d^{\uparrow}s^{\uparrow}u^{\uparrow}\rangle
\end{aligned}
$$

$$
\begin{aligned}
|{}^4\Sigma^8_{5/2}\rangle_{3/2} = \frac{1}{2\sqrt{30}}\,|1\rangle_\alpha\,|\,&u^\uparrow d^\uparrow s^\downarrow + u^\uparrow d^\downarrow s^\uparrow + u^\downarrow d^\uparrow s^\uparrow + u^\uparrow s^\uparrow d^\downarrow + u^\uparrow s^\downarrow d^\uparrow \\
&+ u^\downarrow s^\uparrow d^\uparrow + d^\uparrow u^\uparrow s^\downarrow + d^\uparrow u^\downarrow s^\uparrow + d^\downarrow u^\uparrow s^\uparrow + d^\uparrow s^\uparrow u^\downarrow \\
&+ d^\uparrow s^\downarrow u^\uparrow + d^\downarrow s^\uparrow u^\uparrow - 2s^\uparrow d^\uparrow u^\downarrow - 2s^\uparrow d^\downarrow u^\uparrow \\
&- 2s^\downarrow d^\uparrow u^\uparrow - 2s^\uparrow u^\uparrow d^\downarrow - 2s^\uparrow u^\downarrow d^\uparrow - 2s^\downarrow u^\uparrow d^\uparrow \rangle \\
+ \frac{1}{2\sqrt{15}}\,|0\rangle_\alpha\,|\,&u^\uparrow d^\uparrow s^\uparrow + u^\uparrow s^\uparrow d^\uparrow + d^\uparrow u^\uparrow s^\uparrow + d^\uparrow s^\uparrow u^\uparrow \\
&- 2s^\uparrow u^\uparrow d^\uparrow - 2s^\uparrow d^\uparrow u^\uparrow \rangle \\
+ \frac{1}{2\sqrt{10}}\,|1\rangle_\beta\,|\,&u^\uparrow d^\uparrow s^\downarrow + u^\uparrow d^\downarrow s^\uparrow + u^\downarrow d^\uparrow s^\uparrow + d^\uparrow u^\uparrow s^\downarrow + d^\uparrow u^\downarrow s^\uparrow \\
&+ d^\downarrow u^\uparrow s^\uparrow - d^\uparrow s^\uparrow u^\downarrow - d^\uparrow s^\downarrow u^\uparrow - d^\downarrow s^\uparrow u^\uparrow \\
&- u^\uparrow s^\uparrow d^\downarrow - u^\uparrow s^\downarrow d^\uparrow - u^\downarrow s^\uparrow d^\uparrow \rangle \\
+ \frac{1}{2\sqrt{5}}\,|0\rangle_\beta\,|\,&u^\uparrow d^\uparrow s^\uparrow + d^\uparrow u^\uparrow s^\uparrow - d^\uparrow s^\uparrow u^\uparrow - u^\uparrow s^\uparrow d^\uparrow \rangle
\end{aligned}
$$

$$
\begin{aligned}
|{}^4\Sigma^8_{5/2}\rangle_{1/2} = \frac{1}{4\sqrt{15}}\,|1\rangle_\alpha\,|\,&u^\uparrow d^\downarrow s^\downarrow + u^\downarrow d^\uparrow s^\downarrow + u^\downarrow d^\downarrow s^\uparrow + u^\uparrow s^\downarrow d^\downarrow + u^\downarrow s^\uparrow d^\downarrow \\
&+ u^\downarrow s^\downarrow d^\uparrow + d^\uparrow u^\downarrow s^\downarrow + d^\downarrow u^\uparrow s^\downarrow + d^\downarrow u^\downarrow s^\uparrow + d^\uparrow s^\downarrow u^\downarrow \\
&+ d^\downarrow s^\uparrow u^\downarrow + d^\downarrow s^\downarrow u^\uparrow - 2s^\uparrow u^\downarrow d^\downarrow - 2s^\downarrow u^\uparrow d^\downarrow \\
&- 2s^\downarrow u^\downarrow d^\uparrow - 2s^\uparrow d^\downarrow u^\downarrow - 2s^\downarrow d^\uparrow u^\downarrow - 2s^\downarrow d^\downarrow u^\uparrow \rangle \\
+ \frac{1}{2\sqrt{30}}\,|0\rangle_\alpha\,|\,&u^\uparrow d^\uparrow s^\downarrow + u^\uparrow d^\downarrow s^\uparrow + u^\downarrow d^\uparrow s^\uparrow + u^\uparrow s^\uparrow d^\downarrow + u^\uparrow s^\downarrow d^\uparrow \\
&+ u^\downarrow s^\uparrow d^\uparrow + d^\uparrow u^\uparrow s^\downarrow + d^\uparrow u^\downarrow s^\uparrow + d^\downarrow u^\uparrow s^\uparrow \\
&+ d^\uparrow s^\uparrow u^\downarrow + d^\uparrow s^\downarrow u^\uparrow + d^\downarrow s^\uparrow u^\uparrow - 2s^\uparrow u^\uparrow d^\downarrow \\
&- 2s^\uparrow u^\downarrow d^\uparrow - 2s^\downarrow u^\uparrow d^\uparrow - 2s^\uparrow d^\uparrow u^\downarrow - 2s^\uparrow d^\downarrow u^\uparrow \\
&- 2s^\downarrow d^\uparrow u^\uparrow \rangle \\
+ \frac{1}{4\sqrt{15}}\,|-1\rangle_\alpha\,|\,&u^\uparrow d^\uparrow s^\uparrow + u^\uparrow s^\uparrow d^\uparrow + d^\uparrow u^\uparrow s^\uparrow + d^\uparrow s^\uparrow u^\uparrow \\
&- 2s^\uparrow u^\uparrow d^\uparrow - 2s^\uparrow d^\uparrow u^\uparrow \rangle
\end{aligned}
$$

$+$

$$\begin{aligned}
&+ \frac{1}{4\sqrt{5}}\,|1\rangle_{\beta}\,|u^{\uparrow}d^{\downarrow}s^{\downarrow} + u^{\downarrow}d^{\uparrow}s^{\downarrow} + u^{\downarrow}d^{\downarrow}s^{\uparrow} + d^{\uparrow}u^{\downarrow}s^{\downarrow} + d^{\downarrow}u^{\uparrow}s^{\downarrow} \\
&\qquad + d^{\downarrow}u^{\downarrow}s^{\uparrow} - u^{\uparrow}s^{\downarrow}d^{\downarrow} - u^{\downarrow}s^{\uparrow}d^{\downarrow} - u^{\downarrow}s^{\downarrow}d^{\uparrow} \\
&\qquad - d^{\uparrow}s^{\downarrow}u^{\downarrow} - d^{\downarrow}s^{\uparrow}u^{\downarrow} - d^{\downarrow}s^{\downarrow}u^{\uparrow}\rangle
\end{aligned}$$

$$\begin{aligned}
&+ \frac{1}{2\sqrt{10}}\,|0\rangle_{\beta}\,|u^{\uparrow}d^{\uparrow}s^{\downarrow} + u^{\uparrow}d^{\downarrow}s^{\uparrow} + u^{\downarrow}d^{\uparrow}s^{\uparrow} + d^{\uparrow}u^{\uparrow}s^{\downarrow} + d^{\uparrow}u^{\downarrow}s^{\uparrow} \\
&\qquad + d^{\downarrow}u^{\uparrow}s^{\uparrow} - u^{\uparrow}s^{\uparrow}d^{\downarrow} - u^{\uparrow}s^{\downarrow}d^{\uparrow} - u^{\downarrow}s^{\uparrow}d^{\uparrow} \\
&\qquad - d^{\uparrow}s^{\uparrow}u^{\downarrow} - d^{\uparrow}s^{\downarrow}u^{\uparrow} - d^{\downarrow}s^{\uparrow}u^{\uparrow}\rangle
\end{aligned}$$

$$+ \frac{1}{4\sqrt{5}}\,|-1\rangle_{\beta}\,|u^{\uparrow}d^{\uparrow}s^{\uparrow} + d^{\uparrow}u^{\uparrow}s^{\uparrow} - u^{\uparrow}s^{\uparrow}d^{\uparrow} - d^{\uparrow}s^{\uparrow}u^{\uparrow}\rangle$$

$$\begin{aligned}
|{}^{4}\Sigma^{8}_{3/2}\rangle_{3/2} = &- \frac{1}{6\sqrt{5}}\,|1\rangle_{\alpha}\,|u^{\uparrow}d^{\uparrow}s^{\downarrow} + u^{\uparrow}d^{\downarrow}s^{\uparrow} + u^{\downarrow}d^{\uparrow}s^{\uparrow} + u^{\uparrow}s^{\uparrow}d^{\downarrow} + u^{\uparrow}s^{\downarrow}d^{\uparrow} \\
&\qquad + u^{\downarrow}s^{\uparrow}d^{\uparrow} + d^{\uparrow}u^{\uparrow}s^{\downarrow} + d^{\uparrow}u^{\downarrow}s^{\uparrow} + d^{\downarrow}u^{\uparrow}s^{\uparrow} \\
&\qquad + d^{\uparrow}s^{\uparrow}u^{\downarrow} + d^{\uparrow}s^{\downarrow}u^{\uparrow} + d^{\downarrow}s^{\uparrow}u^{\uparrow} - 2s^{\uparrow}d^{\uparrow}u^{\downarrow} \\
&\qquad - 2s^{\uparrow}d^{\downarrow}u^{\uparrow} - 2s^{\downarrow}d^{\uparrow}u^{\uparrow} - 2s^{\uparrow}u^{\uparrow}d^{\downarrow} - 2s^{\uparrow}u^{\downarrow}d^{\uparrow} \\
&\qquad - 2s^{\downarrow}u^{\uparrow}d^{\uparrow}\rangle \\
&+ \frac{1}{2\sqrt{10}}\,|0\rangle_{\alpha}\,|u^{\uparrow}d^{\uparrow}s^{\uparrow} + u^{\uparrow}s^{\uparrow}d^{\uparrow} + d^{\uparrow}u^{\uparrow}s^{\uparrow} + d^{\uparrow}s^{\uparrow}u^{\uparrow} \\
&\qquad - 2s^{\uparrow}d^{\uparrow}u^{\uparrow} - 2s^{\uparrow}u^{\uparrow}d^{\uparrow}\rangle \\
&- \frac{1}{2\sqrt{15}}\,|1\rangle_{\beta}\,|u^{\uparrow}d^{\uparrow}s^{\downarrow} + u^{\uparrow}d^{\downarrow}s^{\uparrow} + u^{\downarrow}d^{\uparrow}s^{\uparrow} + d^{\uparrow}u^{\uparrow}s^{\downarrow} + d^{\uparrow}u^{\downarrow}s^{\uparrow} \\
&\qquad + d^{\downarrow}u^{\uparrow}s^{\uparrow} - d^{\uparrow}s^{\uparrow}u^{\downarrow} - d^{\uparrow}s^{\downarrow}u^{\uparrow} - d^{\downarrow}s^{\uparrow}u^{\uparrow} \\
&\qquad - u^{\uparrow}s^{\uparrow}d^{\downarrow} - u^{\uparrow}s^{\downarrow}d^{\uparrow} - u^{\downarrow}s^{\uparrow}d^{\uparrow}\rangle \\
&+ \frac{\sqrt{3}}{2\sqrt{10}}\,|0\rangle_{\beta}\,|u^{\uparrow}d^{\uparrow}s^{\uparrow} + d^{\uparrow}u^{\uparrow}s^{\uparrow} - d^{\uparrow}s^{\uparrow}u^{\uparrow} - u^{\uparrow}s^{\uparrow}d^{\uparrow}\rangle
\end{aligned}$$

$$\begin{aligned}
|{}^{4}\Sigma^{8}_{3/2}\rangle_{1/2} = &- \frac{1}{3\sqrt{15}}\,|1\rangle_{\alpha}\,|u^{\uparrow}d^{\downarrow}s^{\downarrow} + u^{\downarrow}d^{\uparrow}s^{\downarrow} + u^{\downarrow}d^{\downarrow}s^{\uparrow} + u^{\uparrow}s^{\downarrow}d^{\downarrow} + u^{\downarrow}s^{\uparrow}d^{\downarrow} \\
&\qquad + u^{\downarrow}s^{\downarrow}d^{\uparrow} + d^{\uparrow}u^{\downarrow}s^{\downarrow} + d^{\downarrow}u^{\uparrow}s^{\downarrow} + d^{\downarrow}u^{\downarrow}s^{\uparrow} \\
&\qquad + d^{\uparrow}s^{\downarrow}u^{\downarrow} + d^{\downarrow}s^{\uparrow}u^{\downarrow} + d^{\downarrow}s^{\downarrow}u^{\uparrow} - 2s^{\uparrow}d^{\downarrow}u^{\downarrow} \\
&\qquad - 2s^{\downarrow}d^{\uparrow}u^{\downarrow} - 2s^{\downarrow}d^{\downarrow}u^{\uparrow} - 2s^{\uparrow}u^{\downarrow}d^{\downarrow} \\
&\qquad - 2s^{\downarrow}u^{\uparrow}d^{\downarrow} - 2s^{\downarrow}u^{\downarrow}d^{\uparrow}\rangle \quad +
\end{aligned}$$

$$+ \frac{1}{6\sqrt{30}} |0>_{\alpha} |u^{\uparrow}d^{\uparrow}s^{\downarrow} + u^{\uparrow}d^{\downarrow}s^{\uparrow} + u^{\downarrow}d^{\uparrow}s^{\uparrow} + u^{\uparrow}s^{\uparrow}d^{\downarrow} + u^{\uparrow}s^{\downarrow}d^{\uparrow}$$
$$+ u^{\downarrow}s^{\uparrow}d^{\uparrow} + d^{\uparrow}u^{\uparrow}s^{\downarrow} + d^{\uparrow}u^{\downarrow}s^{\uparrow} + d^{\downarrow}u^{\uparrow}s^{\uparrow}$$
$$+ d^{\uparrow}s^{\uparrow}u^{\downarrow} + d^{\uparrow}s^{\downarrow}u^{\uparrow} + d^{\downarrow}s^{\uparrow}u^{\uparrow} - 2s^{\uparrow}d^{\uparrow}u^{\downarrow}$$
$$- 2s^{\uparrow}d^{\downarrow}u^{\uparrow} - 2s^{\downarrow}d^{\uparrow}u^{\uparrow} - 2s^{\uparrow}u^{\uparrow}d^{\downarrow}$$
$$- 2s^{\uparrow}u^{\downarrow}d^{\uparrow} - 2s^{\downarrow}u^{\uparrow}d^{\uparrow}>$$

$$+ \frac{1}{2\sqrt{15}} |-1>_{\alpha} |u^{\uparrow}d^{\uparrow}s^{\uparrow} + u^{\uparrow}s^{\uparrow}d^{\uparrow} + d^{\uparrow}u^{\uparrow}s^{\uparrow} + d^{\uparrow}s^{\uparrow}u^{\uparrow}$$
$$- 2s^{\uparrow}u^{\uparrow}d^{\uparrow} - 2s^{\uparrow}d^{\uparrow}u^{\uparrow}>$$

$$+ \frac{1}{2\sqrt{5}} |-1>_{\beta} |u^{\uparrow}d^{\uparrow}s^{\uparrow} - u^{\uparrow}s^{\uparrow}d^{\uparrow} + d^{\uparrow}u^{\uparrow}s^{\uparrow} - d^{\uparrow}s^{\uparrow}u^{\uparrow}>$$

$$+ \frac{1}{6\sqrt{10}} |0>_{\beta} |u^{\uparrow}d^{\uparrow}s^{\downarrow} + u^{\uparrow}d^{\downarrow}s^{\uparrow} + u^{\downarrow}d^{\uparrow}s^{\uparrow} - u^{\uparrow}s^{\uparrow}d^{\downarrow} - u^{\uparrow}s^{\downarrow}d^{\uparrow}$$
$$- u^{\downarrow}s^{\uparrow}d^{\uparrow} + d^{\uparrow}u^{\uparrow}s^{\downarrow} + d^{\uparrow}u^{\downarrow}s^{\uparrow} + d^{\downarrow}u^{\uparrow}s^{\uparrow}$$
$$- d^{\uparrow}s^{\uparrow}u^{\downarrow} - d^{\uparrow}s^{\downarrow}u^{\uparrow} - d^{\downarrow}s^{\uparrow}u^{\uparrow}>$$

$$- \frac{1}{3\sqrt{5}} |1>_{\beta} |u^{\downarrow}d^{\downarrow}s^{\uparrow} + u^{\downarrow}d^{\uparrow}s^{\downarrow} + u^{\uparrow}d^{\downarrow}s^{\downarrow} - u^{\downarrow}s^{\downarrow}d^{\uparrow} - u^{\downarrow}s^{\uparrow}d^{\downarrow}$$
$$- u^{\uparrow}s^{\downarrow}d^{\downarrow} + d^{\downarrow}u^{\downarrow}s^{\uparrow} + d^{\downarrow}u^{\uparrow}s^{\downarrow} + d^{\uparrow}u^{\downarrow}s^{\downarrow}$$
$$- d^{\downarrow}s^{\downarrow}u^{\uparrow} - d^{\downarrow}s^{\uparrow}u^{\downarrow} - d^{\uparrow}s^{\downarrow}u^{\downarrow}>$$

$$|{}^{4}\Sigma^{8}_{1/2}>_{1/2} = \frac{1}{12\sqrt{3}} |1>_{\alpha} |u^{\uparrow}d^{\downarrow}s^{\downarrow} + u^{\downarrow}d^{\uparrow}s^{\downarrow} + u^{\downarrow}d^{\downarrow}s^{\uparrow} + u^{\uparrow}s^{\downarrow}d^{\downarrow} + u^{\downarrow}s^{\uparrow}d^{\downarrow}$$
$$+ u^{\downarrow}s^{\downarrow}d^{\uparrow} + d^{\uparrow}u^{\downarrow}s^{\downarrow} + d^{\downarrow}u^{\uparrow}s^{\downarrow} + d^{\downarrow}u^{\downarrow}s^{\uparrow} + d^{\uparrow}s^{\downarrow}u^{\downarrow}$$
$$+ d^{\downarrow}s^{\uparrow}u^{\downarrow} + d^{\downarrow}s^{\downarrow}u^{\uparrow} - 2s^{\uparrow}d^{\downarrow}u^{\downarrow} - 2s^{\downarrow}d^{\uparrow}u^{\downarrow}$$
$$- 2s^{\downarrow}d^{\downarrow}u^{\uparrow} - 2s^{\uparrow}u^{\downarrow}d^{\downarrow} - 2s^{\downarrow}u^{\uparrow}d^{\downarrow} - 2s^{\downarrow}u^{\downarrow}d^{\uparrow}>$$

$$- \frac{1}{6\sqrt{6}} |0>_{\alpha} |u^{\uparrow}d^{\uparrow}s^{\downarrow} + u^{\uparrow}d^{\downarrow}s^{\uparrow} + u^{\downarrow}d^{\uparrow}s^{\uparrow} + u^{\uparrow}s^{\uparrow}d^{\downarrow} + u^{\uparrow}s^{\downarrow}d^{\uparrow}$$
$$+ u^{\downarrow}s^{\uparrow}d^{\uparrow} + d^{\uparrow}u^{\uparrow}s^{\downarrow} + d^{\uparrow}u^{\downarrow}s^{\uparrow} + d^{\downarrow}u^{\uparrow}s^{\uparrow}$$
$$+ d^{\uparrow}s^{\uparrow}u^{\downarrow} + d^{\uparrow}s^{\downarrow}u^{\uparrow} + d^{\downarrow}s^{\uparrow}u^{\uparrow} - 2s^{\uparrow}d^{\uparrow}u^{\downarrow}$$
$$- 2s^{\uparrow}d^{\downarrow}u^{\uparrow} - 2s^{\downarrow}d^{\uparrow}u^{\uparrow} - 2s^{\uparrow}u^{\uparrow}d^{\downarrow} - 2s^{\uparrow}u^{\downarrow}d^{\uparrow}$$
$$- 2s^{\downarrow}u^{\uparrow}d^{\uparrow}>$$

$$+$$

$$
\begin{aligned}
&+ \frac{1}{4\sqrt{3}}\,|-1>_{\alpha}\,|u^{\uparrow}d^{\uparrow}s^{\uparrow} + u^{\uparrow}s^{\uparrow}d^{\uparrow} + d^{\uparrow}u^{\uparrow}s^{\uparrow} + d^{\uparrow}s^{\uparrow}u^{\uparrow} \\
&\qquad - 2s^{\uparrow}d^{\uparrow}u^{\uparrow} - 2s^{\uparrow}u^{\uparrow}d^{\uparrow}>
\end{aligned}
$$

$$
\begin{aligned}
&+ \frac{1}{12}\,|1>_{\beta}\,|u^{\uparrow}d^{\downarrow}s^{\downarrow} + u^{\downarrow}d^{\uparrow}s^{\downarrow} + u^{\downarrow}d^{\downarrow}s^{\uparrow} + d^{\uparrow}u^{\downarrow}s^{\downarrow} + d^{\downarrow}u^{\uparrow}s^{\downarrow} \\
&\qquad + d^{\downarrow}u^{\downarrow}s^{\uparrow} - d^{\uparrow}s^{\downarrow}u^{\downarrow} - d^{\downarrow}s^{\uparrow}u^{\downarrow} - d^{\downarrow}s^{\downarrow}u^{\uparrow} \\
&\qquad - u^{\uparrow}s^{\downarrow}d^{\downarrow} - u^{\downarrow}s^{\uparrow}d^{\downarrow} - u^{\downarrow}s^{\downarrow}d^{\uparrow}>
\end{aligned}
$$

$$
\begin{aligned}
&- \frac{1}{6\sqrt{2}}\,|0>_{\beta}\,|u^{\uparrow}d^{\uparrow}s^{\downarrow} + u^{\uparrow}d^{\downarrow}s^{\uparrow} + u^{\downarrow}d^{\uparrow}s^{\uparrow} + d^{\uparrow}u^{\uparrow}s^{\downarrow} + d^{\uparrow}u^{\downarrow}s^{\uparrow} \\
&\qquad + d^{\downarrow}u^{\uparrow}s^{\uparrow} - d^{\uparrow}s^{\uparrow}u^{\downarrow} - d^{\uparrow}s^{\downarrow}u^{\uparrow} - d^{\downarrow}s^{\uparrow}u^{\uparrow} \\
&\qquad - u^{\uparrow}s^{\uparrow}d^{\downarrow} - u^{\uparrow}s^{\downarrow}d^{\uparrow} - u^{\downarrow}s^{\uparrow}d^{\uparrow}>
\end{aligned}
$$

$$
+ \frac{1}{4}\,|-1>_{\beta}\,|u^{\uparrow}d^{\uparrow}s^{\uparrow} + d^{\uparrow}u^{\uparrow}s^{\uparrow} - d^{\uparrow}s^{\uparrow}u^{\uparrow} - u^{\uparrow}s^{\uparrow}d^{\uparrow}>
$$

$$
\begin{aligned}
|{}^4\Lambda^8_{5/2}>_{5/2} &= \frac{1}{2\sqrt{2}}\,|1>_{\alpha}\,|d^{\uparrow}u^{\uparrow}s^{\uparrow} - u^{\uparrow}d^{\uparrow}s^{\uparrow} + d^{\uparrow}s^{\uparrow}u^{\uparrow} - u^{\uparrow}s^{\uparrow}d^{\uparrow}> \\
&+ \frac{1}{2\sqrt{6}}\,|1>_{\beta}\,|d^{\uparrow}s^{\uparrow}u^{\uparrow} + u^{\uparrow}d^{\uparrow}s^{\uparrow} - u^{\uparrow}s^{\uparrow}d^{\uparrow} - d^{\uparrow}u^{\uparrow}s^{\uparrow} \\
&\qquad + 2s^{\uparrow}d^{\uparrow}u^{\uparrow} - 2s^{\uparrow}u^{\uparrow}d^{\uparrow}>
\end{aligned}
$$

$$
\begin{aligned}
|{}^4\Lambda^8_{5/2}>_{3/2} &= \frac{1}{2\sqrt{10}}\,|1>_{\alpha}\,|d^{\uparrow}u^{\uparrow}s^{\downarrow} + d^{\uparrow}u^{\downarrow}s^{\uparrow} + d^{\downarrow}u^{\uparrow}s^{\uparrow} + d^{\uparrow}s^{\uparrow}u^{\downarrow} + d^{\uparrow}s^{\downarrow}u^{\uparrow} \\
&\qquad + d^{\downarrow}s^{\uparrow}u^{\uparrow} - u^{\uparrow}d^{\uparrow}s^{\downarrow} - u^{\uparrow}d^{\downarrow}s^{\uparrow} - u^{\downarrow}d^{\uparrow}s^{\uparrow} \\
&\qquad - u^{\uparrow}s^{\uparrow}d^{\downarrow} - u^{\uparrow}s^{\downarrow}d^{\uparrow} - u^{\downarrow}s^{\uparrow}d^{\uparrow}> \\
&+ \frac{1}{2\sqrt{5}}\,|0>_{\alpha}\,|d^{\uparrow}u^{\uparrow}s^{\uparrow} + d^{\uparrow}s^{\uparrow}u^{\uparrow} - u^{\uparrow}d^{\uparrow}s^{\uparrow} - u^{\uparrow}s^{\uparrow}d^{\uparrow}> \\
&- \frac{1}{2\sqrt{30}}\,|1>_{\beta}\,|u^{\uparrow}s^{\uparrow}d^{\downarrow} + u^{\uparrow}s^{\downarrow}d^{\uparrow} + u^{\downarrow}s^{\uparrow}d^{\uparrow} - d^{\uparrow}s^{\uparrow}u^{\downarrow} - d^{\uparrow}s^{\downarrow}u^{\uparrow} \\
&\qquad - d^{\downarrow}s^{\uparrow}u^{\uparrow} - u^{\uparrow}d^{\uparrow}s^{\downarrow} - u^{\uparrow}d^{\downarrow}s^{\uparrow} - u^{\downarrow}d^{\uparrow}s^{\uparrow} \\
&\qquad + d^{\uparrow}u^{\uparrow}s^{\downarrow} + d^{\uparrow}u^{\downarrow}s^{\uparrow} + d^{\downarrow}u^{\uparrow}s^{\uparrow} - 2s^{\uparrow}d^{\uparrow}u^{\downarrow} \\
&\qquad - 2s^{\uparrow}d^{\downarrow}u^{\uparrow} - 2s^{\downarrow}d^{\uparrow}u^{\uparrow} + 2s^{\uparrow}u^{\uparrow}d^{\downarrow} \\
&\qquad + 2s^{\uparrow}u^{\downarrow}d^{\uparrow} + 2s^{\downarrow}u^{\uparrow}d^{\uparrow}>
\end{aligned}
$$

$$+ \frac{1}{2\sqrt{15}} |0>_\beta |d^\uparrow s^\uparrow u^\uparrow + u^\uparrow d^\uparrow s^\uparrow - u^\uparrow s^\uparrow d^\uparrow - d^\uparrow u^\uparrow s^\uparrow + 2s^\uparrow d^\uparrow u^\uparrow - 2s^\uparrow u^\uparrow d^\uparrow >$$

$$|^4\Lambda^8_{5/2}>_{1/2} = \frac{1}{4\sqrt{5}} |1>_\alpha |d^\downarrow u^\downarrow s^\uparrow + d^\downarrow u^\uparrow s^\downarrow + d^\uparrow u^\downarrow s^\downarrow + d^\downarrow s^\downarrow u^\uparrow + d^\downarrow s^\uparrow u^\downarrow + d^\uparrow s^\downarrow u^\downarrow - u^\downarrow d^\downarrow s^\uparrow - u^\downarrow d^\uparrow s^\downarrow - u^\uparrow d^\downarrow s^\downarrow - u^\downarrow s^\downarrow d^\uparrow - u^\downarrow s^\uparrow d^\downarrow - u^\uparrow s^\downarrow d^\downarrow >$$

$$+ \frac{1}{2\sqrt{10}} |0>_\alpha |d^\uparrow u^\uparrow s^\downarrow + d^\uparrow u^\downarrow s^\uparrow + d^\downarrow u^\uparrow s^\uparrow + d^\uparrow s^\uparrow u^\downarrow + d^\uparrow s^\downarrow u^\uparrow + d^\downarrow s^\uparrow u^\uparrow - u^\uparrow d^\uparrow s^\downarrow - u^\uparrow d^\downarrow s^\uparrow - u^\downarrow d^\uparrow s^\uparrow - u^\uparrow s^\uparrow d^\downarrow - u^\uparrow s^\downarrow d^\uparrow - u^\downarrow s^\uparrow d^\uparrow >$$

$$+ \frac{1}{4\sqrt{5}} |-1>_\alpha |d^\uparrow u^\uparrow s^\uparrow + d^\uparrow s^\uparrow u^\uparrow - u^\uparrow d^\uparrow s^\uparrow - u^\uparrow s^\uparrow d^\uparrow >$$

$$+ \frac{1}{4\sqrt{15}} |1>_\beta |d^\uparrow s^\downarrow u^\downarrow + d^\downarrow s^\uparrow u^\downarrow + d^\downarrow s^\downarrow u^\uparrow + u^\uparrow d^\downarrow s^\downarrow + u^\downarrow d^\uparrow s^\downarrow + u^\downarrow d^\downarrow s^\uparrow - u^\uparrow s^\downarrow d^\downarrow - u^\downarrow s^\uparrow d^\downarrow - u^\downarrow s^\downarrow d^\uparrow - d^\uparrow u^\downarrow s^\downarrow - d^\downarrow u^\uparrow s^\downarrow - d^\downarrow u^\downarrow s^\uparrow + 2s^\uparrow d^\downarrow u^\downarrow + 2s^\downarrow d^\uparrow u^\downarrow + 2s^\downarrow d^\downarrow u^\uparrow - 2s^\uparrow u^\downarrow d^\downarrow - 2s^\downarrow u^\uparrow d^\downarrow - 2s^\downarrow u^\downarrow d^\uparrow >$$

$$+ \frac{1}{2\sqrt{30}} |0>_\beta |d^\uparrow s^\uparrow u^\downarrow + d^\uparrow s^\downarrow u^\uparrow + d^\downarrow s^\uparrow u^\uparrow + u^\uparrow d^\uparrow s^\downarrow + u^\uparrow d^\downarrow s^\uparrow + u^\downarrow d^\uparrow s^\uparrow - u^\uparrow s^\uparrow d^\downarrow - u^\uparrow s^\downarrow d^\uparrow - u^\downarrow s^\uparrow d^\uparrow - d^\uparrow u^\uparrow s^\downarrow - d^\uparrow u^\downarrow s^\uparrow - d^\downarrow u^\uparrow s^\uparrow + 2s^\uparrow d^\uparrow u^\downarrow + 2s^\uparrow d^\downarrow u^\uparrow + 2s^\downarrow d^\uparrow u^\uparrow - 2s^\uparrow u^\uparrow d^\downarrow - 2s^\uparrow u^\downarrow d^\uparrow - 2s^\downarrow u^\uparrow d^\uparrow >$$

$$+ \frac{1}{4\sqrt{15}} |-1>_\beta |d^\uparrow s^\uparrow u^\uparrow + u^\uparrow d^\uparrow s^\uparrow - u^\uparrow s^\uparrow d^\uparrow - d^\uparrow u^\uparrow s^\uparrow + 2s^\uparrow d^\uparrow u^\uparrow - 2s^\uparrow u^\uparrow d^\uparrow >$$

$$
\begin{aligned}
|{}^4\Lambda^8_{3/2}\rangle_{3/2} = &-\frac{1}{2\sqrt{15}}\,|1\rangle_\alpha\,|d^\uparrow u^\uparrow s^\downarrow + d^\uparrow u^\downarrow s^\uparrow + d^\downarrow u^\uparrow s^\uparrow + d^\uparrow s^\uparrow u^\downarrow + d^\uparrow s^\downarrow u^\uparrow \\
&\qquad + d^\downarrow s^\uparrow u^\uparrow - u^\uparrow d^\uparrow s^\downarrow - u^\uparrow d^\downarrow s^\uparrow - u^\downarrow d^\uparrow s^\uparrow \\
&\qquad - u^\uparrow s^\uparrow d^\downarrow - u^\uparrow s^\downarrow d^\uparrow - u^\downarrow s^\uparrow d^\uparrow \rangle \\
&+ \frac{\sqrt{3}}{2\sqrt{10}}\,|0\rangle_\alpha\,|d^\uparrow u^\uparrow s^\uparrow + d^\uparrow s^\uparrow u^\uparrow - u^\uparrow d^\uparrow s^\uparrow - u^\uparrow s^\uparrow d^\uparrow \rangle \\
&- \frac{1}{6\sqrt{5}}\,|1\rangle_\beta\,|u^\uparrow d^\uparrow s^\downarrow + u^\uparrow d^\downarrow s^\uparrow + u^\downarrow d^\uparrow s^\uparrow + d^\uparrow s^\uparrow u^\downarrow + d^\uparrow s^\downarrow u^\uparrow \\
&\qquad + d^\downarrow s^\uparrow u^\uparrow - u^\uparrow s^\uparrow d^\downarrow - u^\uparrow s^\downarrow d^\uparrow - u^\downarrow s^\uparrow d^\uparrow \\
&\qquad - d^\uparrow u^\uparrow s^\downarrow - d^\uparrow u^\downarrow s^\uparrow - d^\downarrow u^\uparrow s^\uparrow + 2s^\uparrow d^\uparrow u^\downarrow \\
&\qquad + 2s^\uparrow d^\downarrow u^\uparrow + 2s^\downarrow d^\uparrow u^\uparrow - 2s^\uparrow u^\uparrow d^\downarrow - 2s^\uparrow u^\downarrow d^\uparrow \\
&\qquad - 2s^\downarrow u^\uparrow d^\uparrow \rangle \\
&+ \frac{1}{2\sqrt{10}}\,|0\rangle_\beta\,|u^\uparrow d^\uparrow s^\uparrow + d^\uparrow s^\uparrow u^\uparrow - u^\uparrow s^\uparrow d^\uparrow - d^\uparrow u^\uparrow s^\uparrow \\
&\qquad + 2s^\uparrow d^\uparrow u^\uparrow - 2s^\uparrow u^\uparrow d^\uparrow \rangle
\end{aligned}
$$

$$
\begin{aligned}
|{}^4\Lambda^8_{3/2}\rangle_{1/2} = &-\frac{1}{3\sqrt{5}}\,|1\rangle_\alpha\,|d^\uparrow u^\downarrow s^\downarrow + d^\downarrow u^\uparrow s^\downarrow + d^\downarrow u^\downarrow s^\uparrow + d^\uparrow s^\downarrow u^\downarrow + d^\downarrow s^\uparrow u^\downarrow \\
&\qquad + d^\downarrow s^\downarrow u^\uparrow - u^\uparrow d^\downarrow s^\downarrow - u^\downarrow d^\uparrow s^\downarrow - u^\downarrow d^\downarrow s^\uparrow \\
&\qquad - u^\uparrow s^\downarrow d^\downarrow - u^\downarrow s^\uparrow d^\downarrow - u^\downarrow s^\downarrow d^\uparrow \rangle \\
&+ \frac{1}{6\sqrt{10}}\,|0\rangle_\alpha\,|d^\uparrow u^\uparrow s^\downarrow + d^\uparrow u^\downarrow s^\uparrow + d^\downarrow u^\uparrow s^\uparrow + d^\uparrow s^\uparrow u^\downarrow + d^\uparrow s^\downarrow u^\uparrow \\
&\qquad + d^\downarrow s^\uparrow u^\uparrow - u^\uparrow d^\uparrow s^\downarrow - u^\uparrow d^\downarrow s^\uparrow - u^\downarrow d^\uparrow s^\uparrow \\
&\qquad - u^\uparrow s^\uparrow d^\downarrow - u^\uparrow s^\downarrow d^\uparrow - u^\downarrow s^\uparrow d^\uparrow \rangle \\
&+ \frac{1}{2\sqrt{5}}\,|-1\rangle_\alpha\,|d^\uparrow u^\uparrow s^\uparrow + d^\uparrow s^\uparrow u^\uparrow - u^\uparrow d^\uparrow s^\uparrow - u^\uparrow s^\uparrow d^\uparrow \rangle \\
&- \frac{1}{3\sqrt{15}}\,|1\rangle_\beta\,|d^\downarrow s^\downarrow u^\uparrow + d^\downarrow s^\uparrow u^\downarrow + d^\uparrow s^\downarrow u^\downarrow + u^\downarrow d^\downarrow s^\uparrow + u^\downarrow d^\uparrow s^\downarrow \\
&\qquad + u^\uparrow d^\downarrow s^\downarrow - u^\downarrow s^\downarrow d^\uparrow - u^\downarrow s^\uparrow d^\downarrow - u^\uparrow s^\downarrow d^\downarrow \\
&\qquad - d^\downarrow u^\downarrow s^\uparrow - d^\downarrow u^\uparrow s^\downarrow - d^\uparrow u^\downarrow s^\downarrow + 2s^\downarrow d^\downarrow u^\uparrow \\
&\qquad + 2s^\downarrow d^\uparrow u^\downarrow + 2s^\uparrow d^\downarrow u^\downarrow - 2s^\downarrow u^\downarrow d^\uparrow \\
&\qquad - 2s^\downarrow u^\uparrow d^\downarrow - 2s^\uparrow u^\downarrow d^\downarrow \rangle
\end{aligned}
$$

$$
\begin{aligned}
&+ \frac{1}{6\sqrt{30}}\, |0\rangle_\beta \, |d^{\uparrow}s^{\uparrow}u^{\downarrow} + d^{\uparrow}s^{\downarrow}u^{\uparrow} + d^{\downarrow}s^{\uparrow}u^{\uparrow} + u^{\uparrow}d^{\uparrow}s^{\downarrow} + u^{\uparrow}d^{\downarrow}s^{\uparrow} \\
&\qquad + u^{\downarrow}d^{\uparrow}s^{\uparrow} - u^{\uparrow}s^{\uparrow}d^{\downarrow} - u^{\uparrow}s^{\downarrow}d^{\uparrow} - u^{\downarrow}s^{\uparrow}d^{\uparrow} \\
&\qquad - d^{\uparrow}u^{\uparrow}s^{\downarrow} - d^{\uparrow}u^{\downarrow}s^{\uparrow} - d^{\downarrow}u^{\uparrow}s^{\uparrow} + 2s^{\uparrow}d^{\uparrow}u^{\downarrow} \\
&\qquad + 2s^{\uparrow}d^{\downarrow}u^{\uparrow} + 2s^{\downarrow}d^{\uparrow}u^{\uparrow} - 2s^{\uparrow}u^{\uparrow}d^{\downarrow} \\
&\qquad - 2s^{\uparrow}u^{\downarrow}d^{\uparrow} - 2s^{\downarrow}u^{\uparrow}d^{\uparrow}\rangle
\end{aligned}
$$

$$
\begin{aligned}
&+ \frac{1}{2\sqrt{15}}\, |-1\rangle_\beta \, |d^{\uparrow}s^{\uparrow}u^{\uparrow} + u^{\uparrow}d^{\uparrow}s^{\uparrow} - u^{\uparrow}s^{\uparrow}d^{\uparrow} - d^{\uparrow}u^{\uparrow}s^{\uparrow} \\
&\qquad + 2s^{\uparrow}d^{\uparrow}u^{\uparrow} - 2s^{\uparrow}u^{\uparrow}d^{\uparrow}\rangle
\end{aligned}
$$

$$
\begin{aligned}
|{}^4\Lambda^8_{1/2}\rangle_{1/2} = &\frac{1}{12}\, |1\rangle_\alpha \, |d^{\uparrow}u^{\downarrow}s^{\downarrow} + d^{\downarrow}u^{\uparrow}s^{\downarrow} + d^{\downarrow}u^{\downarrow}s^{\uparrow} - u^{\uparrow}d^{\downarrow}s^{\downarrow} - u^{\downarrow}d^{\uparrow}s^{\downarrow} \\
&\qquad - u^{\downarrow}d^{\downarrow}s^{\uparrow} + d^{\uparrow}s^{\downarrow}u^{\downarrow} + d^{\downarrow}s^{\uparrow}u^{\downarrow} + d^{\downarrow}s^{\downarrow}u^{\uparrow} - u^{\uparrow}s^{\downarrow}d^{\downarrow} \\
&\qquad - u^{\downarrow}s^{\uparrow}d^{\downarrow} - u^{\downarrow}s^{\downarrow}d^{\uparrow}\rangle \\
&- \frac{1}{6\sqrt{2}}\, |0\rangle_\alpha \, |d^{\uparrow}u^{\uparrow}s^{\downarrow} + d^{\uparrow}u^{\downarrow}s^{\uparrow} + d^{\downarrow}u^{\uparrow}s^{\uparrow} - u^{\uparrow}d^{\uparrow}s^{\downarrow} - u^{\uparrow}d^{\downarrow}s^{\uparrow} \\
&\qquad - u^{\downarrow}d^{\uparrow}s^{\uparrow} + d^{\uparrow}s^{\uparrow}u^{\downarrow} + d^{\uparrow}s^{\downarrow}u^{\uparrow} + d^{\downarrow}s^{\uparrow}u^{\uparrow} \\
&\qquad - u^{\uparrow}s^{\uparrow}d^{\downarrow} - u^{\uparrow}s^{\downarrow}d^{\uparrow} - u^{\downarrow}s^{\uparrow}d^{\uparrow}\rangle \\
&+ \frac{1}{4}\, |-1\rangle_\alpha \, |d^{\uparrow}u^{\uparrow}s^{\uparrow} - u^{\uparrow}d^{\uparrow}s^{\uparrow} + d^{\uparrow}s^{\uparrow}u^{\uparrow} - u^{\uparrow}s^{\uparrow}d^{\uparrow}\rangle \\
&+ \frac{1}{12\sqrt{3}}\, |1\rangle_\beta \, |d^{\uparrow}s^{\downarrow}u^{\downarrow} + d^{\downarrow}s^{\uparrow}u^{\downarrow} + d^{\downarrow}s^{\downarrow}u^{\uparrow} + u^{\uparrow}d^{\downarrow}s^{\downarrow} + u^{\downarrow}d^{\uparrow}s^{\downarrow} \\
&\qquad + u^{\downarrow}d^{\downarrow}s^{\uparrow} - u^{\uparrow}s^{\downarrow}d^{\downarrow} - u^{\downarrow}s^{\uparrow}d^{\downarrow} - u^{\downarrow}s^{\downarrow}d^{\uparrow} \\
&\qquad - d^{\uparrow}u^{\downarrow}s^{\downarrow} - d^{\downarrow}u^{\uparrow}s^{\downarrow} - d^{\downarrow}u^{\downarrow}s^{\uparrow} + 2s^{\uparrow}d^{\downarrow}u^{\downarrow} \\
&\qquad + 2s^{\downarrow}d^{\uparrow}u^{\downarrow} + 2s^{\downarrow}d^{\downarrow}u^{\uparrow} - 2s^{\uparrow}u^{\downarrow}d^{\downarrow} \\
&\qquad - 2s^{\downarrow}u^{\uparrow}d^{\downarrow} - 2s^{\downarrow}u^{\downarrow}d^{\uparrow}\rangle \\
&- \frac{1}{6\sqrt{6}}\, |0\rangle_\beta \, |d^{\uparrow}s^{\uparrow}u^{\downarrow} + d^{\uparrow}s^{\downarrow}u^{\uparrow} + d^{\downarrow}s^{\uparrow}u^{\uparrow} + u^{\uparrow}d^{\uparrow}s^{\downarrow} + u^{\uparrow}d^{\downarrow}s^{\uparrow} \\
&\qquad + u^{\downarrow}d^{\uparrow}s^{\uparrow} - u^{\uparrow}s^{\uparrow}d^{\downarrow} - u^{\uparrow}s^{\downarrow}d^{\uparrow} - u^{\downarrow}s^{\uparrow}d^{\uparrow} \\
&\qquad - d^{\uparrow}u^{\uparrow}s^{\downarrow} - d^{\uparrow}u^{\downarrow}s^{\uparrow} - d^{\downarrow}u^{\uparrow}s^{\uparrow} + 2s^{\uparrow}d^{\uparrow}u^{\downarrow} \\
&\qquad + 2s^{\uparrow}d^{\downarrow}u^{\uparrow} + 2s^{\downarrow}d^{\uparrow}u^{\uparrow} - 2s^{\uparrow}u^{\uparrow}d^{\downarrow} - 2s^{\uparrow}u^{\downarrow}d^{\uparrow} \\
&\qquad - 2s^{\downarrow}u^{\uparrow}d^{\uparrow}\rangle \quad +
\end{aligned}
$$

$$+ \frac{1}{4\sqrt{3}} \,|-1>_\beta \,|d^\uparrow s^\uparrow u^\uparrow + u^\uparrow d^\uparrow s^\uparrow - u^\uparrow s^\uparrow d^\uparrow - d^\uparrow u^\uparrow s^\uparrow$$

$$+ 2s^\uparrow d^\uparrow u^\uparrow - 2s^\uparrow u^\uparrow d^\uparrow >$$

$$|{}^2\Sigma^8_{3/2}>_{3/2} = \frac{1}{6\sqrt{2}} \,|1>_\beta \,|u^\uparrow d^\uparrow s^\downarrow + u^\downarrow d^\uparrow s^\uparrow - 2u^\uparrow d^\downarrow s^\uparrow + u^\uparrow s^\downarrow d^\uparrow + u^\downarrow s^\uparrow d^\uparrow$$

$$- 2u^\uparrow s^\uparrow d^\downarrow + d^\uparrow u^\uparrow s^\downarrow + d^\downarrow u^\uparrow s^\uparrow - 2d^\uparrow u^\downarrow s^\uparrow$$

$$+ d^\uparrow s^\downarrow u^\uparrow + d^\downarrow s^\uparrow u^\uparrow - 2d^\uparrow s^\uparrow u^\downarrow + s^\uparrow u^\uparrow d^\downarrow + s^\uparrow u^\downarrow d^\uparrow$$

$$- 2s^\downarrow u^\uparrow d^\uparrow + s^\uparrow d^\uparrow u^\downarrow + s^\uparrow d^\downarrow u^\uparrow - 2s^\downarrow d^\uparrow u^\uparrow >$$

$$+ \frac{1}{2\sqrt{6}} \,|1>_\alpha \,|u^\uparrow d^\uparrow s^\downarrow - u^\downarrow d^\uparrow s^\uparrow + u^\downarrow s^\uparrow d^\uparrow - u^\uparrow s^\downarrow d^\uparrow - d^\uparrow s^\downarrow u^\uparrow$$

$$+ d^\downarrow s^\uparrow u^\uparrow - s^\uparrow u^\uparrow d^\downarrow + s^\uparrow u^\downarrow d^\uparrow - s^\uparrow d^\uparrow u^\downarrow$$

$$+ s^\uparrow d^\downarrow u^\uparrow + d^\uparrow u^\uparrow s^\downarrow - d^\downarrow u^\uparrow s^\uparrow >$$

$$|{}^2\Sigma^8_{3/2}>_{1/2} = - \frac{1}{6\sqrt{6}} \,|1>_\alpha \,|u^\downarrow d^\downarrow s^\uparrow + u^\uparrow d^\downarrow s^\downarrow - 2u^\downarrow d^\uparrow s^\downarrow + u^\downarrow s^\uparrow d^\downarrow + u^\uparrow s^\downarrow d^\downarrow$$

$$- 2u^\downarrow s^\downarrow d^\uparrow + d^\uparrow u^\downarrow s^\downarrow + d^\downarrow u^\downarrow s^\uparrow - 2d^\downarrow u^\uparrow s^\downarrow$$

$$+ d^\uparrow s^\downarrow u^\downarrow + d^\downarrow s^\uparrow u^\downarrow - 2d^\downarrow s^\downarrow u^\uparrow + s^\downarrow u^\downarrow d^\uparrow$$

$$+ s^\downarrow u^\uparrow d^\downarrow - 2s^\uparrow u^\downarrow d^\downarrow + s^\downarrow d^\downarrow u^\uparrow + s^\downarrow d^\uparrow u^\downarrow$$

$$- 2s^\uparrow d^\downarrow u^\downarrow >$$

$$+ \frac{1}{6\sqrt{3}} \,|0>_\alpha \,|u^\downarrow d^\uparrow s^\uparrow + \mu^\uparrow d^\uparrow s^\downarrow - 2u^\uparrow d^\downarrow s^\uparrow + u^\downarrow s^\uparrow d^\uparrow + u^\uparrow s^\downarrow d^\uparrow$$

$$- 2u^\uparrow s^\uparrow d^\downarrow + d^\downarrow u^\uparrow s^\uparrow + d^\uparrow u^\uparrow s^\downarrow - 2d^\uparrow u^\downarrow s^\uparrow$$

$$+ d^\downarrow s^\uparrow u^\uparrow + d^\uparrow s^\downarrow u^\uparrow - 2d^\uparrow s^\uparrow u^\downarrow + s^\uparrow u^\uparrow d^\downarrow$$

$$+ s^\uparrow u^\downarrow d^\uparrow - 2s^\downarrow u^\uparrow d^\uparrow + s^\uparrow d^\uparrow u^\downarrow + s^\uparrow d^\downarrow u^\uparrow$$

$$- 2s^\downarrow d^\uparrow u^\uparrow >$$

$$+ \frac{1}{6\sqrt{2}} \,|1>_\beta \,|u^\uparrow d^\downarrow s^\downarrow - u^\downarrow d^\downarrow s^\uparrow + u^\downarrow s^\uparrow d^\downarrow - u^\uparrow s^\downarrow d^\downarrow + d^\downarrow s^\uparrow u^\downarrow$$

$$- d^\uparrow s^\downarrow u^\downarrow - d^\downarrow u^\downarrow s^\uparrow + d^\uparrow u^\downarrow s^\downarrow - s^\downarrow u^\uparrow d^\downarrow$$

$$+ s^\downarrow u^\downarrow d^\uparrow - s^\downarrow d^\uparrow u^\downarrow + s^\downarrow d^\downarrow u^\uparrow >$$

$$+ \frac{1}{6} \,|0>_\beta \,|u^\uparrow d^\uparrow s^\downarrow - u^\downarrow d^\uparrow s^\uparrow - u^\uparrow s^\downarrow d^\uparrow + u^\downarrow s^\uparrow d^\uparrow - d^\uparrow s^\downarrow u^\uparrow + d^\downarrow s^\uparrow u^\uparrow$$

$$- s^\uparrow u^\uparrow d^\downarrow + s^\uparrow u^\downarrow d^\uparrow - s^\uparrow d^\uparrow u^\downarrow + s^\uparrow d^\downarrow u^\uparrow + d^\uparrow u^\uparrow s^\downarrow - d^\downarrow u^\uparrow s^\uparrow >$$

$$
\begin{aligned}
|{}^{2}\Sigma^{8}_{1/2}\rangle_{1/2} = & -\frac{1}{6\sqrt{3}}\,|1\rangle_{\alpha}\,|u^{\downarrow}d^{\downarrow}s^{\uparrow} + u^{\uparrow}d^{\downarrow}s^{\downarrow} - 2u^{\downarrow}d^{\uparrow}s^{\downarrow} + u^{\downarrow}s^{\uparrow}d^{\downarrow} + u^{\uparrow}s^{\downarrow}d^{\downarrow} \\
& \quad - 2u^{\downarrow}s^{\downarrow}d^{\uparrow} + d^{\downarrow}u^{\downarrow}s^{\uparrow} + d^{\uparrow}u^{\downarrow}s^{\downarrow} - 2d^{\downarrow}u^{\uparrow}s^{\downarrow} \\
& \quad + d^{\downarrow}s^{\uparrow}u^{\downarrow} + d^{\uparrow}s^{\downarrow}u^{\downarrow} - 2d^{\downarrow}s^{\downarrow}u^{\uparrow} + s^{\downarrow}u^{\downarrow}d^{\uparrow} \\
& \quad + s^{\downarrow}u^{\uparrow}d^{\downarrow} - 2s^{\uparrow}u^{\downarrow}d^{\downarrow} + s^{\downarrow}d^{\downarrow}u^{\uparrow} + s^{\downarrow}d^{\uparrow}u^{\downarrow} \\
& \quad - 2s^{\uparrow}d^{\downarrow}u^{\downarrow}\rangle \\
& -\frac{1}{6\sqrt{6}}\,|0\rangle_{\alpha}\,|u^{\uparrow}d^{\uparrow}s^{\downarrow} + u^{\downarrow}d^{\uparrow}s^{\uparrow} - 2u^{\uparrow}d^{\downarrow}s^{\uparrow} + u^{\uparrow}s^{\downarrow}d^{\uparrow} + u^{\downarrow}s^{\uparrow}d^{\uparrow} \\
& \quad - 2u^{\uparrow}s^{\uparrow}d^{\downarrow} + d^{\uparrow}u^{\uparrow}s^{\downarrow} + d^{\downarrow}u^{\uparrow}s^{\uparrow} - 2d^{\uparrow}u^{\downarrow}s^{\uparrow} \\
& \quad + s^{\uparrow}u^{\uparrow}d^{\downarrow} + s^{\uparrow}u^{\downarrow}d^{\uparrow} - 2s^{\downarrow}u^{\uparrow}d^{\uparrow} + s^{\uparrow}d^{\uparrow}u^{\downarrow} \\
& \quad + s^{\uparrow}d^{\downarrow}u^{\uparrow} - 2s^{\downarrow}d^{\uparrow}u^{\uparrow} + d^{\uparrow}s^{\downarrow}u^{\uparrow} + d^{\downarrow}s^{\uparrow}u^{\uparrow} \\
& \quad - 2d^{\uparrow}s^{\uparrow}u^{\downarrow}\rangle \\
& +\frac{1}{6}\,|1\rangle_{\beta}\,|u^{\uparrow}d^{\downarrow}s^{\downarrow} - u^{\downarrow}d^{\downarrow}s^{\uparrow} + u^{\downarrow}s^{\uparrow}d^{\downarrow} - u^{\uparrow}s^{\downarrow}d^{\downarrow} + d^{\downarrow}s^{\uparrow}u^{\downarrow} \\
& \quad - d^{\uparrow}s^{\downarrow}u^{\downarrow} - d^{\downarrow}u^{\downarrow}s^{\uparrow} + d^{\uparrow}u^{\downarrow}s^{\downarrow} - s^{\downarrow}u^{\uparrow}d^{\downarrow} + s^{\downarrow}u^{\downarrow}d^{\uparrow} \\
& \quad - s^{\downarrow}d^{\uparrow}u^{\downarrow} + s^{\downarrow}d^{\downarrow}u^{\uparrow}\rangle \\
& -\frac{1}{6\sqrt{2}}\,|0\rangle_{\beta}\,|u^{\uparrow}d^{\uparrow}s^{\downarrow} - u^{\downarrow}d^{\uparrow}s^{\uparrow} + u^{\downarrow}s^{\uparrow}d^{\uparrow} - u^{\uparrow}s^{\downarrow}d^{\uparrow} + d^{\downarrow}s^{\uparrow}u^{\uparrow} \\
& \quad - d^{\uparrow}s^{\downarrow}u^{\uparrow} + d^{\uparrow}u^{\uparrow}s^{\downarrow} - d^{\downarrow}u^{\uparrow}s^{\uparrow} + s^{\uparrow}u^{\downarrow}d^{\uparrow} \\
& \quad - s^{\uparrow}u^{\uparrow}d^{\downarrow} + s^{\uparrow}d^{\downarrow}u^{\uparrow} - s^{\uparrow}d^{\uparrow}u^{\downarrow}\rangle
\end{aligned}
$$

$$
\begin{aligned}
|{}^{2}\Lambda^{8}_{3/2}\rangle_{3/2} = & \frac{1}{2\sqrt{6}}\,|1\rangle_{\alpha}\,|d^{\downarrow}u^{\uparrow}s^{\uparrow} - d^{\uparrow}u^{\uparrow}s^{\downarrow} + u^{\uparrow}d^{\uparrow}s^{\downarrow} - u^{\downarrow}d^{\uparrow}s^{\uparrow} - d^{\uparrow}s^{\downarrow}u^{\uparrow} \\
& \quad + d^{\downarrow}s^{\uparrow}u^{\uparrow} + u^{\uparrow}s^{\downarrow}d^{\uparrow} - u^{\downarrow}s^{\uparrow}d^{\uparrow} + s^{\uparrow}d^{\uparrow}u^{\downarrow} - s^{\uparrow}d^{\downarrow}u^{\uparrow} \\
& \quad - s^{\uparrow}u^{\uparrow}d^{\downarrow} + s^{\uparrow}u^{\downarrow}d^{\uparrow}\rangle \\
& +\frac{1}{6\sqrt{2}}\,|1\rangle_{\beta}\,|d^{\uparrow}u^{\uparrow}s^{\downarrow} + d^{\downarrow}u^{\uparrow}s^{\uparrow} - 2d^{\uparrow}u^{\downarrow}s^{\uparrow} - u^{\uparrow}d^{\uparrow}s^{\downarrow} - u^{\downarrow}d^{\uparrow}s^{\uparrow} \\
& \quad + 2u^{\uparrow}d^{\downarrow}s^{\uparrow} - d^{\downarrow}s^{\uparrow}u^{\uparrow} - d^{\uparrow}s^{\downarrow}u^{\uparrow} + 2d^{\uparrow}s^{\uparrow}u^{\downarrow} \\
& \quad + u^{\downarrow}s^{\uparrow}d^{\uparrow} + u^{\uparrow}s^{\downarrow}d^{\uparrow} - 2u^{\uparrow}s^{\uparrow}d^{\downarrow} + s^{\uparrow}d^{\uparrow}u^{\downarrow} + s^{\uparrow}d^{\downarrow}u^{\uparrow} \\
& \quad - 2s^{\downarrow}d^{\uparrow}u^{\uparrow} - s^{\uparrow}u^{\uparrow}d^{\downarrow} - s^{\uparrow}u^{\downarrow}d^{\uparrow} + 2s^{\downarrow}u^{\uparrow}d^{\uparrow}\rangle
\end{aligned}
$$

$$
\begin{aligned}
|{}^2\Lambda^8_{3/2}>_{1/2} = \frac{1}{6\sqrt{2}}\,|1>_{\alpha}\,|&d^{\downarrow}u^{\downarrow}s^{\uparrow} - d^{\uparrow}u^{\downarrow}s^{\downarrow} - u^{\downarrow}d^{\downarrow}s^{\uparrow} + u^{\uparrow}d^{\downarrow}s^{\downarrow} + d^{\downarrow}s^{\uparrow}u^{\downarrow}\\
&- d^{\uparrow}s^{\downarrow}u^{\downarrow} - u^{\downarrow}s^{\uparrow}d^{\downarrow} + u^{\uparrow}s^{\downarrow}d^{\downarrow} + s^{\downarrow}d^{\uparrow}u^{\downarrow} - s^{\downarrow}d^{\downarrow}u^{\uparrow}\\
&- s^{\downarrow}u^{\uparrow}d^{\downarrow} + s^{\downarrow}u^{\downarrow}d^{\uparrow}>\\
+ \frac{1}{6}\,|0>_{\alpha}\,|&d^{\downarrow}u^{\uparrow}s^{\uparrow} - d^{\uparrow}u^{\uparrow}s^{\downarrow} + u^{\uparrow}d^{\uparrow}s^{\downarrow} - u^{\downarrow}d^{\uparrow}s^{\uparrow} - d^{\uparrow}s^{\downarrow}u^{\uparrow}\\
&+ d^{\downarrow}s^{\uparrow}u^{\uparrow} + u^{\uparrow}s^{\downarrow}d^{\uparrow} - u^{\downarrow}s^{\uparrow}d^{\uparrow} + s^{\uparrow}d^{\uparrow}u^{\downarrow} - s^{\uparrow}d^{\downarrow}u^{\uparrow}\\
&- s^{\uparrow}u^{\uparrow}d^{\downarrow} + s^{\uparrow}u^{\downarrow}d^{\uparrow}>\\
- \frac{1}{6\sqrt{6}}\,|1>_{\beta}\,|&d^{\downarrow}u^{\downarrow}s^{\uparrow} + d^{\uparrow}u^{\downarrow}s^{\downarrow} - 2d^{\downarrow}u^{\uparrow}s^{\downarrow} - u^{\downarrow}d^{\downarrow}s^{\uparrow} - u^{\uparrow}d^{\downarrow}s^{\downarrow}\\
&+ 2u^{\downarrow}d^{\uparrow}s^{\downarrow} - d^{\downarrow}s^{\uparrow}u^{\downarrow} - d^{\uparrow}s^{\downarrow}u^{\downarrow} + 2d^{\downarrow}s^{\downarrow}u^{\uparrow}\\
&+ u^{\downarrow}s^{\uparrow}d^{\downarrow} + u^{\uparrow}s^{\downarrow}d^{\downarrow} - 2u^{\downarrow}s^{\downarrow}d^{\uparrow} + s^{\downarrow}d^{\downarrow}u^{\uparrow}\\
&+ s^{\downarrow}d^{\uparrow}u^{\downarrow} - 2s^{\uparrow}d^{\downarrow}u^{\downarrow} - s^{\downarrow}u^{\downarrow}d^{\uparrow} - s^{\downarrow}u^{\uparrow}d^{\downarrow}\\
&+ 2s^{\uparrow}u^{\downarrow}d^{\downarrow}>\\
+ \frac{1}{6\sqrt{3}}\,|0>_{\beta}\,|&d^{\uparrow}u^{\uparrow}s^{\downarrow} + d^{\downarrow}u^{\uparrow}s^{\uparrow} - 2d^{\uparrow}u^{\downarrow}s^{\uparrow} - u^{\uparrow}d^{\uparrow}s^{\downarrow} - u^{\downarrow}d^{\uparrow}s^{\uparrow}\\
&+ 2u^{\uparrow}d^{\downarrow}s^{\uparrow} - d^{\uparrow}s^{\downarrow}u^{\uparrow} - d^{\downarrow}s^{\uparrow}u^{\uparrow} + 2d^{\uparrow}s^{\uparrow}u^{\downarrow}\\
&+ u^{\uparrow}s^{\downarrow}d^{\uparrow} + u^{\downarrow}s^{\uparrow}d^{\uparrow} - 2u^{\uparrow}s^{\uparrow}d^{\downarrow} + s^{\uparrow}d^{\uparrow}u^{\downarrow}\\
&+ s^{\uparrow}d^{\downarrow}u^{\uparrow} - 2s^{\downarrow}d^{\uparrow}u^{\uparrow} - s^{\uparrow}u^{\uparrow}d^{\downarrow} - s^{\uparrow}u^{\downarrow}d^{\uparrow}\\
&+ 2s^{\downarrow}u^{\uparrow}d^{\uparrow}>
\end{aligned}
$$

$$
\begin{aligned}
|{}^2\Lambda^8_{1/2}>_{1/2} = \frac{1}{6}\,|1>_{\alpha}\,|&d^{\downarrow}u^{\downarrow}s^{\uparrow} - d^{\uparrow}u^{\downarrow}s^{\downarrow} - u^{\downarrow}d^{\downarrow}s^{\uparrow} + u^{\uparrow}d^{\downarrow}s^{\downarrow} + d^{\downarrow}s^{\uparrow}u^{\downarrow}\\
&- d^{\uparrow}s^{\downarrow}u^{\downarrow} - u^{\downarrow}s^{\uparrow}d^{\downarrow} + u^{\uparrow}s^{\downarrow}d^{\downarrow} + s^{\downarrow}d^{\uparrow}u^{\downarrow} - s^{\downarrow}d^{\downarrow}u^{\uparrow}\\
&- s^{\downarrow}u^{\uparrow}d^{\downarrow} + s^{\downarrow}u^{\downarrow}d^{\uparrow}>\\
- \frac{1}{6\sqrt{2}}\,|0>_{\alpha}\,|&d^{\downarrow}u^{\uparrow}s^{\uparrow} - d^{\uparrow}u^{\uparrow}s^{\downarrow} + u^{\uparrow}d^{\uparrow}s^{\downarrow} - u^{\downarrow}d^{\uparrow}s^{\uparrow} - d^{\uparrow}s^{\downarrow}u^{\uparrow}\\
&+ d^{\downarrow}s^{\uparrow}u^{\uparrow} + u^{\uparrow}s^{\downarrow}d^{\uparrow} - u^{\downarrow}s^{\uparrow}d^{\uparrow} + s^{\uparrow}d^{\uparrow}u^{\downarrow}\\
&- s^{\uparrow}d^{\downarrow}u^{\uparrow} - s^{\uparrow}u^{\uparrow}d^{\downarrow} + s^{\uparrow}u^{\downarrow}d^{\uparrow}>\\
+ &
\end{aligned}
$$

$$- \frac{1}{6\sqrt{3}} |1>_{\beta} |d^{\downarrow}u^{\downarrow}s^{\uparrow} + d^{\uparrow}u^{\downarrow}s^{\downarrow} - 2d^{\downarrow}u^{\uparrow}s^{\downarrow} - u^{\downarrow}d^{\downarrow}s^{\uparrow}$$

$$- u^{\uparrow}d^{\downarrow}s^{\downarrow} + 2u^{\downarrow}d^{\uparrow}s^{\downarrow} - d^{\downarrow}s^{\uparrow}u^{\downarrow} - d^{\uparrow}s^{\downarrow}u^{\downarrow}$$

$$+ 2d^{\downarrow}s^{\downarrow}u^{\uparrow} + u^{\downarrow}s^{\uparrow}d^{\downarrow} + u^{\uparrow}s^{\downarrow}d^{\downarrow} - 2u^{\downarrow}s^{\downarrow}d^{\uparrow}$$

$$+ s^{\downarrow}d^{\downarrow}u^{\uparrow} + s^{\downarrow}d^{\uparrow}u^{\downarrow} - 2s^{\uparrow}d^{\downarrow}u^{\downarrow} - s^{\downarrow}u^{\downarrow}d^{\uparrow}$$

$$- s^{\downarrow}u^{\uparrow}d^{\downarrow} + 2s^{\uparrow}u^{\downarrow}d^{\downarrow} >$$

$$- \frac{1}{6\sqrt{6}} |0>_{\beta} |d^{\uparrow}u^{\uparrow}s^{\downarrow} + d^{\downarrow}u^{\uparrow}s^{\uparrow} - 2d^{\uparrow}u^{\downarrow}s^{\uparrow} - u^{\uparrow}d^{\uparrow}s^{\downarrow} - u^{\downarrow}d^{\uparrow}s^{\uparrow}$$

$$+ 2u^{\uparrow}d^{\downarrow}s^{\uparrow} - d^{\uparrow}s^{\downarrow}u^{\uparrow} - d^{\downarrow}s^{\uparrow}u^{\uparrow}$$

$$+ 2d^{\uparrow}s^{\uparrow}u^{\downarrow} + u^{\uparrow}s^{\downarrow}d^{\uparrow} + u^{\downarrow}s^{\uparrow}d^{\uparrow} - 2u^{\uparrow}s^{\uparrow}d^{\downarrow}$$

$$+ s^{\uparrow}d^{\uparrow}u^{\downarrow} + s^{\uparrow}d^{\downarrow}u^{\uparrow} - 2s^{\downarrow}d^{\uparrow}u^{\uparrow} - s^{\uparrow}u^{\uparrow}d^{\downarrow}$$

$$- s^{\uparrow}u^{\downarrow}d^{\uparrow} + 2s^{\downarrow}u^{\uparrow}d^{\uparrow} >$$

C.4.3 *Singlet states*

$$|^{2}\Lambda^{1}_{3/2}>_{3/2} = - \frac{1}{2\sqrt{6}} |1>_{\alpha} |u^{\uparrow}d^{\downarrow}s^{\uparrow} - u^{\uparrow}d^{\uparrow}s^{\downarrow} + u^{\uparrow}s^{\uparrow}d^{\downarrow} - u^{\uparrow}s^{\downarrow}d^{\uparrow} - d^{\uparrow}s^{\uparrow}u^{\downarrow}$$

$$+ d^{\uparrow}s^{\downarrow}u^{\uparrow} + d^{\uparrow}u^{\uparrow}s^{\downarrow} - d^{\uparrow}u^{\downarrow}s^{\uparrow} - s^{\uparrow}u^{\uparrow}d^{\downarrow}$$

$$+ s^{\uparrow}u^{\downarrow}d^{\uparrow} + s^{\uparrow}d^{\uparrow}u^{\downarrow} - s^{\uparrow}d^{\downarrow}u^{\uparrow} >$$

$$+ \frac{1}{6\sqrt{2}} |1>_{\beta} |u^{\uparrow}s^{\uparrow}d^{\downarrow} + u^{\uparrow}s^{\downarrow}d^{\uparrow} - 2u^{\downarrow}s^{\uparrow}d^{\uparrow} - u^{\uparrow}d^{\uparrow}s^{\downarrow} - u^{\uparrow}d^{\downarrow}s^{\uparrow}$$

$$+ 2u^{\downarrow}d^{\uparrow}s^{\uparrow} - d^{\uparrow}s^{\uparrow}u^{\downarrow} - d^{\uparrow}s^{\downarrow}u^{\uparrow} + 2d^{\downarrow}s^{\uparrow}u^{\uparrow}$$

$$+ d^{\uparrow}u^{\uparrow}s^{\downarrow} + d^{\uparrow}u^{\downarrow}s^{\uparrow} - 2d^{\downarrow}u^{\uparrow}s^{\uparrow} - s^{\uparrow}u^{\uparrow}d^{\downarrow}$$

$$- s^{\uparrow}u^{\downarrow}d^{\uparrow} + 2s^{\downarrow}u^{\uparrow}d^{\uparrow} + s^{\uparrow}d^{\uparrow}u^{\downarrow} + s^{\uparrow}d^{\downarrow}u^{\uparrow}$$

$$- 2s^{\downarrow}d^{\uparrow}u^{\uparrow} >$$

$$|^{2}\Lambda^{1}_{3/2}>_{1/2} = - \frac{1}{6\sqrt{2}} |1>_{\alpha} |u^{\downarrow}d^{\downarrow}s^{\uparrow} - u^{\downarrow}d^{\uparrow}s^{\downarrow} + u^{\downarrow}s^{\uparrow}d^{\downarrow} - u^{\downarrow}s^{\downarrow}d^{\uparrow} + d^{\downarrow}s^{\downarrow}u^{\uparrow}$$

$$- d^{\downarrow}s^{\uparrow}u^{\downarrow} + d^{\downarrow}u^{\uparrow}s^{\downarrow} - d^{\downarrow}u^{\downarrow}s^{\uparrow} + s^{\downarrow}u^{\downarrow}d^{\uparrow}$$

$$- s^{\downarrow}u^{\uparrow}d^{\downarrow} + s^{\downarrow}d^{\uparrow}u^{\downarrow} - s^{\downarrow}d^{\downarrow}u^{\uparrow} >$$

$$+$$

$$+ \frac{1}{6}\,|0>_{\alpha}\,|u^{\uparrow}d^{\downarrow}s^{\uparrow} - u^{\uparrow}d^{\uparrow}s^{\downarrow} + u^{\uparrow}s^{\uparrow}d^{\downarrow} - u^{\uparrow}s^{\downarrow}d^{\uparrow} - d^{\uparrow}s^{\uparrow}u^{\downarrow}$$

$$+ d^{\uparrow}s^{\downarrow}u^{\uparrow} + d^{\uparrow}u^{\uparrow}s^{\downarrow} - d^{\uparrow}u^{\downarrow}s^{\uparrow} - s^{\uparrow}u^{\uparrow}d^{\downarrow} + s^{\uparrow}u^{\downarrow}d^{\uparrow}$$

$$+ s^{\uparrow}d^{\uparrow}u^{\downarrow} - s^{\uparrow}d^{\downarrow}u^{\uparrow}>$$

$$- \frac{1}{6\sqrt{6}}\,|1>_{\beta}\,|u^{\downarrow}d^{\downarrow}s^{\uparrow} + u^{\downarrow}d^{\uparrow}s^{\downarrow} - 2u^{\uparrow}d^{\downarrow}s^{\downarrow} - u^{\downarrow}s^{\downarrow}d^{\uparrow} - u^{\downarrow}s^{\uparrow}d^{\downarrow}$$

$$+ 2u^{\uparrow}s^{\downarrow}d^{\downarrow} + d^{\downarrow}s^{\downarrow}u^{\uparrow} + d^{\downarrow}s^{\uparrow}u^{\downarrow} - 2d^{\uparrow}s^{\downarrow}u^{\downarrow}$$

$$- d^{\downarrow}u^{\downarrow}s^{\uparrow} - d^{\downarrow}u^{\uparrow}s^{\downarrow} + 2d^{\uparrow}u^{\downarrow}s^{\downarrow} + s^{\downarrow}u^{\downarrow}d^{\uparrow}$$

$$+ s^{\downarrow}u^{\uparrow}d^{\downarrow} - 2s^{\uparrow}u^{\downarrow}d^{\downarrow} - s^{\downarrow}d^{\downarrow}u^{\uparrow} - s^{\downarrow}d^{\uparrow}u^{\downarrow}$$

$$+ 2s^{\uparrow}u^{\downarrow}d^{\downarrow}>$$

$$+ \frac{1}{6\sqrt{3}}\,|0>_{\beta}\,|u^{\uparrow}s^{\uparrow}d^{\downarrow} + u^{\uparrow}s^{\downarrow}d^{\uparrow} - 2u^{\downarrow}s^{\uparrow}d^{\uparrow} - u^{\uparrow}d^{\uparrow}s^{\downarrow} - u^{\uparrow}d^{\downarrow}s^{\uparrow}$$

$$+ 2u^{\downarrow}d^{\uparrow}s^{\uparrow} - d^{\uparrow}s^{\uparrow}u^{\downarrow} - d^{\uparrow}s^{\downarrow}u^{\uparrow} + 2d^{\downarrow}s^{\uparrow}u^{\uparrow}$$

$$+ d^{\uparrow}u^{\uparrow}s^{\downarrow} + d^{\uparrow}u^{\downarrow}s^{\uparrow} - 2d^{\downarrow}u^{\uparrow}s^{\uparrow} - s^{\uparrow}u^{\uparrow}d^{\downarrow}$$

$$- s^{\uparrow}u^{\downarrow}d^{\uparrow} + 2s^{\downarrow}u^{\uparrow}d^{\uparrow} + s^{\uparrow}d^{\uparrow}u^{\downarrow} + s^{\uparrow}d^{\downarrow}u^{\uparrow}$$

$$- 2s^{\downarrow}d^{\uparrow}u^{\uparrow}>$$

$$|{}^{2}\Lambda^{1}_{1/2}>_{1/2} = - \frac{1}{6}\,|1>_{\alpha}\,|u^{\downarrow}d^{\downarrow}s^{\uparrow} - u^{\downarrow}d^{\uparrow}s^{\downarrow} + u^{\downarrow}s^{\uparrow}d^{\downarrow} - u^{\downarrow}s^{\downarrow}d^{\uparrow} + d^{\downarrow}s^{\downarrow}u^{\uparrow}$$

$$- d^{\downarrow}s^{\uparrow}u^{\downarrow} + d^{\downarrow}u^{\uparrow}s^{\downarrow} - d^{\downarrow}u^{\downarrow}s^{\uparrow} + s^{\downarrow}u^{\downarrow}d^{\uparrow} - s^{\downarrow}u^{\uparrow}d^{\downarrow}$$

$$+ s^{\downarrow}d^{\uparrow}u^{\downarrow} - s^{\downarrow}d^{\downarrow}u^{\uparrow}>$$

$$- \frac{1}{6\sqrt{2}}\,|0>_{\alpha}\,|u^{\uparrow}d^{\downarrow}s^{\uparrow} - u^{\uparrow}d^{\uparrow}s^{\downarrow} + u^{\uparrow}s^{\uparrow}d^{\downarrow} - u^{\uparrow}s^{\downarrow}d^{\uparrow} - d^{\uparrow}s^{\downarrow}u^{\uparrow}$$

$$+ d^{\uparrow}s^{\uparrow}u^{\downarrow} + d^{\uparrow}u^{\uparrow}s^{\downarrow} - d^{\uparrow}u^{\downarrow}s^{\uparrow} - s^{\uparrow}u^{\downarrow}d^{\uparrow}$$

$$+ s^{\uparrow}u^{\uparrow}d^{\downarrow} + s^{\uparrow}d^{\uparrow}u^{\downarrow} - s^{\uparrow}d^{\downarrow}u^{\uparrow}>$$

$$+ \frac{1}{6\sqrt{3}}\,|1>_{\beta}\,|u^{\downarrow}d^{\downarrow}s^{\uparrow} + u^{\downarrow}d^{\uparrow}s^{\downarrow} - 2u^{\uparrow}d^{\downarrow}s^{\downarrow} - u^{\downarrow}s^{\downarrow}d^{\uparrow} - u^{\downarrow}s^{\uparrow}d^{\downarrow}$$

$$+ 2u^{\uparrow}s^{\downarrow}d^{\downarrow} + d^{\downarrow}s^{\downarrow}u^{\uparrow} + d^{\downarrow}s^{\uparrow}u^{\downarrow} - 2d^{\uparrow}s^{\downarrow}u^{\downarrow}$$

$$- d^{\downarrow}u^{\downarrow}s^{\uparrow} - d^{\downarrow}u^{\uparrow}s^{\downarrow} + 2d^{\uparrow}u^{\downarrow}s^{\downarrow} + s^{\downarrow}u^{\downarrow}d^{\uparrow}$$

$$+ s^{\downarrow}u^{\uparrow}d^{\downarrow} - 2s^{\uparrow}u^{\downarrow}d^{\downarrow} - s^{\downarrow}d^{\downarrow}u^{\uparrow} - s^{\downarrow}d^{\uparrow}u^{\downarrow}$$

$$+ 2s^{\uparrow}d^{\downarrow}u^{\downarrow}>$$

$$+$$

$$- \frac{1}{6\sqrt{6}} |0>_\beta |u^\uparrow s^\uparrow d^\downarrow + u^\uparrow s^\downarrow d^\uparrow - 2u^\downarrow s^\uparrow d^\uparrow - u^\uparrow d^\uparrow s^\downarrow - u^\uparrow d^\downarrow s^\uparrow$$
$$+ 2u^\downarrow d^\uparrow s^\uparrow - d^\uparrow s^\uparrow u^\downarrow - d^\uparrow s^\downarrow u^\uparrow + 2d^\downarrow s^\uparrow u^\uparrow$$
$$+ d^\uparrow u^\uparrow s^\downarrow + d^\uparrow u^\downarrow s^\uparrow - 2d^\downarrow u^\uparrow s^\uparrow - s^\uparrow u^\uparrow d^\downarrow$$
$$- s^\uparrow u^\downarrow d^\uparrow + 2s^\downarrow u^\uparrow d^\uparrow + s^\uparrow d^\uparrow u^\downarrow + s^\uparrow d^\downarrow u^\uparrow$$
$$- 2s^\downarrow d^\uparrow u^\uparrow >$$

C.4.4 *Mixing angles and decay schemes*

The observed P-wave $J^P = 5/2^-$ baryons are pure states of the $^48_{5/2}$ octet; $3/2^-$ and $1/2^-$ Δ, Ω states are also pure states of the corresponding decouplets. The remaining observed 70-plet baryons are mixed states of different SU(3) multiplets. The experimental data for the mixing angles are rather scarce and sometimes inconsistent; thus in our calculations in Chap. 7 we used the values of masses, mixing angles and partial decay widths for 70-plet baryons, quoted in the Ref. 3. The missing decay probabilities of any 70-plet baryon B_{70} were always ascribed to the channel $B_{70} \to B_{56}\pi\pi$, where B_{56} is the corresponding 56-plet baryon, and the $\pi\pi$ pair is assumed to be in an isoscalar S-wave state. As a result we get:

$$|N(1670)> = |^4N_{5/2}> \to 0.372N\pi + 0.441\Delta\pi + 0.047N\eta + 0.14N\pi\pi$$

$$|\Lambda(1800)> = |^4\Lambda^8_{5/2}> \to 0.447\Sigma\pi + 0.258\Sigma^*\pi + 0.02\Lambda\eta + 0.275\Lambda\pi\pi$$

$$|\Sigma(1770)> = |^4\Sigma^8_{5/2}> \to 0.066\Sigma\pi + 0.174\Lambda\pi + 0.032\Sigma^*\pi + 0.388N\bar{K} + 0.34\Sigma\pi\pi$$

$$|\Xi(\ ?\ )> = |^4\Xi^8_{5/2}> \to 0.588\Xi\pi + 0.167\Lambda\bar{K} + 0.098\Sigma\bar{K} + 0.108\Xi\pi + 0.039\Sigma^*\bar{K}$$

$$|N(1725)> = 0.98|^4N_{3/2}> + 0.18|^2N_{3/2}> \to 0.123N\pi + 0.653\Delta\pi + 0.224N\pi\pi$$

$$|\Lambda(1815)> = -0.96|^4\Lambda^8_{3/2}> + 0.26|^2\Lambda^8_{3/2}> - 0.07|^2\Lambda^1_{3/2}> \to 0.908\Sigma^*\pi + 0.07N\bar{K} + 0.022\Lambda\pi\pi$$

$$|\Sigma(1580)\rangle = 0.3|{}^4\Sigma^8_{3/2}\rangle + 0.28|{}^2\Sigma^8_{3/2}\rangle - 0.91|{}^2\Sigma^{10}_{3/2}\rangle$$

$$\rightarrow 0.33\Lambda\pi + 0.25\Sigma^*\pi + 0.035N\bar{K} + 0.003\Sigma\pi + 0.382\Sigma\pi\pi$$

$$|\Xi(\ ?\ )\rangle = |{}^4\Xi^8_{3/2}\rangle \rightarrow 0.2\Xi\pi + 0.06\Lambda\bar{K} + 0.03\Sigma\bar{K} + 0.44\Xi^*\pi + 0.27\Sigma^*\bar{K}$$

$$|N(1520)\rangle = -0.18|{}^4N_{3/2}\rangle + 0.98|{}^2N_{3/2}\rangle \rightarrow 0.59N\pi + 0.115\Delta\pi + 0.295N\pi\pi$$

$$|\Lambda(1690)\rangle = -0.27|{}^4\Lambda^8_{3/2}\rangle - 0.89|{}^2\Lambda^8_{3/2}\rangle + 0.37|{}^2\Lambda^1_{3/2}\rangle$$

$$\rightarrow 0.396\Sigma\pi + 0.422\Sigma^*\pi + 0.182N\bar{K}$$

$$|\Sigma(1675)\rangle = -0.29|{}^4\Sigma^8_{3/2}\rangle + 0.94|{}^2\Sigma^8_{3/2}\rangle + 0.19|{}^2\Sigma^{10}_{3/2}\rangle$$

$$\rightarrow 0.806\Sigma\pi + 0.028\Lambda\pi + 0.094 + 0.052N\bar{K} + 0.044\Sigma\pi\pi$$

$$|\Xi(\ ?\ )\rangle = |{}^2\Xi^8_{3/2}\rangle \rightarrow 0.052\Xi\pi + 0.227\Lambda\bar{K} + 0.543\Sigma\bar{K} + 0.048\Xi\eta$$

$$+ 0.084\Xi^*\pi + 0.052\Sigma^*\bar{K}$$

$$|N(1690)\rangle = 0.79|{}^4N_{1/2}\rangle - 0.61|{}^2N_{1/2}\rangle$$

$$\rightarrow 0.50N\pi + 0.30\Delta\pi + 0.043\Lambda K + 0.018\Sigma K + 0.139N\pi\pi$$

$$|\Lambda(1825)\rangle = 0.39|{}^4\Lambda^8_{1/2}\rangle + 0.92|{}^2\Lambda^8_{1/2}\rangle - 0.03|{}^2\Lambda^1_{1/2}\rangle$$

$$\rightarrow 0.105\Sigma^*\pi + 0.342N\bar{K} + 0.057\ K + 0.503\Lambda\pi\pi$$

$$|\Sigma(1675)\rangle = 0.23|{}^4\Sigma^8_{1/2}\rangle + 0.71|{}^2\Sigma^8_{1/2}\rangle + 0.66|{}^2\Sigma^{10}_{1/2}\rangle$$

$$\rightarrow 0.94\Sigma\pi + 0.04N\bar{K} + 0.01\Lambda\pi + 0.01\Sigma^*\pi$$

$$|\Xi(\ ?\ )\rangle = |{}^4\Xi^8_{1/2}\rangle \rightarrow 0.62\Xi\pi + 0.16\Lambda\bar{K} + 0.14\Sigma\bar{K} + 0.05\Xi^*\pi + 0.03\Sigma^*\bar{K}$$

$$|N(1515)\rangle = 0.61|{}^4N_{1/2}\rangle + 0.79|{}^2N_{1/2}\rangle \rightarrow 0.37N\pi + 0.022\Delta\pi$$

$$+ 0.238N\eta + 0.37N\pi\pi$$

$$|\Lambda(1670)\rangle = -0.92|{}^4\Lambda^8_{1/2}\rangle + 0.39|{}^2\Lambda^8_{1/2}\rangle \to 0.76\Sigma\pi + 0.04\Sigma^*\pi + 0.04\Lambda\eta + 0.15N\bar{K} + 0.01\Lambda\pi\pi$$

$$|\Sigma(1770)\rangle = -0.62|{}^4\Sigma^8_{1/2}\rangle + 0.63|{}^2\Sigma^8_{1/2}\rangle - 0.47|{}^2\Sigma^{10}_{1/2}\rangle \to 0.07\Sigma\pi + 0.31\Lambda\pi + 0.03\Sigma\eta + 0.07N\bar{K} + 0.04\Delta\bar{K} + 0.48\Sigma\pi\pi$$

$$|\Xi(\ ?\ )\rangle = |{}^2\Xi^8_{1/2}\rangle \to 0.033\Xi\pi + 0.128\Lambda\bar{K} + 0.447\Sigma\bar{K} + 0.176\Xi\eta + 0.153\Xi^*\pi + 0.063\Sigma^*\bar{K}$$

$$|\Delta(1675)\rangle = |{}^2\Delta_{3/2}\rangle \to 0.142N\pi + 0.371\Delta\pi + 0.487\Delta\pi\pi$$

$$|\Sigma(1940)\rangle = 0.91|{}^4\Sigma^8_{1/2}\rangle + 0.21|{}^2\Sigma^8_{1/2}\rangle + 0.37|{}^2\Sigma^{10}_{1/2}\rangle \to 0.036\Sigma\pi + 0.02\Lambda\pi + 0.243\Sigma^*\pi + 0.129N\bar{K} + 0.015\Sigma\eta + 0.403\Delta\bar{K} + 0.154\Sigma\pi\pi$$

$$|\Xi(\ ?\ )\rangle = |{}^2\Xi^{10}_{3/2}\rangle \to 0.12\Xi\pi + 0.13\Lambda\bar{K} + 0.84\Sigma\bar{K} + 0.0009\Xi\eta + 0.195\Sigma^*\pi + 0.462\Sigma^*\bar{K}$$

$$|\Delta(1615)\rangle = |{}^2\Delta_{1/2}\rangle \to 0.264N\pi + 0.705\Delta\pi + 0.031\Delta\pi\pi$$

$$|\Sigma(1930)\rangle = -0.75|{}^4\Sigma^8_{1/2}\rangle - 0.30|{}^2\Sigma^8_{1/2}\rangle + 0.59|{}^2\Sigma^{10}_{1/2}\rangle \to 0.142\Sigma\pi + 0.015\Lambda\pi + 0.281\Sigma^*\pi + 0.389N\bar{K} + 0.077 + 0.096\Delta\bar{K}$$

$$|\Xi(\ ?\ )\rangle = |{}^2\Xi^{10}_{1/2}\rangle \to 0.063\Xi\pi + 0.063\Lambda\bar{K} + 0.063\Sigma\bar{K} + 0.031\Xi\eta + 0.312\Xi^*\pi + 0.468\Sigma^*\bar{K}$$

$$|\Lambda(1520)\rangle = 0.03|{}^4\Lambda^8_{3/2}\rangle + 0.37|{}^2\Lambda^8_{3/2}\rangle + 0.93|{}^2\Lambda^1_{3/2}\rangle \to 0.43\Sigma\pi + 0.03\Sigma^*\pi + 0.48N\bar{K} + 0.06\Lambda\pi\pi$$

$$|\Lambda(1402)\rangle = 0.01|{}^4\Lambda^8_{1/2}\rangle + 0.03|{}^2\Lambda^8_{1/2}\rangle + 0.99|{}^2\Lambda^1_{1/2}\rangle \rightarrow \Sigma\pi$$

We did not consider decays of the P-wave Ω^- states, which have not been observed experimentally.

References

1. Particle data group. Rev. Mod. Phys. 56, No. 2, part 2 (1984).

2. V.M. Shekhter, L.M. Scheglova, Yad. Fiz. 27, 1070 (1978); Sov. J. Nucl. Phys. 27, 567 (1978).

3. P.L. Litchfield, R.J. Cashmore, A.J.G. Hey, $SU(6)_W$ and Decays of Baryon Resonances - an update, in Proc. of the Topical Confer. on Baryon Resonances, Oxford, 1976, p. 477.

Appendix D

THE HADRON CONTENT OF MESON AND BARYON ENSEMBLES

Here we present the coefficients $\mu_{h(L)}$ and $\beta_{h(L)}$ of the decompositions

$$M_i(L) = \sum_{h_M(L)} \mu_{h_M(L)}(i)h_M(L) \quad ,$$

$$M(L) = \sum_{h_M(L)} \mu_{h_M(L)}h_M(L) \quad ,$$

$$B_{ij}(L) = \sum_{h_B(L)} \beta_{h_B(L)}(ij)h_B(L) \quad ,$$

$$B_i(L) = \sum_{h_B(L)} \beta_{h_B(L)}(i)h_B(L) \quad ,$$

$$B(L) = \sum_{h_B(L)} \beta_{h_B(L)}h_B(L) \quad ,$$

(see also Eqs. (7.31)-(7.33) and (7.39)) which are calculated with the help of the rules of quark statistics. The calculations of the leakage and of the sea meson content in hadron-hadron collisions are also considered in detail and all explicit expressions are given (Tables D.1 to D.4).

As it was explained in Sec. 7.4, sea mesons carry a fraction of the quantum numbers of the initial quarks due to the leakage effect. Let us show, how it is possible to calculate the content of the sea meson ensemble M explicitly. We shall start with the case of a baryon beam.

In order to carry oat this calculation it is convenient to present M in the form

$$M = \sum_q C_q M_q \tag{D.1}$$

where $q = u,d,s,\bar{u},\bar{d},\bar{s}$. In (D.1) every $q\bar{q}$ state is counted twice: once in the M_q ensemble, the second time in the $M_{\bar{q}}$ ensemble. The coefficients C_i in (D.1) depend on the beam quark content, so for the baryon B_{ijk} beam we write $C_q = C_q(ijk)$. The obvious normalization condition is

$$\sum_q C_q(ijk) = 1 \quad . \tag{D.2}$$

Moreover, the symmetry between the M_q and $M_{\bar{q}}$ ensembles assumes that

$$\sum_{q=u,d,s} C_q(ijk) = \sum_{q=\bar{u},\bar{d},\bar{s}} C_q(ijk) \quad . \tag{D.3}$$

When the projectile particle is a baryon, only quark (not antiquark) quantum numbers "leak" into the sea, so it is natural to require that for $q = \bar{u},\bar{d},\bar{s}$ the coefficients C_q in (D.1) should be the same as in the absence of any leakage

$$\begin{aligned} C_{\bar{u}}(ijk) &= C_{\bar{d}}(ijk) \\ &= \frac{1}{\lambda} C_{\bar{s}}(ijk) \quad . \end{aligned} \tag{D.4}$$

Similarly for the antibaryon beam (D.4) holds for C_u, C_d, C_s. Two more equations for C_q express the conservation of average quantum numbers (charge and strangeness) in the process of nondiffractive inelastic collisions

$$\frac{\Delta Q(ijk)}{d \cdot N(s)} = \frac{1+\lambda}{2+\lambda}\,[C_u(ijk) - C_{\bar{u}}(ijk)]$$

$$- \frac{1}{2+\lambda}\,[C_d(ijk) - C_{\bar{d}}(ijk) + C_s(ijk) - C_{\bar{s}}(ijk)] \quad ,$$

$$\frac{\Delta S(ijk)}{d \cdot N(s)} = \frac{\lambda}{2+\lambda}\,[C_u(ijk) - C_{\bar{u}}(ijk) + C_d(ijk) - C_{\bar{d}}(ijk)]$$

$$- \frac{2}{2+\lambda}\,[C_s(ijk) - C_{\bar{s}}(ijk)] \quad . \qquad \text{(D.5)}$$

Here $\Delta Q(ijk)$ and $\Delta S(ijk)$ are the loss of charge and strangeness in the process of hadronization of the spectator and tagged quarks:

$$\Delta Q(ijk) =. Q(B_{ijk}) - \tilde{\xi}_1 Q(B_{ij}) - \tilde{\xi}_2(Q(B_i) + Q(B_j)) - \tilde{\xi}_3(Q(M_i) + Q(M_j))$$

$$- \tilde{\xi}_4 Q(B_k) - \tilde{\xi}_5 Q(M_k)$$

$$\Delta S(ijk) = S(B_{ijk}) - \tilde{\xi}_1 S(B_{ij}) - \tilde{\xi}_2 S(B_j) - \tilde{\xi}_3(S(M_i) + S(M_j))$$

$$- \tilde{\xi}_4 S(B_k) - \tilde{\xi}_5 S(M_k) \quad . \qquad \text{(D.6)}$$

In (D.6) $Q(B_{ijk})$ and $S(B_{ijk})$ are the charge and strangeness of the beam particle, $Q(B_{ij})$, $Q(B_i)$, $Q(M_i)$, etc. — average values of the charge and strangeness of the corresponding hadron ensembles; they are presented in Table D.5. Since the hadron contents of (B_i, B_j, B_k) and (M_i, M_j, M_k) are the same, according to Eqs. (7.2), (7.63)-(7.66) we can write for the case of a proton beam

$$\Delta Q(p) = 1 - \xi_1 \left[\frac{1}{3} Q(B_{uu}) + \frac{1}{9} Q(B_{ud_1}) + \frac{5}{9} Q(B_{ud_{op}})\right] \qquad \text{(D.7)}$$

$$- (1 - \xi_1)\left[\frac{1}{3} Q(b_d) + \frac{2}{3} Q(B_u)\right] - (2 - \xi_1)\left[\frac{1}{3} Q(M_d) + \frac{2}{3} Q(M_u)\right]$$

and a similar expression for $\Delta S(p)$.

Equations (D.2)-(D.5) for $C_q(ijk)$ may be easily solved; the obtained solution is displayed in Table D.6.

A similar procedure of calculation of the coefficients $C_q(ij)$ is valid for the case of meson beam. The only difference is that it is necessary to consider the leakage of quark and antiquark quantum numbers independently, so here we shall write instead of two Eqs. (D.4), two more equations like (D.5): four equations of the type (D.5) will express the conservation of the quantum numbers of the initial quark $Q(q_i)$ and $S(q_i)$ and antiquark $Q(\bar{q}_{\bar{j}})$ and $S(\bar{q}_{\bar{j}})$. The solutions for $C_q(i\bar{j})$ are also presented in the Table D.6.

When inclusive spectra of secondary sea mesons are calculated, one should take instead of one universal function $\Phi(x)$ normalized by the condition (7.18) six functions $\Phi_q(x)$, normalized by the conditions

$$\int_0^1 \Phi_q(x)dx = C_q N(s) \quad . \tag{D.8}$$

Thus the normalization factors for different sea meson distributions slightly differ by energy-independent items. This means that these distributions are of the same behaviour at $x \to 0$ (which in fact determines $N(s)$) and of slightly different behaviour at $x \gtrsim 0.05 - 0.1$. As an illustration, in Fig. D.1 we present prompt sea distributions of π^+ and π^- mesons in pp collisions. In the absence of leakage these two spectra would coincide; leakage makes them different by approximately 10 - 15%.

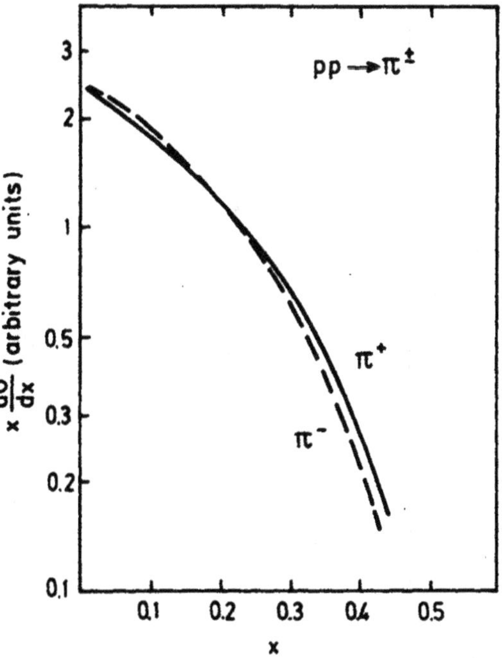

Fig. D.1 Prompt sea pion spectra in pp collisions. π^+ and π^- spectra are slightly different at $x \gtrsim 0.05$ due to the leakage effect.

Table D.1 Coefficients $\mu_{h(L)}$ (i) and $\mu_{h(L)}$ of the decomposition of ensembles M_i and M in terms of S-wave and P-wave meson multiplets. Here $n_f^J = (2J+1) \cdot (1+\lambda_f/2)^{-1}$, $n^J = (2J+1)(1+\lambda/2)^{-2}$.

$^hM(0)$		L = 0			
J = 0	J = 1	$\mu_h(u)$	$\mu_h(d)$	$\mu_h(s)$	μ_h
π^+	ρ^+	$\frac{1}{8} n_f^J$			$\frac{1}{16} n^J$
π^0	ρ^0	$\frac{1}{16} n_f^J$	$\frac{1}{16} n_f^J$		$\frac{1}{16} n^J$
π^-	ρ^-		$\frac{1}{8} n_f^J$		$\frac{1}{16} n^J$
K^+	K^{*+}	$\frac{\lambda_f}{8} n_f^J$			$\frac{\lambda}{16} n^J$
K^0	K^{*0}		$\frac{\lambda_f}{8} n_f^J$		$\frac{\lambda}{16} n^J$
$\bar{K}^0$	$\bar{K}^{*0}$			$\frac{1}{8} n_f^J$	$\frac{\lambda}{16} n^J$
K^-	K^{*-}			$\frac{1}{8} n_f^J$	$\frac{\lambda}{16} n^J$
	ω	$\frac{1}{16} n_f^J$	$\frac{1}{16} n_f^J$		$\frac{1}{16} n^J$
	ϕ			$\frac{\lambda_f}{8} n_f^J$	$\frac{\lambda^2}{16} n^J$
η^8		$\frac{1}{48} n_f^J$	$\frac{1}{48} n_f^J$	$\frac{\lambda_f}{12} n_f^J$	$\frac{1+2\lambda^2}{48} n^J$
η^1		$\frac{1}{24} n_f^J$	$\frac{1}{24} n_f^J$	$\frac{\lambda_f}{24} n_f^J$	$\frac{2+\lambda^2}{48} n^J$

Table D.1 Coefficients $\mu_{h(L)}$ (i) and $\mu_{h(L)}$ of the decomposition of ensembles M_i and M in terms of S-wave and P-wave meson multiplets. Here $n_f^J = (2J+1) \cdot (1+\lambda_f/2)^{-1}$, $n^J = (2J+1)(1+\lambda/2)^{-2}$.

$h_{M(1)}$				L = 1			
J = 0	J = 1	J = 1	J = 2	$\mu_h(u)$	$\mu_h(d)$	$\mu_h(s)$	μ_h
δ^+	A_1^+	B^+	A_2^+	$\frac{1}{24} n_f^J$			$\frac{1}{48} n^J$
δ^0	A_1^0	B^0	A_2^0	$\frac{1}{48} n_f^J$	$\frac{1}{48} n_f^J$		$\frac{1}{48} n^J$
δ^-	A_1^-	B^-	A_2^-		$\frac{1}{24} n_f^J$		$\frac{1}{48} n^J$
κ^+	Q_A^+	Q_B^+	K^{**+}	$\frac{\lambda_f}{24} n_f^J$			$\frac{\lambda}{48} n^J$
κ^0	Q_A^0	Q_B^0	K^{**0}		$\frac{\lambda_f}{24} n_f^J$		$\frac{\lambda}{48} n^J$
$\bar{\kappa}^0$	$\bar{Q}_A^0$	$\bar{Q}_B^0$	$\bar{K}^{**0}$			$\frac{1}{24} n_f^J$	$\frac{\lambda}{48} n^J$
κ^-	Q_A^-	Q_B^-	K^{**-}			$\frac{1}{24} n_f^J$	$\frac{\lambda}{48} n^J$
ϵ	D	H	f	$\frac{1}{48} n_f^J$	$\frac{1}{48} n_f^J$		$\frac{1}{48} n^J$
S^*	E	?	f'			$\frac{\lambda_f}{24} n_f^J$	$\frac{\lambda^2}{48} n^J$

Table D.2 Coefficients $\beta_{h(o)}(ij)$, $\beta_{h(o)}(i)$ and $\beta_{h(o)}$ of the decomposition of ensembles B_{ij}, B_i and B in terms of the 56-plet baryons. Here $\{ud_1\} = \{u^\uparrow d^\uparrow\}$, $\{us_o\} = \{u^\uparrow s^\downarrow\}$, etc. $\{ud_p\} = 1/\sqrt{5}\,(2\{u^\uparrow d^\downarrow\} - \{u^\downarrow d^\uparrow\})$, $\{ud_\Lambda\} = 1/\sqrt{2}(\{u^\uparrow d^\downarrow\} - \{u^\downarrow d^\uparrow\})$, $\{us_\Sigma\} = 1/\sqrt{5}(2\{u^\uparrow s^\downarrow\} - \{u^\downarrow s^\uparrow\})$. $n_{qq} = (1+\lambda_f/3)^{-1}$, $n_{qs} = (1+3\lambda_f/5)^{-1}$, $n_{ss} = (1+\lambda_f)^{-1}$, $n_q = (1+2\lambda_f/3+\lambda_f^2/5)^{-1}$, $n_s = (1+6\lambda_f/5+3\lambda_f^2/5)^{-1}$, $n_o = (1+\lambda+3\lambda^2/5+\lambda^3/5)^{-1}$.

		B_{ij}								B_i			B
	h_B	$\beta_h(uu)$	$\beta_h(ud_1)$	$\beta_h(ud_p)$	$\beta_h(ud_\Lambda)$	$\beta_h(us_1)$	$\beta_h(us_o)$	$\beta_h(us_\Sigma)$	$\beta_h(ss)$	$\beta_h(u)$	$\beta_h(d)$	$\beta_h(s)$	β_h
{8}	P	$\frac{1}{9}n_{qq}$	$\frac{1}{18}n_{qq}$	$\frac{41}{90}n_{qq}$	$\frac{1}{2}n_{qq}$					$\frac{2}{15}n_q$	$\frac{1}{15}n_q$		$\frac{1}{10}n_o$
	n		$\frac{1}{18}n_{qq}$	$\frac{41}{90}n_{qq}$	$\frac{1}{2}n_{qq}$					$\frac{1}{15}n_q$	$\frac{2}{15}n_q$		$\frac{1}{10}n_o$
	Λ			$\frac{3\lambda_f}{10}n_{qq}$	$\frac{\lambda_f}{3}n_{qq}$	$\frac{1}{10}n_{qs}$	$\frac{1}{10}n_{qs}$	$\frac{1}{10}n_{qs}$		$\frac{\lambda_f}{15}n_q$	$\frac{\lambda_f}{15}n_q$	$\frac{1}{10}n_s$	$\frac{\lambda}{10}n_o$
	Σ^+	$\frac{\lambda_f}{9}n_{qq}$				$\frac{1}{15}n_{qs}$	$\frac{1}{3}n_{qs}$	$\frac{41}{75}n_{qs}$		$\frac{2\lambda_f}{15}n_q$		$\frac{1}{10}n_s$	$\frac{\lambda}{10}n_o$
	Σ^0		$\frac{\lambda_f}{9}n_{qq}$	$\frac{\lambda_f}{90}n_{qq}$		$\frac{1}{30}n_{qs}$	$\frac{1}{6}n_{qs}$	$\frac{41}{150}n_{qs}$		$\frac{\lambda_f}{15}n_q$	$\frac{\lambda_f}{15}n_q$	$\frac{1}{10}n_s$	$\frac{\lambda}{10}n_o$
	Σ^-										$\frac{2\lambda_f}{15}n_q$	$\frac{1}{10}n_s$	$\frac{\lambda}{10}n_o$
	Ξ^0					$\frac{\lambda_f}{15}n_{qs}$	$\frac{\lambda_f}{3}n_{qs}$	$\frac{41\lambda_f}{75}n_{qs}$	$\frac{1}{6}n_{ss}$	$\frac{\lambda_f^2}{15}n_q$		$\frac{\lambda_f}{5}n_s$	$\frac{\lambda^2}{10}n_o$
	Ξ^-								$\frac{1}{6}n_{ss}$		$\frac{\lambda_f^2}{15}n_q$	$\frac{\lambda_f}{15}n_s$	$\frac{\lambda^2}{10}n_o$
{1}	Ω^-								$\lambda_f n_{ss}$			$\frac{3\lambda_f^2}{5}n_s$	$\frac{\lambda^3}{5}n_o$

Table D.2 Coefficients $\beta_{h(o)}$ (ij), $\beta_{h(o)}$ (i) and $\beta_{h(o)}$ of the decomposition of ensembles B_{ij}, B_i and B in terms of the 56-plet baryons. Here $\{ud_1\} = \{u^{\uparrow}d^{\uparrow}\} \cdot \{us_0\} = \{u^{\uparrow}s^{\downarrow}\}$, etc. $\{ud_p\} = 1/\sqrt{5}\ (2\{u^{\uparrow}d^{\downarrow}\} - \{u^{\downarrow}d^{\uparrow}\})$, $\{ud_{\Lambda}\} = 1/\sqrt{2}(\{u^{\uparrow}d^{\downarrow}\} - \{u^{\downarrow}d^{\uparrow}\})$, $\{us_{\Sigma}\} = 1/\sqrt{5}(2\{u^{\uparrow}s^{\downarrow}\} - \{u^{\downarrow}s^{\uparrow}\})$. $n_{qq} = (1 + \lambda_f/3)^{-1}$, $n_{qs} = (1 + 3\lambda_f/5)^{-1}$, $n_{ss} = (1 + \lambda_f)^{-1}$, $n_q = (1 + 2\lambda_f/3 + \lambda_f^2/5)^{-1}$, $n_s = (1 + 6\lambda_f/5 + 3\lambda_f^2/5)^{-1}$, $n_o = (1 + \lambda + 3\lambda^2/5 + \lambda^3/5)^{-1}$.

	h_B	B_{ij}								B_i			B
		$\beta_h(uu)$	$\beta_h(ud_1)$	$\beta_h(ud_p)$	$\beta_h(ud_\Lambda)$	$\beta_h(us_1)$	$\beta_h(us_0)$	$\beta_h(us_\Sigma)$	$\beta_h(ss)$	$\beta_h(u)$	$\beta_h(d)$	$\beta_h(s)$	β_h
	Δ^{++}	$\frac{2}{3}n_{qq}$								$\frac{2}{5}n_q$			$\frac{1}{5}n_o$
	Δ^{+}	$\frac{2}{9}n_{qq}$	$\frac{4}{9}n_{qq}$	$\frac{2}{45}n_{qq}$						$\frac{4}{15}n_q$	$\frac{2}{15}n_q$		$\frac{1}{5}n_o$
	Δ^{0}		$\frac{4}{9}n_{qq}$	$\frac{2}{45}n_{qq}$						$\frac{2}{15}n_q$	$\frac{4}{15}n_q$		$\frac{1}{5}n_o$
	Δ^{-}										$\frac{2}{5}n_q$		$\frac{1}{5}n_o$
{10}	Σ^{*+}	$\frac{2\lambda_f}{9}n_{qq}$				$\frac{8}{15}n_{qs}$	$\frac{4}{15}n_{qs}$	$\frac{4}{75}n_{qs}$		$\frac{4\lambda_f}{15}n_q$		$\frac{1}{5}n_s$	$\frac{\lambda}{5}n_o$
	Σ^{*0}		$\frac{2\lambda_f}{9}n_{qq}$	$\frac{\lambda_f}{45}n_{qq}$		$\frac{4}{15}n_{qs}$	$\frac{2}{15}n_{qs}$	$\frac{2}{75}n_{qs}$		$\frac{2\lambda_f}{15}n_q$	$\frac{2\lambda_f}{15}n_q$	$\frac{1}{5}n_s$	$\frac{\lambda}{5}n_o$
	Σ^{*-}										$\frac{4\lambda_f}{15}n_q$	$\frac{1}{5}n_s$	$\frac{\lambda}{5}n_o$
	Ξ^{*0}					$\frac{8\lambda_f}{15}n_{qs}$	$\frac{4\lambda_f}{15}n_{qs}$	$\frac{4\lambda_f}{15}n_{qs}$	$\frac{1}{3}n_{ss}$	$\frac{2\lambda_f^2}{15}n_q$		$\frac{2\lambda_f}{5}n_s$	$\frac{\lambda^2}{5}n_o$
	Ξ^{*-}								$\frac{1}{3}n_{ss}$		$\frac{2\lambda_f^2}{15}n_q$	$\frac{2\lambda_f}{5}n_s$	$\frac{\lambda^2}{5}n_o$

Table D.3 Coefficients of decomposition of ensembles B_{ij}, B_i and B in terms of the baryons of the 70-plet. Here $N_1 = 1 + 2\lambda_f/3$.

h_B		B(uu)					$B\left(\frac{u^\uparrow d^\uparrow + d^\uparrow u^\uparrow}{\sqrt{2}}\right)$				
		${}^4h_{5/2}$	${}^4h_{3/2}$	${}^4h_{1/2}$	${}^2h_{3/2}$	${}^2h_{1/2}$	${}^4h_{5/2}$	${}^4h_{3/2}$	${}^4h_{1/2}$	${}^2h_{3/2}$	${}^2h_{1/2}$
{8}	N^+	$\frac{2}{9N_1}$	$\frac{4}{27N_1}$	$\frac{2}{27N_1}$	$\frac{2}{27N_1}$	$\frac{1}{27N_1}$	$\frac{1}{9N_1}$	$\frac{2}{27N_1}$	$\frac{1}{27N_1}$	$\frac{1}{27N_1}$	$\frac{1}{54N_1}$
	N^0						$\frac{1}{9N_1}$	$\frac{2}{27N_1}$	$\frac{1}{27N_1}$	$\frac{1}{27N_1}$	$\frac{1}{54N_1}$
	Λ									$\frac{\lambda_f}{5N_1}$	$\frac{\lambda_f}{10N_1}$
	Σ^+	$\frac{2\lambda_f}{9N_1}$	$\frac{4\lambda_f}{27N_1}$	$\frac{2\lambda_f}{27N_1}$	$\frac{2\lambda_f}{27N_1}$	$\frac{\lambda_f}{27N_1}$					
	Σ^0						$\frac{2\lambda_f}{9N_1}$	$\frac{4\lambda_f}{27N_1}$	$\frac{2\lambda_f}{27N_1}$	$\frac{2\lambda_f}{27N_1}$	$\frac{\lambda_f}{27N_1}$
	Σ^-										
	Ξ^0										
	Ξ^-										
{1}	Λ										

Table D.3 Coefficients of decomposition of ensembles B_{ij}, B_i and B in terms of the baryons of the 70-plet. Here $N_1 = 1 + 2\lambda_f/3$.

h_B		B(uu)					$B\left(\frac{u^\uparrow d^\uparrow + d^\uparrow u^\uparrow}{\sqrt{2}}\right)$				
		$^4h_{5/2}$	$^4h_{3/2}$	$^4h_{1/2}$	$^2h_{3/2}$	$^2h_{1/2}$	$^4h_{5/2}$	$^4h_{3/2}$	$^4h_{1/2}$	$^2h_{3/2}$	$^2h_{1/2}$
{10}	Δ^{++}				$\frac{2}{9N_1}$	$\frac{1}{9N_1}$					
	Δ^+				$\frac{2}{27N_1}$	$\frac{1}{27N_1}$				$\frac{4}{27N_1}$	$\frac{2}{27N_1}$
	Δ^0									$\frac{4}{27N_1}$	$\frac{2}{27N_1}$
	Δ^-										
	Σ^+				$\frac{2\lambda_f}{27N_1}$	$\frac{\lambda_f}{27N_1}$					
	Σ^0									$\frac{2\lambda_f}{27N_1}$	$\frac{\lambda_f}{27N_1}$
	Σ^-										
	Ξ^0										
	Ξ^-										
	Ω										

Table D.3 (Cont'd) — Here $N_1 = 1 + 2\lambda_f/3$.

h_B		$B\left(\frac{2\{u^{\uparrow}d^{\downarrow}\} - \{u^{\downarrow}d^{\uparrow}\}}{\sqrt{5}}\right)$					$B\left(\frac{\{u^{\uparrow}d^{\downarrow}\} - \{u^{\downarrow}d^{\uparrow}\}}{\sqrt{2}}\right)$				
		$^4h_{5/2}$	$^4h_{3/2}$	$^4h_{1/2}$	$^2h_{3/2}$	$^2h_{1/2}$	$^4h_{5/2}$	$^4h_{3/2}$	$^4h_{1/2}$	$^2h_{3/2}$	$^2h_{1/2}$
{8}	N^+	$\frac{1}{90N_1}$	$\frac{1}{135N_1}$	$\frac{1}{270N_1}$	$\frac{41}{135N_1}$	$\frac{41}{270N_1}$				$\frac{1}{3N_1}$	$\frac{1}{6N_1}$
	N^0	$\frac{1}{90N_1}$	$\frac{1}{135N_1}$	$\frac{1}{270N_1}$	$\frac{41}{135N_1}$	$\frac{41}{270N_1}$				$\frac{1}{3N_1}$	$\frac{1}{6N_1}$
	Λ				$\frac{\lambda_f}{5N_1}$	$\frac{\lambda_f}{10N_1}$				$\frac{2\lambda_f}{9N_1}$	$\frac{\lambda_f}{9N_1}$
	Σ^+										
	Σ^0	$\frac{\lambda_f}{45N_1}$	$\frac{2\lambda_f}{135N_1}$	$\frac{\lambda_f}{135N_1}$	$\frac{\lambda_f}{135N_1}$	$\frac{\lambda_f}{270N_1}$					
	Σ^-										
	Ξ^0										
	Ξ^-										
{1}	Λ				$\frac{\lambda_f}{5N_1}$	$\frac{\lambda_f}{10N_1}$				$\frac{2\lambda_f}{9N_1}$	$\frac{\lambda_f}{9N_1}$

Table D.3 (Cont'd) — Here $N_1 = 1 + 2\lambda_f/3$.

h_B		$B\left(\frac{2\{u^\uparrow d^\downarrow\} - \{u^\downarrow d^\uparrow\}}{\sqrt{5}}\right)$					$B\left(\frac{\{u^\uparrow d^\downarrow\} - \{u^\downarrow d^\uparrow\}}{\sqrt{2}}\right)$				
		$^4h_{5/2}$	$^4h_{3/2}$	$^4h_{1/2}$	$^2h_{3/2}$	$^2h_{1/2}$	$^4h_{5/2}$	$^4h_{3/2}$	$^4h_{1/2}$	$^2h_{3/2}$	$^2h_{1/2}$
{10}	Δ^{++}										
	Δ^+				$\frac{2}{135N_1}$	$\frac{1}{135N_1}$					
	Δ^0				$\frac{2}{135N_1}$	$\frac{1}{135N_1}$					
	Δ^-										
	Σ^+										
	Σ^0				$\frac{\lambda_f}{135N_1}$	$\frac{\lambda_f}{270N_1}$					
	Σ^-										
	Ξ^0										
	Ξ^-										
	Ω										

Table D.3 (Cont'd) — Here $N_2 = 1 + 3\lambda_f/7$.

h_B		$B\left(\frac{u^\uparrow s^\uparrow + s^\uparrow u^\uparrow}{\sqrt{2}}\right)$					$B\left(\frac{u^\downarrow s^\uparrow + s^\uparrow u^\downarrow}{\sqrt{2}}\right)$				
		${}^4h_{5/2}$	${}^4h_{3/2}$	${}^4h_{1/2}$	${}^2h_{3/2}$	${}^2h_{1/2}$	${}^4h_{5/2}$	${}^4h_{3/2}$	${}^4h_{1/2}$	${}^2h_{3/2}$	${}^2h_{1/2}$
{8}	N^+										
	N^0										
	Λ	$\frac{1}{7N_2}$	$\frac{2}{21N_2}$	$\frac{1}{21N_2}$	$\frac{1}{21N_2}$	$\frac{1}{42N_2}$	$\frac{1}{14N_2}$	$\frac{1}{21N_2}$	$\frac{1}{42N_2}$	$\frac{1}{21N_2}$	$\frac{1}{42N_2}$
	Σ^+	$\frac{2}{21N_2}$	$\frac{4}{63N_2}$	$\frac{2}{63N_2}$	$\frac{2}{63N_2}$	$\frac{1}{63N_2}$	$\frac{1}{21N_2}$	$\frac{2}{63N_2}$	$\frac{1}{63N_2}$	$\frac{10}{63N_2}$	$\frac{5}{63N_2}$
	Σ^0	$\frac{1}{21N_2}$	$\frac{2}{63N_2}$	$\frac{1}{63N_2}$	$\frac{1}{63N_2}$	$\frac{1}{126N_2}$	$\frac{1}{42N_2}$	$\frac{1}{63N_2}$	$\frac{1}{126N_2}$	$\frac{5}{63N_2}$	$\frac{5}{126N_2}$
	Σ^-										
	Ξ^0	$\frac{2\lambda_f}{21N_2}$	$\frac{4\lambda_f}{63N_2}$	$\frac{2\lambda_f}{63N_2}$	$\frac{2\lambda_f}{63N_2}$	$\frac{\lambda_f}{63N_2}$	$\frac{\lambda_f}{21N_2}$	$\frac{2\lambda_f}{63N_2}$	$\frac{\lambda_f}{63N_2}$	$\frac{10\lambda_f}{63N_2}$	$\frac{5\lambda_f}{63N_2}$
	Ξ^-										
{1}	Λ									$\frac{2}{21N_2}$	$\frac{1}{21N_2}$

Table D.3 (Cont'd) — Here $N_2 = 1 + 3\lambda_f/7$.

h_B		$B\left(\frac{u^{\uparrow}s^{\uparrow}+s^{\uparrow}u^{\uparrow}}{\sqrt{2}}\right)$					$B\left(\frac{u^{\downarrow}s^{\uparrow}+s^{\uparrow}u^{\downarrow}}{\sqrt{2}}\right)$				
		${}^4h_{5/2}$	${}^4h_{3/2}$	${}^4h_{1/2}$	${}^2h_{3/2}$	${}^2h_{1/2}$	${}^4h_{5/2}$	${}^4h_{3/2}$	${}^4h_{1/2}$	${}^2h_{3/2}$	${}^2h_{1/2}$
{10}	Δ^{++}										
	Δ^{+}										
	Δ^{0}										
	Δ^{-}										
	Σ^{+}				$\frac{8}{63N_2}$	$\frac{4}{63N_2}$				$\frac{4}{63N_2}$	$\frac{2}{63N_2}$
	Σ^{0}				$\frac{4}{63N_2}$	$\frac{2}{63N_2}$				$\frac{2}{63N_2}$	$\frac{1}{63N_2}$
	Σ^{-}										
	Ξ^{0}				$\frac{8\lambda_f}{63N_2}$	$\frac{4\lambda_f}{63N_2}$				$\frac{4\lambda_f}{63N_2}$	$\frac{2\lambda_f}{63N_2}$
	Ξ^{-}										
	Ω										

Table D.3 (Cont'd) — Here $N_2 = 1 + 3\lambda_f/7$, $N = 30 + 32\lambda_f + 8\lambda_f^2$.

h_B		$B\left(\frac{2\{u^\uparrow s^\downarrow\} - \{u^\downarrow s^\uparrow\}}{\sqrt{5}}\right)$					$B(u)$				
		$^4h_{5/2}$	$^4h_{3/2}$	$^4h_{1/2}$	$^2h_{3/2}$	$^2h_{1/2}$	$^4h_{5/2}$	$^4h_{3/2}$	$^4h_{1/2}$	$^2h_{3/2}$	$^2h_{1/2}$
{8}	N^+						$\frac{4}{N}$	$\frac{8}{3N}$	$\frac{4}{3N}$	$\frac{8}{3N}$	$\frac{4}{3N}$
	N^0						$\frac{2}{N}$	$\frac{4}{3N}$	$\frac{2}{3N}$	$\frac{4}{3N}$	$\frac{2}{3N}$
	Λ	$\frac{1}{70N_2}$	$\frac{1}{105N_2}$	$\frac{1}{210N_2}$	$\frac{1}{21N_2}$	$\frac{1}{42N_2}$	$\frac{2\lambda_f}{N}$	$\frac{4\lambda_f}{3N}$	$\frac{2\lambda_f}{3N}$	$\frac{4\lambda_f}{3N}$	$\frac{2\lambda_f}{3N}$
	Σ^+	$\frac{1}{105N_2}$	$\frac{2}{315N_2}$	$\frac{1}{315N_2}$	$\frac{82}{315N_2}$	$\frac{41}{315N_2}$	$\frac{4\lambda_f}{N}$	$\frac{8\lambda_f}{3N}$	$\frac{4\lambda_f}{3N}$	$\frac{8\lambda_f}{3N}$	$\frac{4\lambda_f}{3N}$
	Σ^0	$\frac{1}{210N_2}$	$\frac{1}{315N_2}$	$\frac{1}{630N_2}$	$\frac{41}{315N_2}$	$\frac{41}{630N_2}$	$\frac{2\lambda_f}{N}$	$\frac{4\lambda_f}{3N}$	$\frac{2\lambda_f}{3N}$	$\frac{4\lambda_f}{3N}$	$\frac{2\lambda_f}{3N}$
	Σ^-										
	Ξ^0	$\frac{\lambda_f}{105N_2}$	$\frac{2\lambda_f}{315N_2}$	$\frac{\lambda_f}{315N_2}$	$\frac{82\lambda_f}{315N_2}$	$\frac{41\lambda_f}{315N_2}$	$\frac{2\lambda_f^2}{N}$	$\frac{4\lambda_f^2}{3N}$	$\frac{2\lambda_f^2}{3N}$	$\frac{4\lambda_f^2}{3N}$	$\frac{2\lambda_f^2}{3N}$
	Ξ^-										
{1}	Λ				$\frac{6}{35N_2}$	$\frac{3}{35N_2}$				$\frac{4\lambda_f}{3N}$	$\frac{2\lambda_f}{3N}$

Table D.3 (Cont'd) — Here $N_2 = 1 + 3\lambda_f/7$, $N = 30 + 32\lambda_f + 8\lambda_f^2$.

h_B		$B\left(\frac{2\{u^\uparrow s^\downarrow\} - \{u^\downarrow s^\uparrow\}}{\sqrt{5}}\right)$					$B(u)$				
		$^4h_{5/2}$	$^4h_{3/2}$	$^4h_{1/2}$	$^2h_{3/2}$	$^2h_{1/2}$	$^4h_{5/2}$	$^4h_{3/2}$	$^4h_{1/2}$	$^2h_{3/2}$	$^2h_{1/2}$
	Δ^{++}									$\frac{4}{N}$	$\frac{2}{N}$
	Δ^{+}									$\frac{8}{3N}$	$\frac{4}{3N}$
	Δ^{0}									$\frac{4}{3N}$	$\frac{2}{3N}$
	Δ^{-}										
	Σ^{+}				$\frac{4}{315N_2}$	$\frac{2}{315N_2}$				$\frac{8\lambda_f}{3N}$	$\frac{4\lambda_f}{3N}$
{10}	Σ^{0}				$\frac{2}{315N_2}$	$\frac{1}{315N_2}$				$\frac{4\lambda_f}{3N}$	$\frac{2\lambda_f}{3N}$
	Σ^{-}										
	Ξ^{0}				$\frac{4\lambda_f}{315N_2}$	$\frac{2\lambda_f}{315N_2}$				$\frac{4\lambda_f^2}{3N}$	$\frac{2\lambda_f^2}{3N}$
	Ξ^{-}										
	Ω										

Table D.3 (Cont'd) — Here $N_3 = 1 + \lambda_f/4$, $N_s = 32 + 32\lambda_f + 6\lambda_f^2$.

h_B		B(s)					B(ss)				
		$^4h_{5/2}$	$^4h_{3/2}$	$^4h_{1/2}$	$^2h_{3/2}$	$^2h_{1/2}$	$^4h_{5/2}$	$^4h_{3/2}$	$^4h_{1/2}$	$^2h_{3/2}$	$^2h_{1/2}$
{8}	N^+										
	N^0										
	Λ	$\frac{2}{N_s}$	$\frac{4}{3N_s}$	$\frac{2}{3N_s}$	$\frac{4}{3N_s}$	$\frac{2}{3N_s}$					
	Σ^+	$\frac{2}{N_s}$	$\frac{4}{3N_s}$	$\frac{2}{3N_s}$	$\frac{4}{3N_s}$	$\frac{2}{3N_s}$					
	Σ^0	$\frac{2}{N_s}$	$\frac{4}{3N_s}$	$\frac{2}{3N_s}$	$\frac{4}{3N_s}$	$\frac{2}{3N_s}$					
	Σ^-	$\frac{2}{N_s}$	$\frac{4}{3N_s}$	$\frac{2}{3N_s}$	$\frac{4}{3N_s}$	$\frac{2}{3N_s}$					
	Ξ^0	$\frac{4\lambda_f}{N_s}$	$\frac{8\lambda_f}{3N_s}$	$\frac{4\lambda_f}{3N_s}$	$\frac{8\lambda_f}{3N_s}$	$\frac{4\lambda_f}{3N_s}$	$\frac{1}{6N_3}$	$\frac{1}{9N_3}$	$\frac{1}{18N_3}$	$\frac{1}{18N_3}$	$\frac{1}{36N_3}$
	Ξ^-	$\frac{4\lambda_f}{N_s}$	$\frac{8\lambda_f}{3N_s}$	$\frac{4\lambda_f}{3N_s}$	$\frac{8\lambda_f}{3N_s}$	$\frac{4\lambda_f}{3N_s}$	$\frac{1}{6N_3}$	$\frac{1}{9N_3}$	$\frac{1}{18N_3}$	$\frac{1}{18N_3}$	$\frac{1}{36N_3}$
{1}	Λ				$\frac{4}{3N_s}$	$\frac{2}{3N_s}$					

Table D.3 (Cont'd) — Here $N_3 = 1 + \lambda_f/4$, $N_s = 32 + 32\lambda_f + 6\lambda_f^2$.

h_B		B(s)					B(ss)				
		$^4h_{5/2}$	$^4h_{3/2}$	$^4h_{1/2}$	$^2h_{3/2}$	$^2h_{1/2}$	$^4h_{5/2}$	$^4h_{3/2}$	$^4h_{1/2}$	$^2h_{3/2}$	$^2h_{1/2}$
{10}	Δ^{++}										
	Δ^{+}										
	Δ^{0}										
	Δ^{-}										
	Σ^{+}				$\frac{4}{3N_s}$	$\frac{2}{3N_s}$					
	Σ^{0}				$\frac{4}{3N_s}$	$\frac{2}{3N_s}$					
	Σ^{-}				$\frac{4}{3N_s}$	$\frac{2}{3N_s}$					
	Ξ^{0}				$\frac{8\lambda_f}{3N_s}$	$\frac{4\lambda_f}{3N_s}$				$\frac{1}{18N_3}$	$\frac{1}{36N_3}$
	Ξ^{-}				$\frac{8\lambda_f}{3N_s}$	$\frac{4\lambda_f}{3N_s}$				$\frac{1}{18N_3}$	$\frac{1}{36N_3}$
	Ω				$\frac{4\lambda_f^2}{N_s}$	$\frac{2\lambda_f^2}{N_s}$				$\frac{\lambda_f}{6N_3}$	$\frac{\lambda_f}{12N_3}$

Table D.3 (Cont'd) — Here $N' = (1 + 8\lambda/3 + 4\lambda^2/3 + \lambda^3/6)^{-1}$.

h_B		B				
		${}^4h_{5/2}$	${}^4h_{3/2}$	${}^4h_{1/2}$	${}^2h_{3/2}$	${}^2h_{1/2}$
{8}	N^+	$\frac{N'}{10}$	$\frac{N'}{15}$	$\frac{N'}{30}$	$\frac{N'}{15}$	$\frac{N'}{30}$
	N^0	$\frac{N'}{10}$	$\frac{N'}{15}$	$\frac{N'}{30}$	$\frac{N'}{15}$	$\frac{N'}{30}$
	Λ	$\frac{\lambda N'}{10}$	$\frac{\lambda N'}{15}$	$\frac{\lambda N'}{30}$	$\frac{\lambda N'}{15}$	$\frac{\lambda N'}{30}$
	Σ^+	$\frac{\lambda N'}{10}$	$\frac{\lambda N'}{15}$	$\frac{\lambda N'}{30}$	$\frac{\lambda N'}{15}$	$\frac{\lambda N'}{30}$
	Σ^0	$\frac{\lambda N'}{10}$	$\frac{\lambda N'}{15}$	$\frac{\lambda N'}{30}$	$\frac{\lambda N'}{15}$	$\frac{\lambda N'}{30}$
	Σ^-	$\frac{\lambda N'}{10}$	$\frac{\lambda N'}{15}$	$\frac{\lambda N'}{30}$	$\frac{\lambda N'}{15}$	$\frac{\lambda N'}{30}$
	Ξ^0	$\frac{\lambda^2 N'}{10}$	$\frac{\lambda^2 N'}{15}$	$\frac{\lambda^2 N'}{30}$	$\frac{\lambda^2 N'}{15}$	$\frac{\lambda^2 N'}{30}$
	Ξ^-	$\frac{\lambda^2 N'}{10}$	$\frac{\lambda^2 N'}{15}$	$\frac{\lambda^2 N'}{30}$	$\frac{\lambda^2 N'}{15}$	$\frac{\lambda^2 N'}{30}$
{1}	Λ				$\frac{\lambda N'}{15}$	$\frac{\lambda N'}{30}$

Table D.3 (Cont'd) — Here $N' = (1 + 8\lambda/3 + 4\lambda^2/3 + \lambda^3/6)^{-1}$.

h_B		B				
		$^4h_{5/2}$	$^4h_{3/2}$	$^4h_{1/2}$	$^2h_{3/2}$	$^2h_{1/2}$
{10}	Δ^{++}				$\frac{N'}{15}$	$\frac{N'}{30}$
	Δ^{+}				$\frac{N'}{15}$	$\frac{N'}{30}$
	Δ^{0}				$\frac{N'}{15}$	$\frac{N'}{30}$
	Δ^{-}				$\frac{N'}{15}$	$\frac{N'}{30}$
	Σ^{+}				$\frac{\lambda N'}{15}$	$\frac{\lambda N'}{30}$
	Σ^{0}				$\frac{\lambda N'}{15}$	$\frac{\lambda N'}{30}$
	Σ^{-}				$\frac{\lambda N'}{15}$	$\frac{\lambda N'}{30}$
	Ξ^{0}				$\frac{\lambda^2 N'}{15}$	$\frac{\lambda^2 N'}{30}$
	Ξ^{-}				$\frac{\lambda^2 N'}{15}$	$\frac{\lambda^2 N'}{30}$
	Ω				$\frac{\lambda^3 N'}{15}$	$\frac{\lambda^3 N'}{30}$

Table D.4 Coefficients of decomposition of ensembles M_c and B_c in terms of charmed hadrons.

$M = c\bar{q}$ $(\bar{q} = \bar{u}, \bar{d}, \bar{s})$	$\mu_{h(0)}{}^{(c)}$	$B = cqq$ $(q = u, d)$	$\beta_{h(0)}{}^{(c)}$	$B = csq$, $B = css$ $(q = u, d)$	$\beta_{h(0)}{}^{(c)}$
D^+	$\frac{n}{8}$	Σ_c^{++}	$\frac{N}{10}$	Ξ_{c1}^+	$\frac{\lambda_f N}{10}$
D^0	$\frac{n}{8}$	Σ_c^+	$\frac{N}{10}$	Ξ_{c1}^0	$\frac{\lambda_f N}{10}$
D^{*+}	$\frac{3n}{8}$	Σ_c^0	$\frac{N}{10}$	Ξ_{c2}^+	$\frac{\lambda_f N}{10}$
D^{*0}	$\frac{3n}{8}$	Λ_c^+	$\frac{N}{10}$	Ξ_{c2}^0	$\frac{\lambda_f N}{10}$
F^+	$\frac{\lambda_f n}{8}$	Σ_c^{*++}	$\frac{N}{5}$	Ξ_c^{*+}	$\frac{\lambda_f N}{5}$
F^{*+}	$\frac{3\lambda_f n}{8}$	Σ_c^{*+}	$\frac{N}{5}$	Ξ_c^{*0}	$\frac{\lambda_f N}{5}$
		Σ_c^{*0}	$\frac{N}{5}$	Ω_c	$\frac{\lambda_f^2 N}{10}$
				Ω_c^*	$\frac{\lambda_f^2 N}{5}$

Table D.5* Average values of charge $Q(H_I)$ and strangeness $S(H_I)$ of the hadron ensembles H_I. As $Q(B_{ij})$ and $S(B_{ij})$ depend only on the quark content of the initial diquark q_iq_j, not on the detailed form of the wave function, we indicate only quark content of the diquark in the subscript $B_{ij}: Q(B_{ud}) = Q(B_{ud_0}) = Q(B_{ud_1}) = Q(B_{ud})$ etc.

H_I	$Q(H_I)$
M_u	$n_0\left(\frac{1}{2}+\frac{\lambda_f}{2}\right)$
M_d	$-n_0\frac{\lambda_f}{2}$
M_s	$-n_0\frac{\lambda_f}{2}$
B_u	$\alpha_i^B(0)n_q^{56}\left(\frac{6}{5}+\frac{2\lambda_f}{5}\right)+\alpha_i^B(1)n_q^{70}\left(\frac{14}{15}+\frac{8\lambda_f}{15}\right)$
B_d	$-\alpha_i^B(0)n_q^{56}\left(\frac{1}{5}+\frac{2\lambda_f}{5}+\frac{\lambda_f^2}{5}\right)+\alpha_i^B(1)n_q^{70}\left(\frac{1}{15}-\frac{8\lambda_f}{15}-\frac{4\lambda_f^2}{15}\right)$
B_s	$-\alpha_i^B(0)n_s^{56}\left(\frac{3\lambda_f}{5}+\frac{3\lambda_f^2}{5}\right)-\alpha_i^B(1)n_s^{70}\left(\frac{1}{2}+\frac{3\lambda_f^2}{16}\right)$
B_{uu}	$\alpha_{ij}^B(0)n_{qq}^{56}\left(\frac{5}{3}+\frac{\lambda_f}{3}\right)+\alpha_{ij}^B(1)n_{qq}^{70}\left(\frac{4}{3}+\frac{2\lambda_f}{3}\right)$
B_{ud}	$\alpha_{ij}^B(0)n_{qq}^{56}\frac{1}{2}+\alpha_{ij}^B(1)n_{qq}^{70}\frac{1}{2}$
B_{us}	$\alpha_{ij}^B(0)n_{qs}^{56}\frac{3}{5}+\alpha_{ij}^B(1)n_{qs}^{70}\frac{3}{7}$
B_{dd}	$-\alpha_{ij}^B(0)n_{qq}^{56}\left(\frac{2}{3}+\frac{\lambda_f}{3}\right)-\alpha_{ij}^B(1)n_{qq}^{70}\left(\frac{1}{3}+\frac{2\lambda_f}{3}\right)$
B_{ds}	$-\alpha_{ij}^B(0)n_{qs}^{56}\left(\frac{3}{5}+\frac{3\lambda_f}{5}\right)-\alpha_{ij}^B(1)n_{qs}^{70}\left(\frac{3}{7}+\frac{3\lambda_f}{7}\right)$
B_{ss}	$-\alpha_{ij}^B(0)n_{ss}^{56}\left(\frac{1}{2}+\lambda_f\right)-\alpha_{ij}^B(1)n_{ss}^{70}\left(\frac{1}{2}+\frac{\lambda_f}{4}\right)$

Table D.5 (cont'd.)

H_I	$S(H_I)$
M_u	$n_0 \dfrac{\lambda_f}{2}$
M_d	$n_0 \dfrac{\lambda_f}{2}$
M_s	$-n_0$
B_u	$-\alpha_i^B(0)n_q^{56}\left(\dfrac{2\lambda_f}{3}+\dfrac{2\lambda_f^2}{5}\right)-\alpha_i^B(1)n_q^{70}\left(\dfrac{16\lambda_f}{15}+\dfrac{2\lambda_f^2}{15}\right)$
B_d	$-\alpha_i^B(0)n_q^{56}\left(\dfrac{2\lambda_f}{3}+\dfrac{2\lambda_f^2}{5}\right)-\alpha_i^B(1)n_q^{70}\left(\dfrac{16\lambda_f}{15}+\dfrac{2\lambda_f^2}{15}\right)$
B_s	$-\alpha_i^B(0)n_s^{56}\left(1+\dfrac{12\lambda_f}{5}+\dfrac{9\lambda_f^2}{5}\right)-\alpha_i^B(1)n_s^{70}\left(\dfrac{16}{15}+2\lambda_f+\dfrac{9\lambda_f^2}{16}\right)$
B_{uu}	$-\alpha_{ij}^B(0)n_{qq}^{56}\dfrac{\lambda_f}{3}-\alpha_{ij}^B(1)n_{qq}^{70}\dfrac{2\lambda_f}{3}$
B_{ud}	$-\alpha_{ij}^B(0)n_{qq}^{56}\dfrac{\lambda_f}{3}-\alpha_{ij}^B(1)n_{qq}^{70}\dfrac{2\lambda_f}{3}$
B_{us}	$-\alpha_{ij}^B(0)n_{qs}^{56}\left(1+\dfrac{6\lambda_f}{5}\right)-\alpha_{ij}^B(1)n_{qs}^{70}\left(1+\dfrac{6\lambda_f}{7}\right)$
B_{dd}	$-\alpha_{ij}^B(0)n_{qq}^{56}\dfrac{\lambda_f}{3}-\alpha_{ij}^B(1)n_{qq}^{70}\left(1+\dfrac{6\lambda_f}{7}\right)$
B_{ds}	$-\alpha_{ij}^B(0)n_{qs}^{56}\left(1+\dfrac{6\lambda_f}{5}\right)-\alpha_{ij}^B(1)n_{qs}^{70}\left(1+\dfrac{6\lambda_f}{7}\right)$
B_{ss}	$-\alpha_{ij}^B(0)n_{ss}^{56}(2+3\lambda_f)-\alpha_{ij}^B(1)n_{ss}^{70}\left(2+\dfrac{3\lambda_f}{4}\right)$

*For Table D.5 $n_0=(1+\lambda_f/2)^{-1}$, $n_q^{56}=(1+2\lambda_f/3+\lambda_f^2/5)^{-1}$, $n_s^{56}=(1+6\lambda_f/5+3\lambda_f^2/5)^{-1}$. $n_{qq}^{56}=(1+\lambda_f/3)^{-1}$, $n_{qs}^{56}=(1+3\lambda_f/5)^{-1}$, $n_{ss}^{56}=(1+3\lambda_f)^{-1}$, $n_q^{70}=(1+16\lambda_f/15+4\lambda_f^2/15)^{-1}$, $n_s^{70}=(1+\lambda_f+3\lambda_f^2/16)^{-1}$, $n_{qq}^{70}=(1+2\lambda_f/3)^{-1}$, $n_{qs}^{70}=(1+3\lambda_f/7)^{-1}$, $n_{ss}^{70}=(1+\lambda_f/4)^{-1}$.

Table D.6 Solutions for C_q in the decomposition (D – 1) for baryon, antibaryon and meson beams. $n = (4 + 2\lambda)^{-1}$.

Beam hadron	B_{ijk}	$\bar{B}_{\bar{i}\bar{j}\bar{k}}$	$M_{i\bar{j}}$
C_u	$n + \dfrac{\Delta Q(ijk)}{dN(S)}$	n	$n + \dfrac{\Delta Q(i)}{dN(S)}$
C_d	$n + \dfrac{\Delta S(ijk) - \Delta Q(ijk)}{dN(S)}$	n	$n + \dfrac{\Delta S(i) - \Delta Q(i)}{dN(S)}$
C_s	$\lambda n - \dfrac{\Delta S(ijk)}{dN(S)}$	λn	$\lambda n - \dfrac{\Delta S(i)}{dN(S)}$
$C_{\bar{u}}$	n	$n - \dfrac{\Delta Q(\bar{i}\bar{j}\bar{k})}{dN(S)}$	$n - \dfrac{\Delta Q(\bar{j})}{dN(S)}$
$C_{\bar{d}}$	n	$n + \dfrac{\Delta Q(\bar{i}\bar{j}\bar{k}) - \Delta S(\bar{i}\bar{j}\bar{k})}{dN(S)}$	$n + \dfrac{\Delta Q(\bar{j}) - \Delta S(\bar{j})}{dN(S)}$
$C_{\bar{s}}$	λn	$\lambda n + \dfrac{\Delta S(\bar{i}\bar{j}\bar{k})}{dN(S)}$	$\lambda n + \dfrac{\Delta S(\bar{j})}{dN(S)}$

Appendix E

SPECTRA OF THE RESONANCE DECAY PRODUCTS

Let us consider the production and subsequent decay of the resonance R in the inclusive reaction $a+b \to R+\text{anything}$. Let then the resonance R take the decay mode $R \to c_1 + c_2 + \ldots c_n$. Here we shall calculate the inclusive spectra of the secondary particle, say c_1, assuming that the resonance distribution

$$\frac{d^2\sigma_R}{dy_R dp_{RT}^2}(a+b \to R+X) = f(y_R,\ p_{RT}^2)$$

is given.

The phase space element for particle c_1 in the rest frame of the resonance may be written in the form

$$d\Phi^* = \frac{dp_L^*}{2p^*} \cdot \frac{d\varphi^*}{2\pi} \tag{E.1}$$

where

$$p^* = 1/(2m_R) \cdot \sqrt{(m_R+m+W)(m_R-m+W)(m_R+m-W)(m_R-m-W)} \tag{E.2}$$

is the momentum of particle c_1 in the resonance rest frame, m_R is the mass of the resonance, m is that of the particle c_1, W is the invariant mass of the other particles $c_2, c_3, \ldots, c_n$. Hereafter all starred values refer to the resonance rest frame; p_L^* is the longitudinal component of c_1 momentum along the beam axis; φ^* is the azimuthal angle (see Fig. E.1).

In the simplest case the resonance R is not polarized and not aligned in any direction. Then the decay probability does not depend on p_L^* and φ^* (i.e. the decay is isotropic in the resonance rest frame) and may be written as

$$d\Gamma = g(W^2)dW^2 \frac{dp_L^*}{2p^*} \frac{d\varphi^*}{2\pi} \quad , \tag{E.3}$$

where $g(W^2)$ is the square of the matrix element of the considered decay, integrated over the (n - 1)-particle phase space:

$$g(W^2) \sim \int |M|^2 \prod_{i=2}^{n} \frac{d^3p_i^*}{2E_i^*} \delta^3(\sum_{i=1}^{n} \vec{p}_i^*)\delta(\Sigma E_i^* - m_R)\delta((\sum_{i=2}^{n} p_i^*)^2 - W^2) \tag{E.4}$$

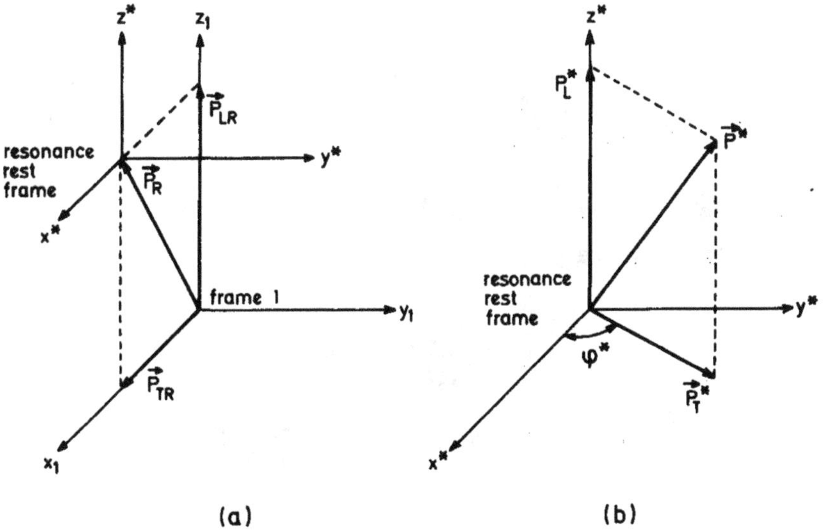

Fig. E.1 (a) Resonance decay in an arbitrary reference frame and
(b) In its rest frame.

and normalized so that

$$\int_{(\sum_{i=2}^{n} m_i)^2}^{(m_R - m)^2} g(W^2)dW^2 = b \tag{E.5}$$

and b is the partial fraction of the decay mode considered.

In order to calculate the inclusive spectra of the particle c_1 it is necessary to integrate the right-hand side of Eq. (E.3) over all possible values of W^2, p_{RT} and y_R with the weight function $f(y_R, p_{RT}^2)$ — the resonance production cross section. However, (E.3) is written in the resonance rest frame, while we need it in the frame where $f(y_R, p_{RT}^2)$ is given (frame 1). The relative velocity of these two frames is just $-\vec{\beta}_R = -\vec{p}_R/E_R$; so for the energy and momentum of the particle c_1 in frame 1 we get

$$E = \frac{E^* E_R + p^* p_R \cos\varphi^*}{m_R}$$

$$\vec{p} = \vec{p}^* + \vec{p}_R \left(\frac{E^*}{m_R} + \frac{p^* p_R \cos\varphi^*}{m_R(E_R + m_R)} \right) \tag{E.6}$$

(non-starred values E, $\vec{p}$, E_R, $\vec{p}_R$ refer to the frame 1).

Reversing (E.6) we get

$$p_L^* = p_L - p_{RL} \frac{E + E^*}{E_R + m_R} ,$$

$$p_T^{*2} = p^{*2} - p_L^{*2} = E^{*2} - m^2 - \left(p_L - p_{RL} \frac{E + E^*}{E_R + m_R}\right)^2 ,$$

$$\cos\varphi^* = \frac{E m_R - E^* E_R - p_L p_{RL} + p_{RL}^2 \frac{E + E^*}{E_R + m_R}}{p_{RT} p_T^*} . \tag{E.7}$$

Using (E.7) it is easy to rewrite (E.3) in the variables y, p_T^2 for the particle c_1 in the frame 1:

$$d\Gamma = dy\,dp_T^2\,dW^2 g(W^2)\,\frac{1}{4p^*}\,\frac{m_R}{\sqrt{p_{RT}^2 p_T^2 - [m_R E^* - m_{RT} m_T \mathrm{ch}(y-y_R)]^2}} \quad . \tag{E.8}$$

Here $m_T^2 = p_T^2 + m^2$, $m_{RT}^2 = p_{RT}^2 + m_R^2$. The last factor in the right-hand side of (E.8) is just the Jacobian of the transformation from the variables p^*, φ^* to y, p_T^2. Thus for the inclusive distribution of the particle c_1, $f(y, p_T^2)$, we get

$$f(y, p_T^2) = \frac{d^2\sigma}{dy\,dp_T^2}\,(a + b \to R + X) \quad R \to c_1 + (c_2 + \ldots + c_n)$$

$$= \frac{m_R}{4\pi}\int_{(\sum_{i=2}^{n} m_i)^2}^{(m_R - m)^2} dW^2 g(W^2)\,\frac{1}{p^*}\int_{y_R^{(-)}}^{y_R^{(+)}} dy_R \int_{p_{RT}^{2(-)}}^{p_{RT}^{2(+)}} dp_{RT}^2$$

$$\times \frac{f(y_R, p_{RT}^2)}{\sqrt{p_{RT}^2 p_T^2 - [m_R E^* - m_{RT} m_T \mathrm{ch}(y-y_R)]^2}} \quad . \tag{E.9}$$

The limits of integration in (E.9) are

$$p_{RT}^{2(\pm)} = m_{RT}^{2(\pm)} - m_R^2 \quad ; \quad p_{RT}^{2(\pm)} \geq 0$$

$$m_{RT}^{(\pm)} = \frac{m_R\left[E^* m_T \mathrm{ch}(y-y_R) \pm p_T\sqrt{E^{*2} + p_T^2 - m_T^2 \mathrm{ch}^2(y-y_R)}\,\right]}{m_T^2 \mathrm{ch}^2(y-y_R) - p_T^2} \quad ;$$

$$y_R^{(\pm)} = y \pm \ln\left(\frac{\sqrt{E^{*2} + p_T^2} + p^*}{m_T}\right) \quad ; \quad y_{R\min} \leq y_R^{(\pm)} \leq y_{R\max} \tag{E.10}$$

are simply the zeroes of the expression under the square root sign in the denominator of the last integrand in (E.9), so that this expression is non-negative in the whole space of integration.

In the simplest case of two-particle decay $R \to c_1 + c_2$ the first integration in (E.9) disappears:

$$f(y, p_T^2) = \frac{m_R}{4\pi p^*} b \int_{y_R^{(-)}}^{y_R^{(+)}} dy_R \int_{p_{RT}^{2(-)}}^{p_{RT}^{2(+)}} dp_{RT}^2 \times \frac{f(y_R, p_{RT}^2)}{\sqrt{p_{RT}^2 p_T^2 - [m_R E^* - m_{RT} m_T \mathrm{ch}(y - y_R)]^2}} \quad \text{(E.11)}$$

The experimental data for fast secondaries are often presented as the distributions over the scaling variable $x = 2p_L/\sqrt{s}$, so it is useful to rewrite (E.9) and (E.11) in x and p_T^2. If the total center-of-mass energy $\sqrt{s}$ is high enough, we may neglect all particle masses and transverse momenta compared with the longitudinal momenta p_L, p_{RL}. Then (E.7) may be written as

$$p_L^* = m_R x/x_R - E^* ,$$

$$p_T^{*2} = p^{*2} - p_L^{*2} = (2x/x_R)m_R E^* - (x^2/x_R^2)m_R^2 - m^2 ,$$

$$\cos\varphi^* = \frac{p_T^2 - p_T^{*2} - p_{RT}^2 \cdot x^2/x_R^2}{2p_{RT}p_T^2} \cdot \frac{x_R}{x} , \quad \text{(E.12)}$$

where $x = 2p_L/\sqrt{s}$, $x_R = 2p_{RL}/\sqrt{s}$; p_L, $p_{RL} >> m$, m_R, p_T, p_{RT}. Now, if $f(x_R, p_{RT}^2) = (x_R/\pi)d^2\sigma/dx_R dp_{RT}^2$ is the resonance production invariant cross section, then for the secondary particle c_1

$$f(x, p_T^2) = \frac{x}{\pi} \cdot \frac{d^2\sigma}{dx\, dp_T^2} (a + b \to R + X) \quad \hookrightarrow c_1 + (c_2 + \ldots + c_n)$$

$$= \frac{m_R x}{2\pi} \int_{(\sum_{i=2}^{n} m_i)^2}^{(m_R - m)^2} \frac{g(W^2)dW^2}{p^*} \int_{x_R^{(-)}}^{x_R^{(+)}} \frac{dx_R}{x_R^2} \int_{p_{RT}^{2(-)}}^{p_{RT}^{2(+)}} dp_{RT}^2$$

$$\times \frac{f(x_R, p_{RT}^2)}{\sqrt{[p_{RT}^2 \frac{x^2}{x_R^2} - (p_T - p_T^*)^2] \cdot [(p_T + p_T^*)^2 - p_{RT}^2 \frac{x^2}{x_R^2}]}} \quad . \tag{E.13}$$

Here the limits of integration are (assuming $0 < x_1 < 1$)

$$p_{RT}^{2(\pm)} = \frac{(p_T \pm p_T^*)^2 x_R^2}{x^2}$$

$$x_R^{(-)} = \frac{m_R x_1}{E^* + p^*} \quad ; \quad x_R^{(+)} = \min\left\{1, \frac{m_R x}{E^* - p^*}\right\} \quad . \tag{E.14}$$

For the two-particle decays $R \to c_1 + c_2$:

$$f(x, p_T^2) = \frac{m_R x}{2\pi p^*} \int_{x_R^{(-)}}^{x_R^{(+)}} \frac{dx_R}{x_R^2} \int_{p_{RT}^{2(-)}}^{p_{RT}^{2(+)}} dp_{RT}^2$$

$$\times \frac{f(x_R, p_{RT}^2)}{\sqrt{[p_{RT}^2 \cdot \frac{x^2}{x_R^2} - (p_T - p_T^*)^2][(p_T + p_T^*)^2 - p_{RT}^2 \frac{x^2}{x_R^2}]}} \quad . \tag{E.15}$$

The expression obtained ((E.13) or (E.15)) is illustrated by the simple example in Fig. E.2. We have chosen an artificial distribution for prompt particles (resonances) $f(x_R, p_{RT}^2) \sim (1 - x_R) \cdot \exp(-7p_{RT}^2)$. The x-distributions of the secondary pions from the decay $\rho \to 2\pi$ are shown in Fig. E.2a for fixed p_T value and integrated over all p_T. The same distributions over p_T^2 for the fixed x-value (and integrated over x) are shown in Fig. E.2b. One can see that the initial distri-

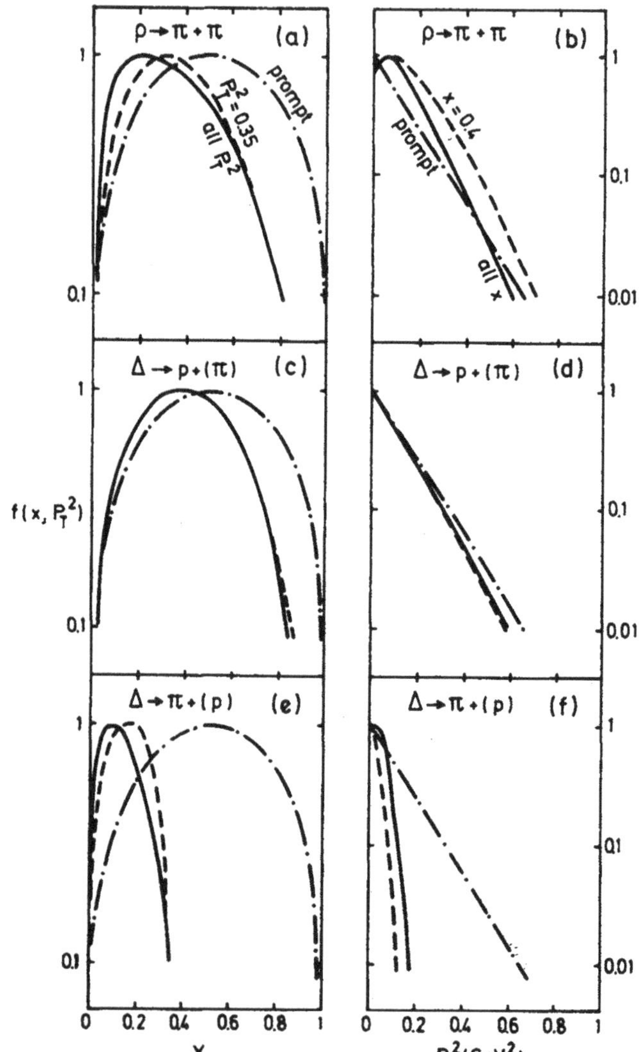

Fig. E.2 x and p_T^2 distribution of secondary pions and protons produced as a result of the decays $\rho \to \pi\pi$ and $\Delta \to p\pi$. The prompt resonance distribution is taken to be $f(x_R, p_{RT}^2) \sim (1 - x_R) \cdot \exp(-7p_{RT}^2)$.

bution is distorted rather significantly. In Figs. E.2c, E.2d we show the distributions for the decay protons from the decay $\Delta^{++} \to p\pi^{+}$; here the difference between the initial and final distributions is less because of the large mass difference between the decay products. Indeed, a proton carries away almost all Δ^{++}-momentum. Thus the distributions of the secondary pions from the same decay $\Delta^{++} \to p\pi^{+}$ (Figs. E.2e, 2f) are considerably softened as compared to those for prompt distribution.

www.ingramcontent.com/pod-product-compliance
Lightning Source LLC
Chambersburg PA
CBHW050633190326
41458CB00008B/2248
9789971966683